The Unofficial

MINECRAFT
COLORING MATH BOOK
CRACKING FRACTIONS

A Complete Guide to Master 2500+ Fractions Tasks and Word Problems with Test Prep, Word Search, Mazes, Coloring, and More!

STEM mindset, LLC 1603 Capitol Ave. Suite 310 A293 Cheyenne, WY 82001 USA
www.stemmindset.com info@stemmindset.com

First published in the USA 2019. ISBN 9781948737579

Table of Contents

Table of Contents

Table of Contents

Table of Contents

Table of Contents

Table of Contents

EVERYBODY'S A CRITIC -
And if you're one, too, we want You!

Speak up, write on, and let your voice be heard! We want to know what you, parents and users, really think – share your feedback on info@stemmindset.com or www.amazon.com!

Kids learn and stay engaged, thanks to puzzles, coloring, mazes, word search tasks, along with challenging math problems. These methods help kids understand math concepts, master math skills, and even help struggling students gain the confidence to improve their math comprehension and testing.

These activities are perfect for daily practice, morning work, homework, math centers, early finishers, test preparation, assessment, or for high school students struggling with fractions.

They will work great in 4th Grade to challenge students. They are perfect in 5th-6th Grades to practice fractions, but they also might work in 7th-9th Grades as a review or for those who are struggling with math.

Hi. I'm Sunny. For me, everything is an adventure. I am ready to try anything, take chances, see what happens - and help you try, too! I like to think I'm confident, caring and have an open mind. I will cheer for your success and encourage everyone! I'm ready to be a really good friend!

I've got a problem. Well, I've always got a problem. And I don't like it. It makes me cranky, and grumpy, impatient and the truth is, I got a bad attitude. There. I said it. I admit it. And the reason I feel this way? Math! I don't get it and it bums me out. Grrrr!

Not trying to brag, but I am the smartest Brainer that ever lived - and I'm a brilliant shade of blue. That's why they call me Smarty. I love to solve problems and I'm always happy to explain how things work - to help any Brainer out there! To me, work is fun, and math is a blast!

I scare easily. Like, even just a little ...Boo! Oh wow, I've scared myself! Anyway, they call me Pickles because I turn a little green when I get panicky. Especially with new stuff. Eek! And big complicated problems. Really any problem. Eek! There, I did it again.

Hi! Name's Pepper. I have what you call a positive outlook. I just think being alive is exciting! And you know something? By being friendly, kind and maybe even wise, you can have a pretty awesome day every day on this amazing planet.

A famous movie star once said, "I want to be alone." Well, I do too! I'm best when I'm dreaming, thinking, and in my own world. And so, I resist! Yes, I resist anything new, and only do things my way or quit. The rest of the Brainers have math, but I'd rather have a headache and complain. Or pout.

1. Read and write the missing numbers.

*The **numerator** is the numeral that is written above the bar.*

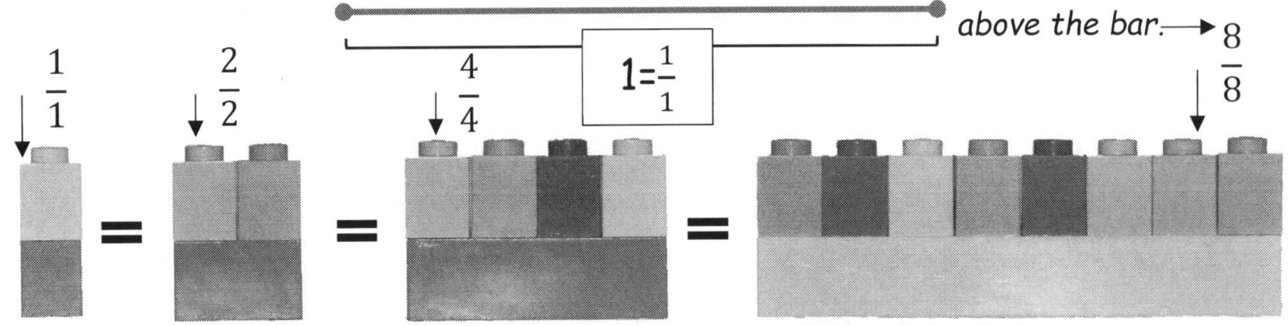

The numerator represents how many equal PARTS are being identified:

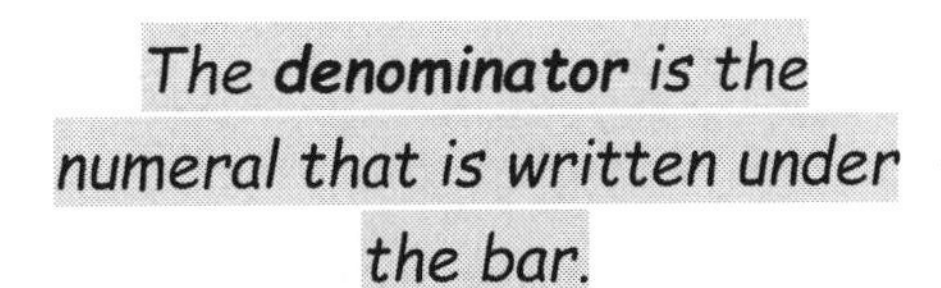

*The **denominator** is the numeral that is written under the bar.*

$$\text{Whole} = 1 = \frac{2}{2} = \frac{4}{4} = \frac{8}{8}$$

The denominator represents the number of equal parts in the WHOLE (the whole is divided into the two, four, or eight equal parts)

*The **numerator** 1 tells that 1 part of the whole is being counted.*

A half = $\frac{1}{2}$ ← *The **denominator** 2 tells that a whole was divided into 2 equal parts.*

$\frac{1}{2}$

The numerator shows you one (brown block) of the two equal parts (green blocks).

The denominator shows you the number of the equal parts (green blocks) in the whole.

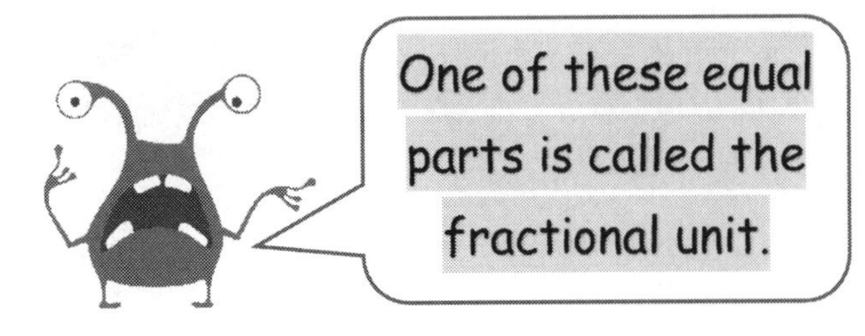

One of these equal parts is called the fractional unit.

The fractional unit for this Minecraft block is

$\frac{1}{4}$

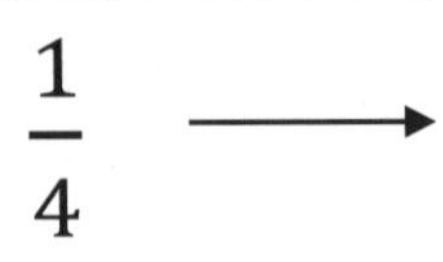

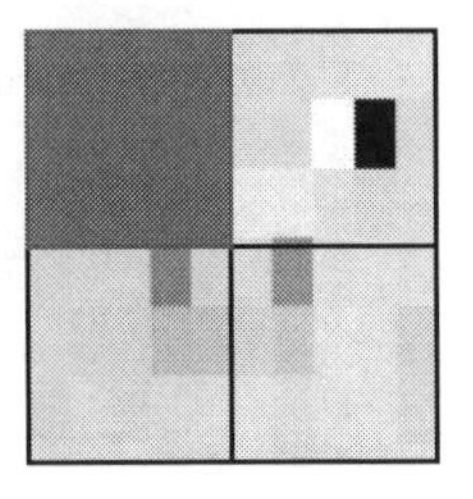

Aha… I start with multiplying the fractions, right? So, to multiply the fractions I need to multiply out the numerators and then, I need to multiply out the denominators.

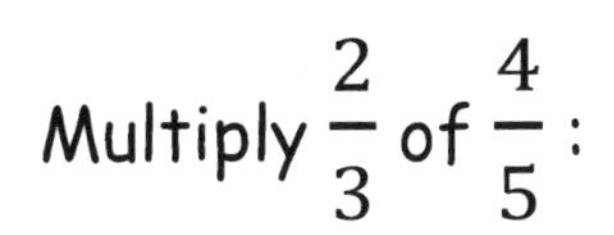

Multiply $\frac{2}{3}$ of $\frac{4}{5}$:

Add 4 two times: 4 + 4 ⟶
Add 5 three times: 5 + 5 + 5 ⟶

$$\frac{2 \times 4}{3 \times 5} = \frac{8}{15}$$

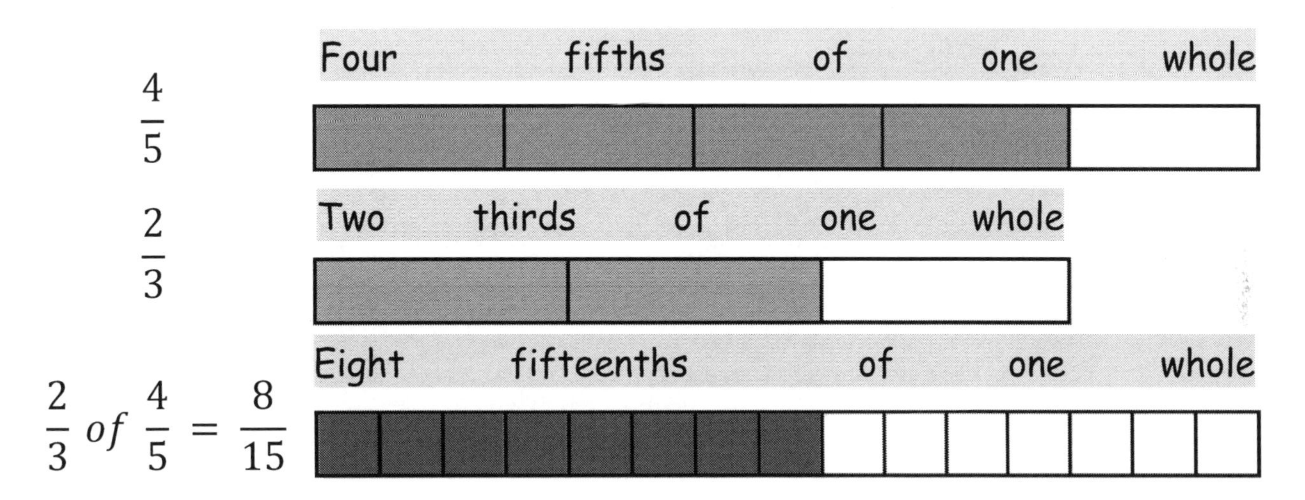

$$\frac{2}{3}\ of\ \frac{4}{5} = \frac{8}{15}$$

Multiply $\frac{3}{5}$ of $\frac{1}{4}$:

$$\frac{add\ 1\ three\ times: 1+1+1}{add\ 4\ five\ times: 4+4+4+4+4} = \frac{3 \times 1}{5 \times 4} = \frac{3}{20}$$

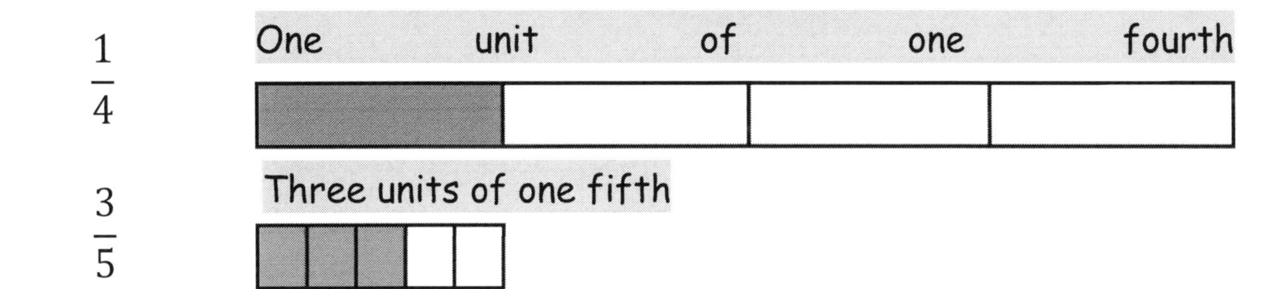

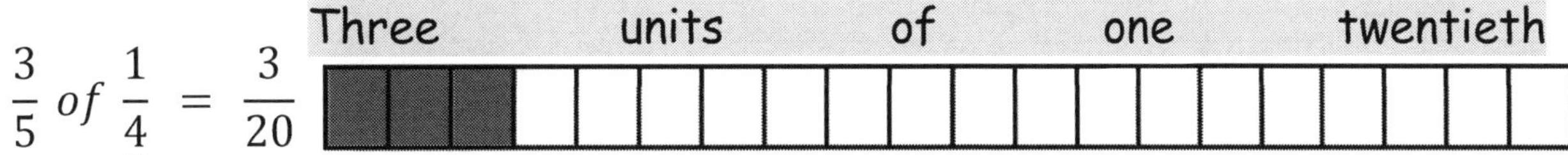

$$\frac{3}{5}\ of\ \frac{1}{4} = \frac{3}{20}$$

Multiplying is easy!

Yes, a common denominator is not needed!

1. Multiply fractions. Color the fractions bars. Write the missing numbers and words.

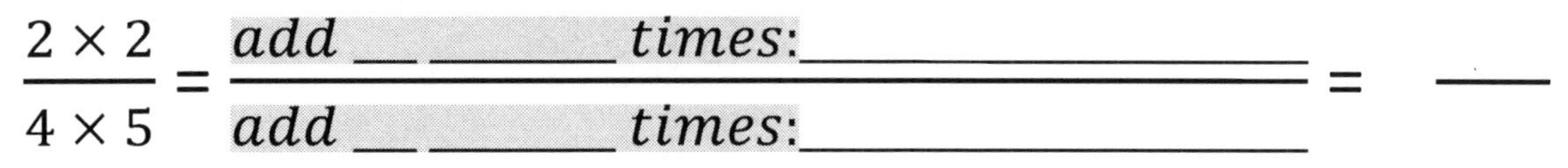

$$\frac{2 \times 2}{4 \times 5} = \frac{add\ \_\_\ \_\_\_\_\_\_\ times: \_\_\_\_\_\_\_\_\_\_\_\_}{add\ \_\_\ \_\_\_\_\_\_\ times: \_\_\_\_\_\_\_\_\_\_\_\_} = \frac{\quad}{\quad}$$

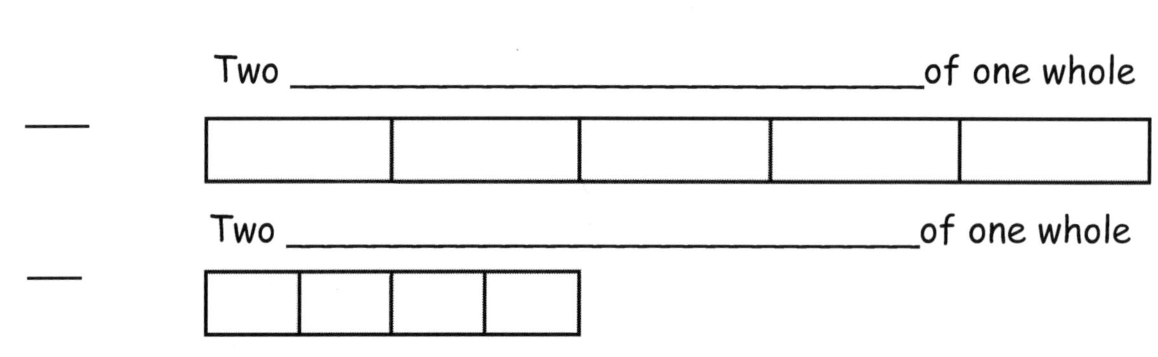

Two ______________________________ of one whole

$\frac{\quad}{\quad}$

Two ______________________________ of one whole

$\frac{\quad}{\quad}$

$\frac{\quad}{\quad}\ of\ \frac{\quad}{\quad} =$ ______________________________ of one whole

$= \frac{\quad}{\quad}$

1. Multiply fractions. Draw and color the fractions bars. Write the missing numbers.

$$\frac{5 \times 4}{6 \times 5} = \frac{add\ \_\_\ \_\_\_\_\_\_\ times: \_\_\_\_\_\_\_\_\_\_\_\_}{add\ \_\_\ \_\_\_\_\_\_\ times: \_\_\_\_\_\_\_\_\_\_\_\_} = \frac{\quad}{\quad}$$

Four units of ______________________________

$\frac{\quad}{\quad}$

Five units of ______________________________

$\frac{\quad}{\quad}$

$\frac{\quad}{\quad}\ of\ \frac{\quad}{\quad} =$ ____________ units of ______________________

$= \frac{\quad}{\quad}$

1. Equivalent fractions. Answer the questions and write the missing numbers.

Equivalent fractions? Oh, boy! What's that?

How can I explain? Equivalent fractions show the same value of the whole.

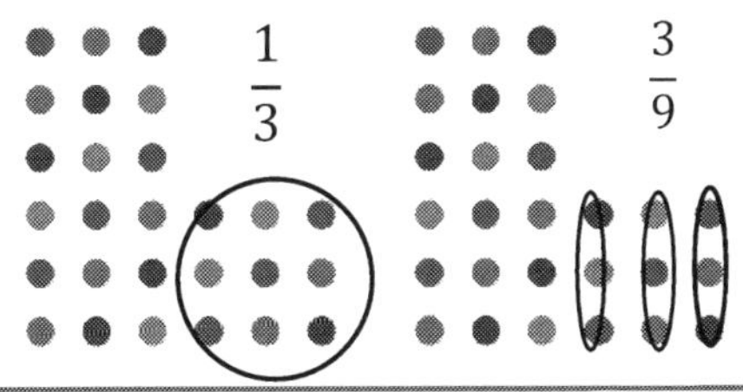

I circled one third black. The whole is divided into 3 equal parts.

I circled three ninths black. The whole is divided into 9 equal parts. I circled 9 dots in both pictures.

The numerators and denominators are different, but the value is the same. Take a half of the block: $\frac{1}{2}$ (a half). If you multiply the numerator and the denominator BY THE SAME NUMBER called the COMMON FACTOR, say, 2, you'll get a different fraction - $\frac{2}{4}$ (two fourths), but their value will stay the same:

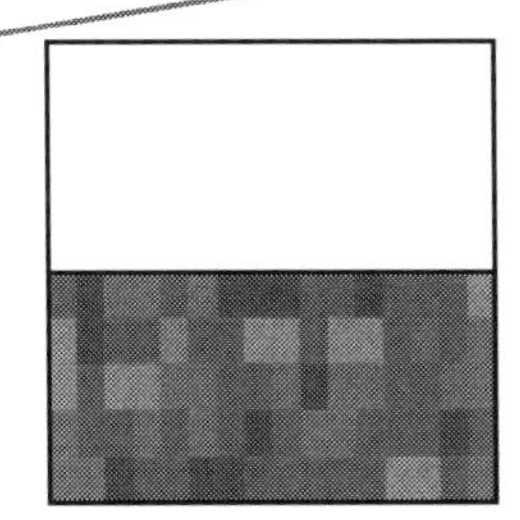

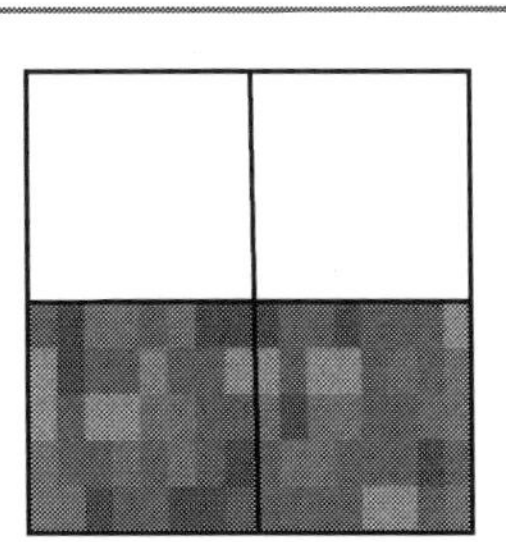

2 is the ***common factor***

$$\frac{1 \times 2}{2 \times 2} = \frac{2}{4}$$

2 is the ***common factor***

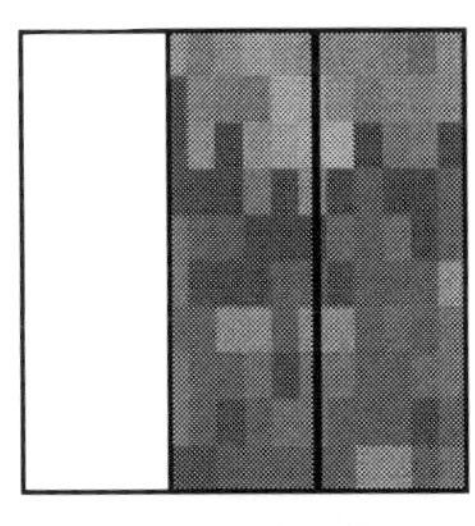

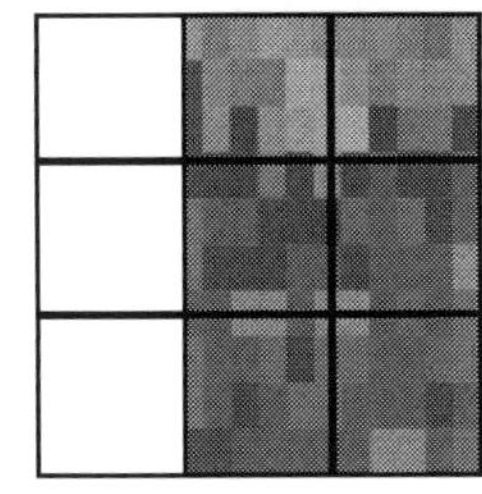

To find an equivalent fraction of $\frac{1}{3}$ I need to multiply the numerator and the denominator BY THE SAME NUMBER called the COMMON FACTOR (say, 3): $\frac{1 \times 3}{3 \times 3} = \frac{3}{9}$.

3 is the ***common factor***

$$\frac{1 \times 3}{3 \times 3} = \frac{3}{9}$$

3 is the ***common factor***

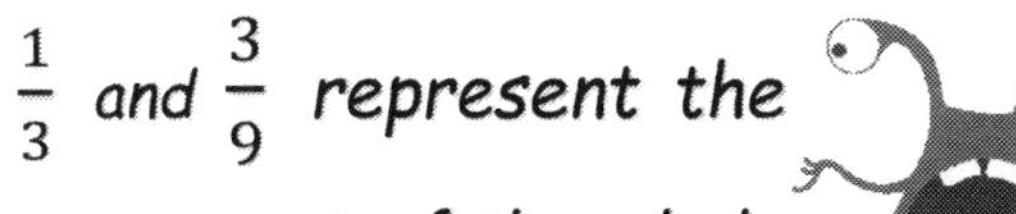

$\frac{1}{3}$ and $\frac{3}{9}$ represent the
same part of the whole

1. Fill in the missing numbers using the fractions strips.

$$\frac{1\times 2}{2\times 2} = \frac{\ }{\ }$$

$$\frac{1\times 4}{2\times 4} = \frac{\ }{\ }$$

$$\frac{1\times 2}{3\times 2} = \frac{\ }{\ }$$

$$\frac{1\times 2}{5\times 2} = \frac{\ }{\ }$$

$$\frac{1\times 2}{4\times 2} = \frac{\ }{\ }$$

1/2									
1/4									
1/5									
1/6									
1/8									
1/10									

1in

$$\frac{1\times}{3\times 2} = \frac{\ }{\ } \qquad \frac{1\times}{5\times 2} = \frac{\ }{\ } \qquad \frac{1\times}{4\times 2} = \frac{\ }{\ }$$

$$\frac{1\times}{2\times 5} = \frac{\ }{\ } \qquad \frac{1\times}{3\times 3} = \frac{\ }{\ } \qquad \frac{1\times}{2\times 4} = \frac{\ }{\ }$$

$$\frac{1\times}{\times 2} = \frac{\ }{8} \qquad \frac{1\times}{\times 3} = \frac{\ }{6} \qquad \frac{1\times}{\times 5} = \frac{\ }{10}$$

$$\frac{1\times}{\times 4} = \frac{\ }{8} \qquad \frac{1\times}{\times 3} = \frac{\ }{9} \qquad \frac{1\times}{\times 5} = \frac{\ }{5}$$

$$\frac{1\times}{\times 6} = \frac{\ }{12} \qquad \frac{1\times}{\times 5} = \frac{\ }{15} \qquad \frac{1\times}{\times 4} = \frac{\ }{12}$$

1. Find an equivalent fraction with the given denominator in higher terms and color the fraction of each shape.

Red Blue Green Yellow Grey Orange Purple Pink

$\frac{1}{2} = \frac{}{4} = \frac{}{8} = \frac{}{10} = \frac{}{14} = \frac{}{16} = \frac{}{20} = \frac{}{40}$

How to create an ***equivalent fraction*** *in higher terms?*

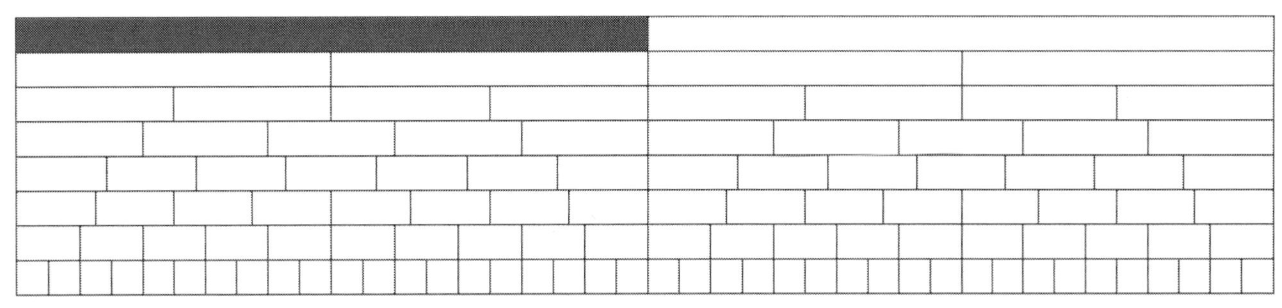

Red Blue Green Yellow Grey Brown Purple Pink

$\frac{1}{3} = \frac{}{6} = \frac{}{9} = \frac{}{12} = \frac{}{18} = \frac{}{21} = \frac{}{24} = \frac{}{36}$

! Multiply both the ***numerator*** *and the* ***denominator*** *by the* ***same whole number****.*

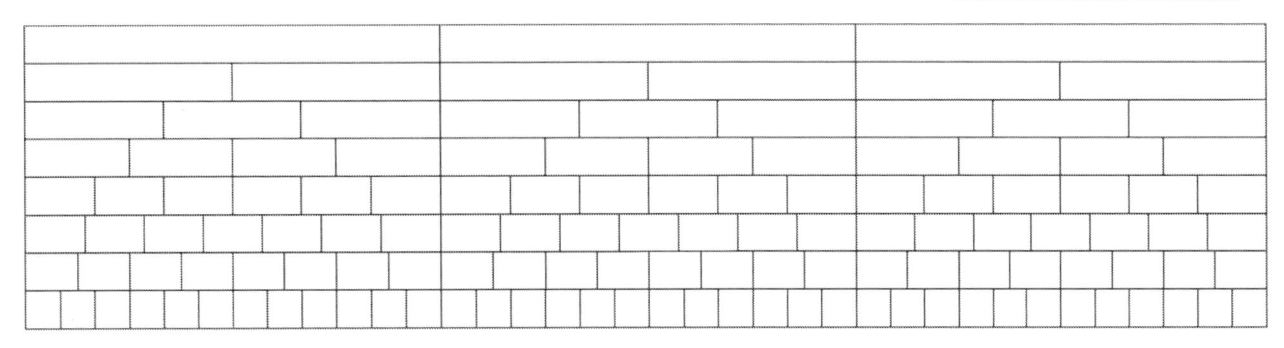

Red Blue Green Yellow Grey Brown Orange Purple White Pink

$\frac{1}{4} = \frac{}{8} = \frac{}{12} = \frac{}{16} = \frac{}{20} = \frac{}{24} = \frac{}{28} = \frac{}{32} = \frac{}{36} = \frac{}{40}$

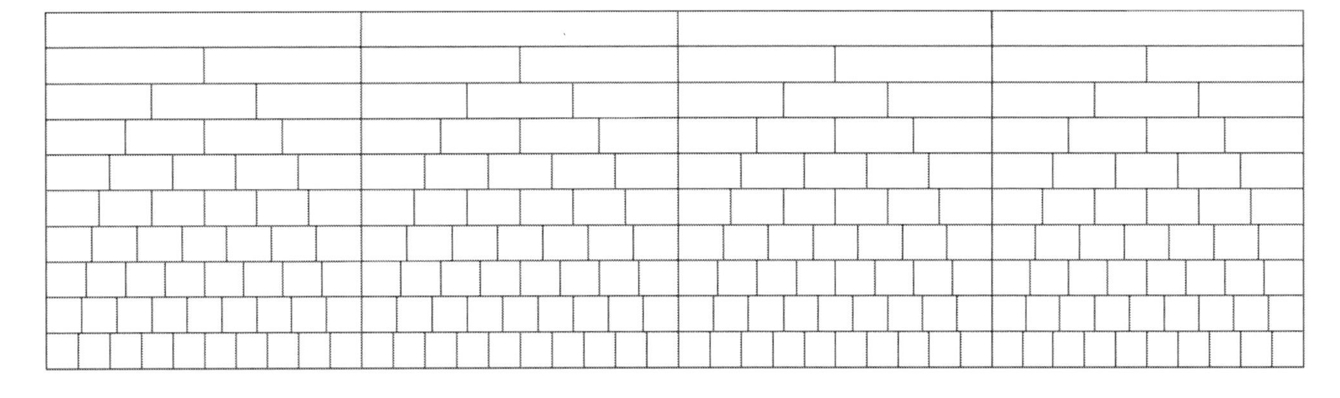

Smarty, can you help me to draw and color $\frac{3}{8}$ of this square?
Grey
$\frac{3}{8}$
Good question! Start with the denominator! How can you get 8 by using multiplication?
That's easy!
1 × 8 = 8; 2 × 4 = 8; 8 × 1 = 8; 4 × 2 = 8
Hmm... So, you can draw:
(1×8: 1 row by 8 columns)
or you can draw:
(2×4: 2 rows by 4 columns)
1
8
2
4
(8×1: 8 rows by 1 column)
8
or:
1
or:
4
2
(4×2: 4 rows by 2 columns)

1. Draw and color the fraction of each shape.

$\frac{4}{6}$

$\frac{3}{5}$

$\frac{2}{7}$

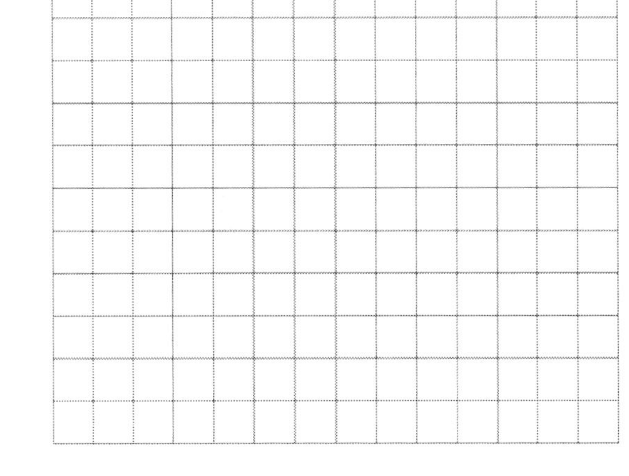

$\frac{6}{9}$

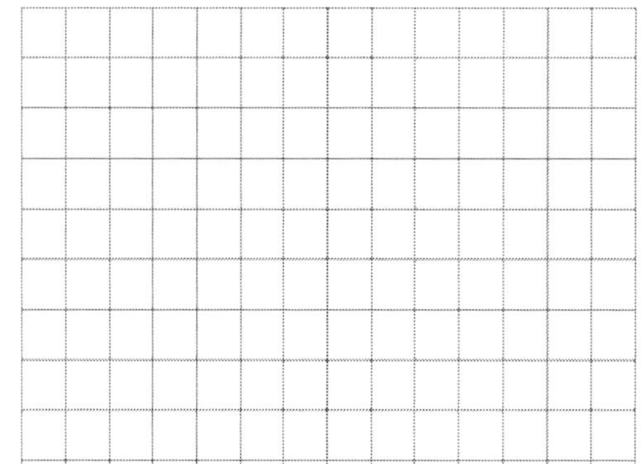

$\frac{1}{4}$

$\frac{5}{12}$

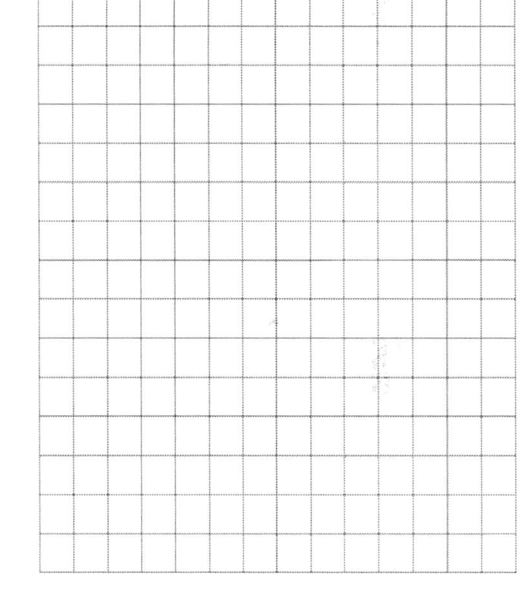

2. I found 26 emeralds. I traded a $\frac{1}{2}$ of them. How many emeralds were left?

Answer: ____________________.

1. Find an equivalent fraction with the given numerator.

Step 1: Use the given fraction $(\frac{1}{2})$.

Step 2: The numerator 1 has been multiplied by 2 to get 2.

Step 3: To get an equivalent fraction, the denominator 2 must also be multiplied by 2 to get 4.

Therefore $\frac{1}{2} = \frac{2}{4}$.

$$\frac{1}{2} = \frac{2}{4} = \frac{11}{\quad} = \frac{4}{\quad} = \frac{9}{\quad} = \frac{6}{\quad} = \frac{10}{\quad} = \frac{8}{\quad} = \frac{5}{\quad} = \frac{7}{\quad} = \frac{3}{\quad} = \frac{12}{\quad}$$

$$\frac{1}{3} = \frac{2}{6} = \frac{12}{\quad} = \frac{4}{\quad} = \frac{8}{\quad} = \frac{11}{\quad} = \frac{7}{\quad} = \frac{5}{\quad} = \frac{9}{\quad} = \frac{10}{\quad} = \frac{6}{\quad} = \frac{3}{\quad}$$

$$\frac{1}{4} = \frac{9}{\quad} = \frac{3}{\quad} = \frac{12}{\quad} = \frac{5}{\quad} = \frac{8}{\quad} = \frac{11}{\quad} = \frac{6}{\quad} = \frac{2}{\quad} = \frac{10}{\quad} = \frac{7}{\quad} = \frac{4}{\quad}$$

$$\frac{1}{5} = \frac{8}{\quad} = \frac{3}{\quad} = \frac{12}{\quad} = \frac{2}{\quad} = \frac{10}{\quad} = \frac{4}{\quad} = \frac{5}{\quad} = \frac{9}{\quad} = \frac{6}{\quad} = \frac{11}{\quad} = \frac{7}{\quad}$$

2. Write the fraction name under each picture.

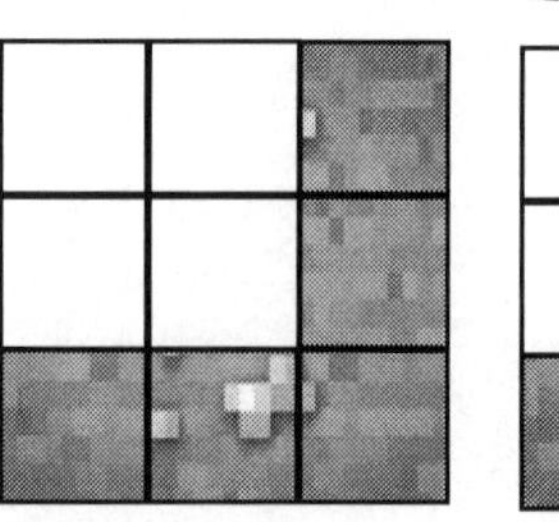

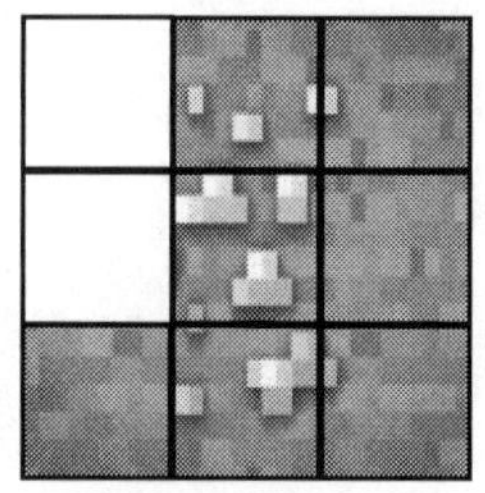

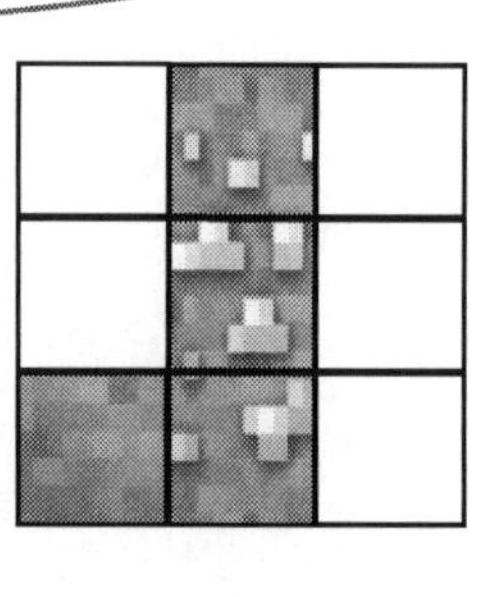

$$\frac{\text{white blocks}}{\text{blocks in total}} \qquad \frac{\dots}{\dots} \qquad \frac{\dots}{\dots} \qquad \frac{\dots}{\dots}$$

1. Find an equivalent fraction with the given denominator.

Step 1: Use the given fraction ($\frac{1}{2}$).

Step 2: The denominator 2 has been multiplied by 9 to get 18.

Step 3: To get an equivalent fraction, the numerator 1 must also be multiplied by 9 to get 9. Therefore $\frac{1}{2} = \frac{9}{18}$.

$$\frac{1}{2} = \frac{9}{18} = \frac{\_\_}{4} = \frac{\_\_}{12} = \frac{\_\_}{22} = \frac{\_\_}{16} = \frac{\_\_}{8} = \frac{\_\_}{28} = \frac{\_\_}{40} = \frac{\_\_}{6} = \frac{\_\_}{10} = \frac{\_\_}{34}$$

$$\frac{1}{3} = \frac{7}{21} = \frac{\_\_}{9} = \frac{\_\_}{60} = \frac{\_\_}{15} = \frac{\_\_}{27} = \frac{\_\_}{12} = \frac{\_\_}{6} = \frac{\_\_}{18} = \frac{\_\_}{36} = \frac{\_\_}{48} = \frac{\_\_}{24}$$

$$\frac{1}{4} = \frac{\_\_}{24} = \frac{\_\_}{40} = \frac{\_\_}{16} = \frac{\_\_}{32} = \frac{\_\_}{12} = \frac{\_\_}{48} = \frac{\_\_}{8} = \frac{\_\_}{36} = \frac{\_\_}{20} = \frac{\_\_}{28} = \frac{\_\_}{56}$$

$$\frac{1}{5} = \frac{\_\_}{35} = \frac{\_\_}{15} = \frac{\_\_}{45} = \frac{\_\_}{20} = \frac{\_\_}{30} = \frac{\_\_}{65} = \frac{\_\_}{25} = \frac{\_\_}{70} = \frac{\_\_}{50} = \frac{\_\_}{10} = \frac{\_\_}{40}$$

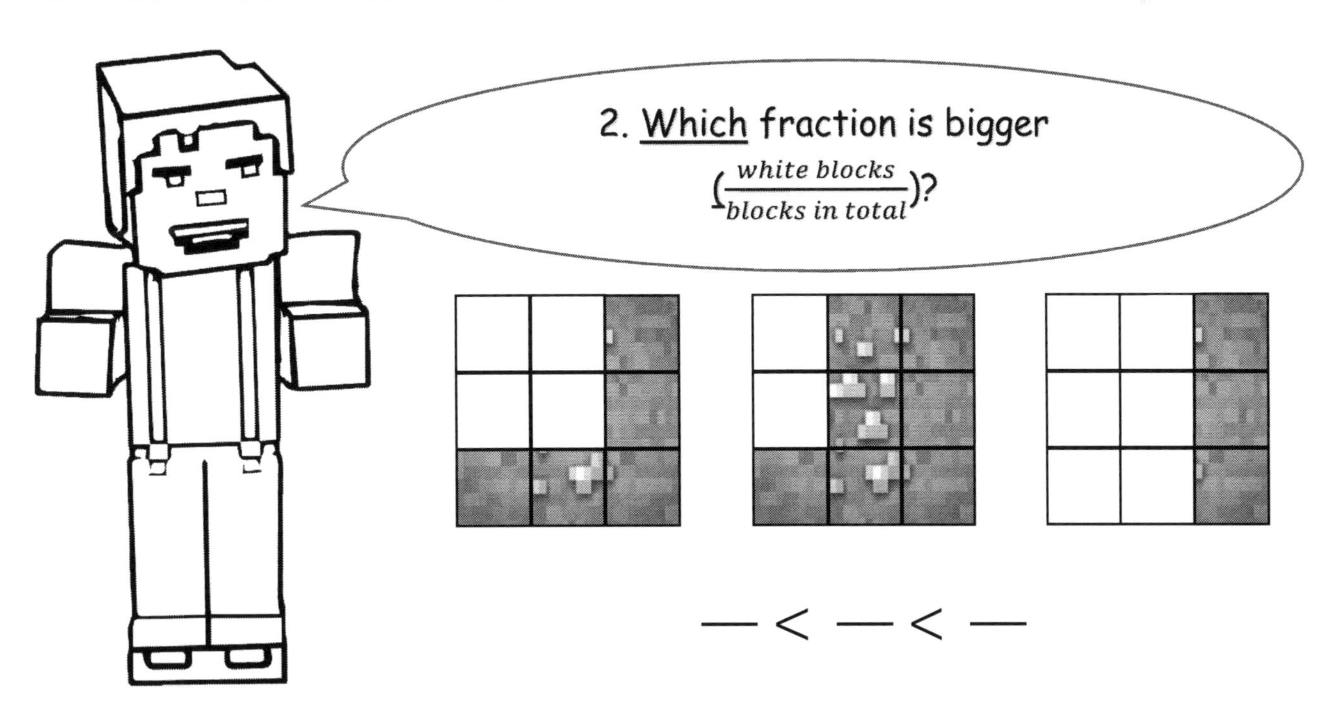

2. Which fraction is bigger
($\frac{white\ blocks}{blocks\ in\ total}$)?
__ < __ < __

1. Find an equivalent fraction with the given numerator.

Step 1: Use the given fraction ($\frac{1}{6}$).

Step 2: The numerator 1 has been multiplied by 2 to get 2.

Step 3: To get an equivalent fraction, the denominator 6 must also be multiplied by 2 to get 12.

Therefore $\frac{1}{6} = \frac{2}{12}$.

$$\frac{1}{6} = \frac{2}{12} = \frac{9}{} = \frac{4}{} = \frac{12}{} = \frac{6}{} = \frac{10}{} = \frac{8}{} = \frac{3}{} = \frac{7}{} = \frac{11}{} = \frac{5}{}$$

$$\frac{1}{7} = \frac{2}{} = \frac{7}{} = \frac{12}{} = \frac{5}{} = \frac{9}{} = \frac{3}{} = \frac{8}{} = \frac{6}{} = \frac{10}{} = \frac{11}{} = \frac{4}{}$$

$$\frac{1}{8} = \frac{2}{} = \frac{6}{} = \frac{4}{} = \frac{9}{} = \frac{3}{} = \frac{11}{} = \frac{8}{} = \frac{5}{} = \frac{10}{} = \frac{7}{} = \frac{12}{}$$

$$\frac{1}{9} = \frac{5}{} = \frac{3}{} = \frac{9}{} = \frac{2}{} = \frac{11}{} = \frac{7}{} = \frac{12}{} = \frac{4}{} = \frac{10}{} = \frac{6}{} = \frac{8}{}$$

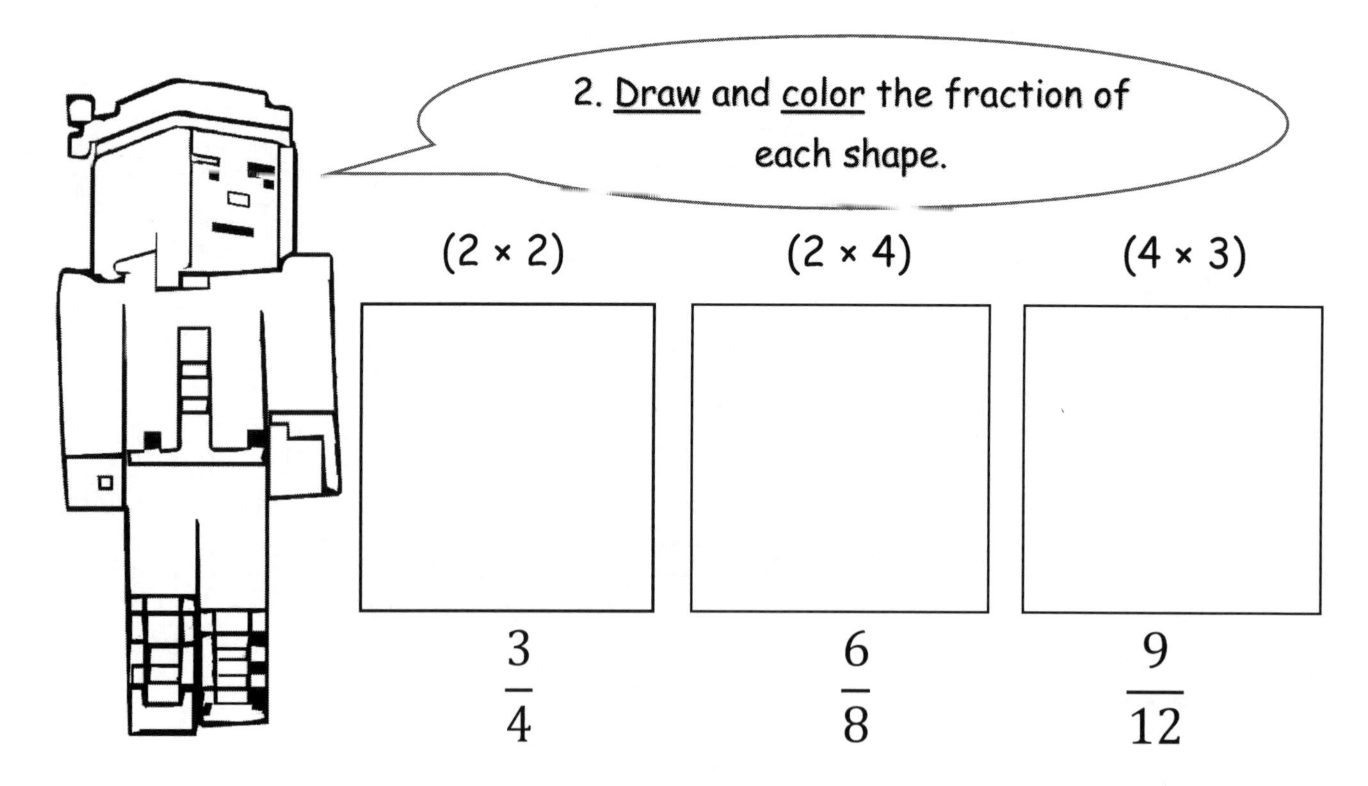

2. Draw and color the fraction of each shape.
(2 × 2)
(2 × 4)
(4 × 3)
3/4
6/8
9/12

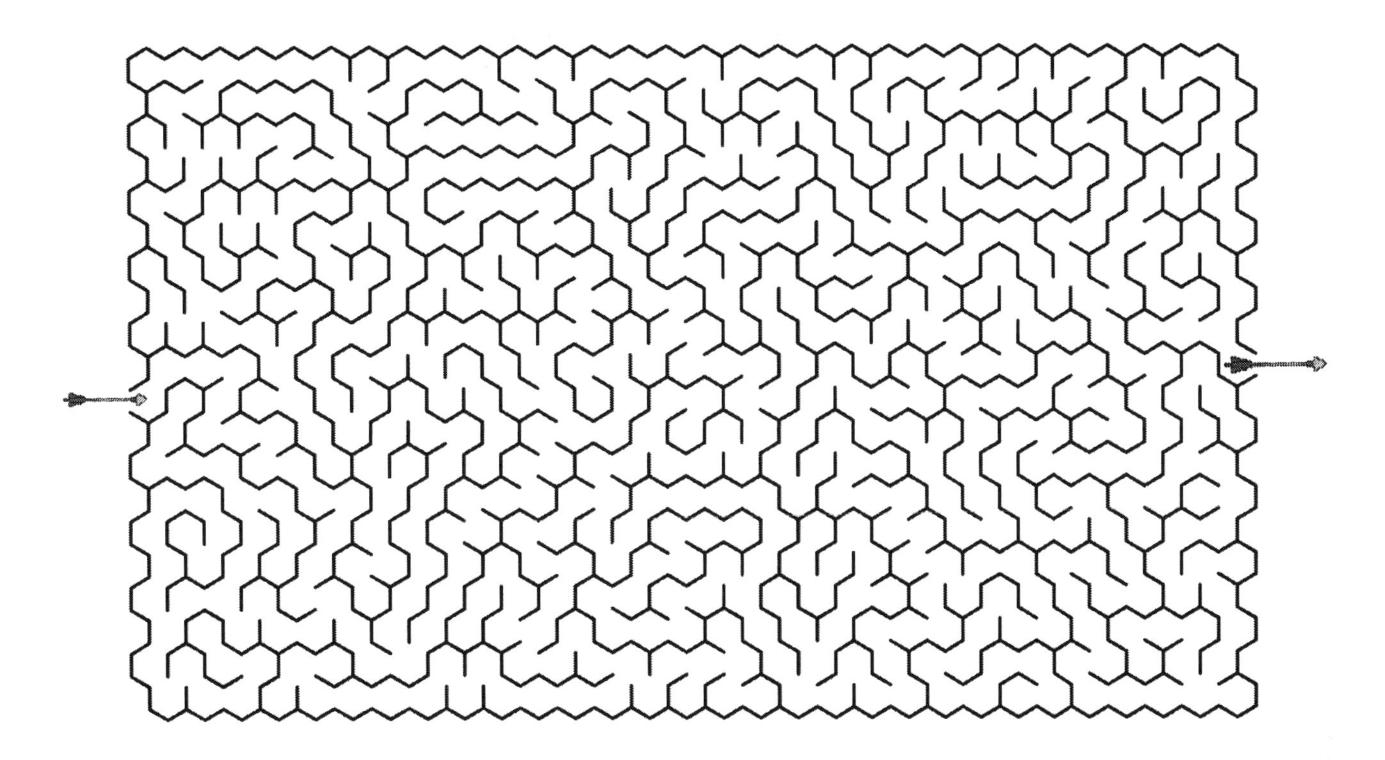

1. Find and circle or cross out the words to find out more about Minecraft.

C A E A W S F D Q D D P R
L O W V F J J G E K O B E
A T N A I M Z T S S E G Q
V H C S C T A I S R G X U
I F G B T L A E M W J Z I
V H M P O R S E O H W R R
R Y L S P S U D R P K H E
U Q I U I A X C J C X D M
S S D O P L X K T L Z C E
T E N T R A N C E I W L N
P S M A E G N G W G O M T
E N O T S E L B B O C N S
R C L E V V T G C Y V X S

CREATIVE
POSSESSIONS
SURVIVAL
REQUIREMENTS
ISOLATED
CONSTRUCTIONS
ENTRANCE
COBBLESTONE

1. Find an equivalent fraction with the given denominator.

Step 1: Use the given fraction $(\frac{1}{6})$.

Step 2: The denominator 6 has been multiplied by 3 to get 18.

Step 3: To get an equivalent fraction, the numerator 1 must also be multiplied by 3 to get 3.

Therefore $\frac{1}{6} = \frac{3}{18}$.

$$\frac{1}{6} = \frac{3}{18} = \frac{\quad}{36} = \frac{\quad}{12} = \frac{\quad}{48} = \frac{\quad}{66} = \frac{\quad}{24} = \frac{\quad}{78} = \frac{\quad}{42} = \frac{\quad}{60} = \frac{\quad}{30} = \frac{\quad}{54}$$

$$\frac{1}{7} = \frac{\quad}{21} = \frac{\quad}{49} = \frac{\quad}{70} = \frac{\quad}{14} = \frac{\quad}{77} = \frac{\quad}{42} = \frac{\quad}{28} = \frac{\quad}{63} = \frac{\quad}{35} = \frac{\quad}{84} = \frac{\quad}{56}$$

$$\frac{1}{8} = \frac{\quad}{24} = \frac{\quad}{40} = \frac{\quad}{16} = \frac{\quad}{88} = \frac{\quad}{72} = \frac{\quad}{48} = \frac{\quad}{64} = \frac{\quad}{80} = \frac{\quad}{96} = \frac{\quad}{32} = \frac{\quad}{56}$$

$$\frac{1}{9} = \frac{\quad}{36} = \frac{\quad}{18} = \frac{\quad}{72} = \frac{\quad}{27} = \frac{\quad}{90} = \frac{\quad}{45} = \frac{\quad}{63} = \frac{\quad}{99} = \frac{\quad}{54} = \frac{\quad}{81} = \frac{\quad}{117}$$

2. Draw and color the fraction of each shape.

(3 × 1)

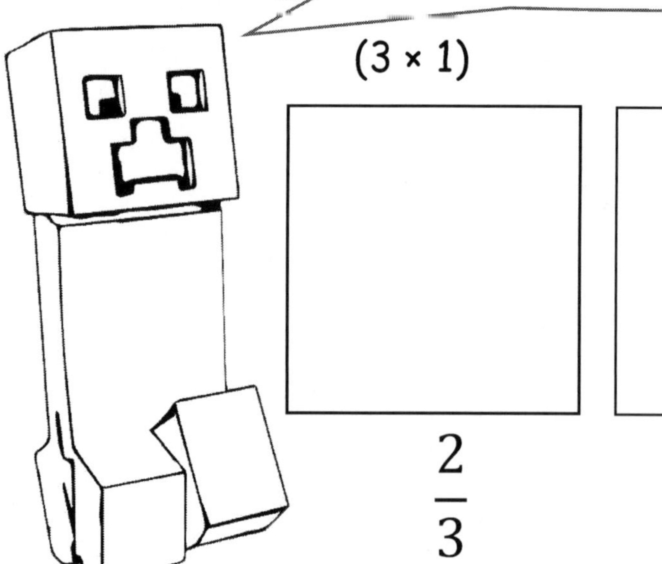

$\frac{2}{3}$

(3 × 3)

$\frac{6}{9}$

(2 × 6)

$\frac{12}{18}$

About simplifying... When you divide both the numerator and denominator by the same number, you simplify the fraction, or write it in lower terms, or reduce it to its lowest terms.

I colored white the same relative amount of my block ($\frac{1}{4}$). Though the sizes of the equal parts my blocks are divided into are different.

$$\frac{1}{4} = \frac{1 \times 1}{4 \times 1}$$

We divided $\frac{1}{4}$ by 2

$$\frac{2}{8} = \frac{1 \times 2}{4 \times 2}$$

We divided $\frac{1}{4}$ by 3

$$\frac{3}{12} = \frac{1 \times 3}{4 \times 3}$$

We divided $\frac{1}{4}$ by 4

$$\frac{4}{16} = \frac{1 \times 4}{4 \times 4}$$

All these fractions represent the same relative amount - one fourth.

The fraction $\frac{4}{16}$ is in higher terms, while $\frac{1}{4}$ is in its lowest terms. So $\frac{4}{16}$was simplified or reduced to its lowest terms.

To simplify a fraction, divide both the numerator and the denominator by the same number and write a fraction in its simplest form:

$$\frac{2}{8} = \frac{(2 \div 2)}{(4 \div 2)} = \frac{1}{4}$$

$$\frac{3}{12} = \frac{(3 \div 3)}{(12 \div 3)} = \frac{1}{4}$$

$$\frac{4}{16} = \frac{(4 \div 4)}{(16 \div 4)} = \frac{1}{4}$$

The greatest common factor is 2

The greatest common factor is 3

The greatest common factor is 4

Find the greatest factor both the numerator and the denominator can be divided evenly without any remainder.

1. Reduce each fraction to simplest form.

Well, to reduce a fraction to its lowest form (or term) means that I need to divide the numerator (the top number) and the denominator (the bottom number) by the SAME GREATEST number.

$$\frac{(3 \div 3)}{(12 \div 3)} = \frac{1}{4}$$

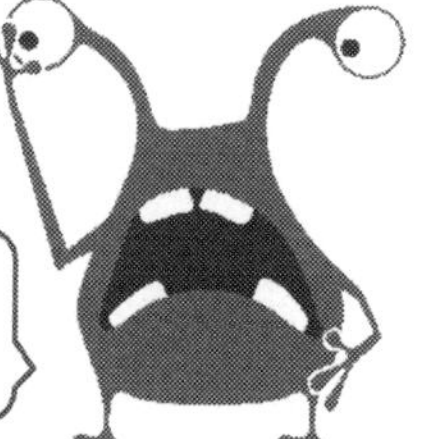

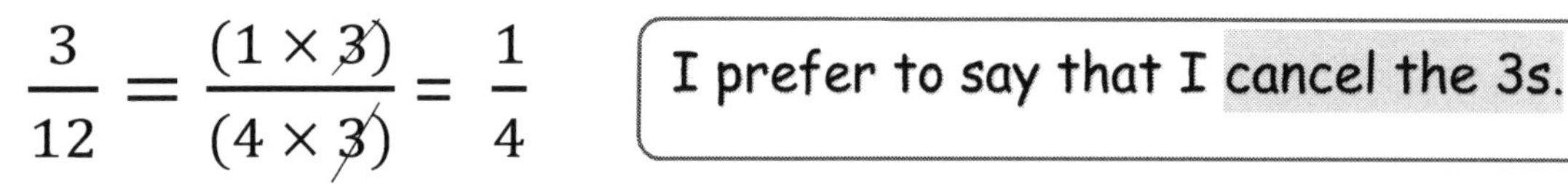

$$\frac{3}{12} = \frac{(1 \times \not{3})}{(4 \times \not{3})} = \frac{1}{4}$$

I prefer to say that I cancel the 3s.

So, you reduce the fraction to make it look as simple as possible. What is the GREATEST number both 2 and 4 can be divided into? Right, they both can be divided by 2 (the ***greatest common factor***): $\frac{2}{4} = \frac{(2 \div 2)}{(4 \div 2)} = \frac{1}{2}$.

*To write an **equivalent fraction in lower terms**, divide both the numerator and denominator by the same number - **common factor (2)**.*

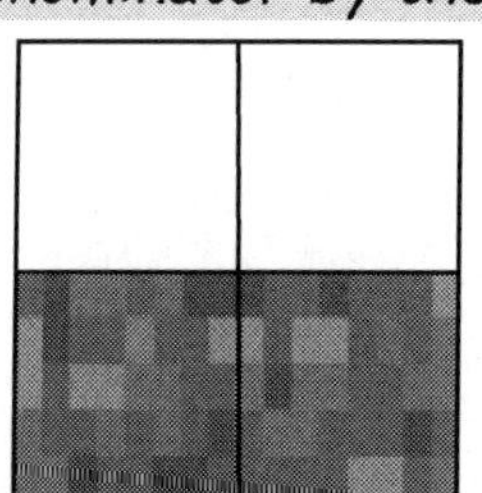

$$\frac{2}{4} = \frac{1}{2}$$

$$\frac{(2 \div 2)}{(4 \div 2)} = \frac{1}{2}$$

I can divide both the numerator and the denominator by 2, 3, 4, 5, 6, and etc. I need to reduce $\frac{18}{42}$ to its lowest terms: $\frac{(18 \div 2)}{(42 \div 2)} = \frac{(9 \div 3)}{(21 \div 3)} = \frac{3}{7}$.

I don't change the VALUE of the fraction.

$$\frac{(16 \div 2)}{(40 \div 2)} = \frac{(8 \div 2)}{(20 \div 2)} = \frac{(4 \div 2)}{(10 \div 2)} = \frac{2}{5}$$

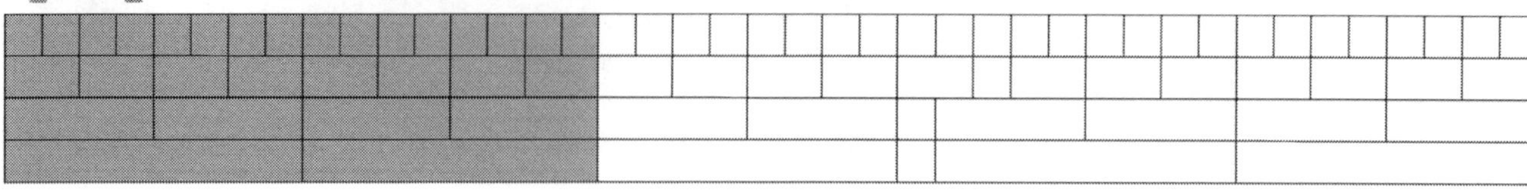

1. Write each fraction in its simplest form. Write the missing numbers.

$\frac{(3 \div 3)}{(6 \div 3)} = \frac{}{}$

The Greatest Common factor = 3

$\frac{(3 \div \ )}{(12 \div \ )} = \frac{}{}$

The Greatest Common factor = 3

$\frac{(2 \div \ )}{(14 \div \ )} = \frac{}{}$

The Greatest Common factor = ____

$\frac{(2 \div \ )}{(4 \div \ )} = \frac{}{}$

The Greatest Common factor = ____

$\frac{(5 \div \ )}{(20 \div \ )} = \frac{}{}$

The Greatest Common factor = ____

$\frac{(5 \div \ )}{(35 \div \ )} = \frac{}{}$

The Greatest Common factor = ____

$\frac{(4 \div \ )}{(24 \div \ )} = \frac{}{}$

The Greatest Common factor = ____

$\frac{(4 \div \ )}{(32 \div \ )} = \frac{}{}$

The Greatest Common factor = ____

$\frac{(6 \div \ )}{(18 \div \ )} = \frac{}{}$

The Greatest Common factor = ____

$\frac{(6 \div \ )}{(42 \div \ )} = \frac{}{}$

The Greatest Common factor = ____

$\frac{(7 \div \ )}{(14 \div \ )} = \frac{}{}$

The Greatest Common factor = ____

$\frac{(7 \div \ )}{(28 \div \ )} = \frac{}{}$

The Greatest Common factor = ____

$\frac{(8 \div \ )}{(48 \div \ )} = \frac{}{}$

The Greatest Common factor = ____

$\frac{(8 \div \ )}{(32 \div \ )} = \frac{}{}$

The Greatest Common factor = ____

2. Compare fractions using ">", "<", or "=".

$\frac{3}{8} \ \dots \ \frac{5}{8}$ $\qquad \frac{4}{12} \ \dots \ \frac{9}{12}$

$\frac{5}{7} \ \dots \ \frac{2}{7}$ $\qquad \frac{3}{9} \ \dots \ \frac{7}{9}$

1. <u>Reduce</u> each fraction to its lowest terms and <u>color</u> the fraction of each shape. <u>Write</u> the missing numbers.

Blue *Green* *Red*

$$\frac{16}{28} = \frac{(16 \div 2)}{(28 \div 2)} = \frac{(\quad \div 2)}{(\quad \div 2)} = \frac{\quad}{\quad}$$

The Greatest common factor is the greatest factor that divides both the numerator and the denominator evenly into a number, and there is no remainder.

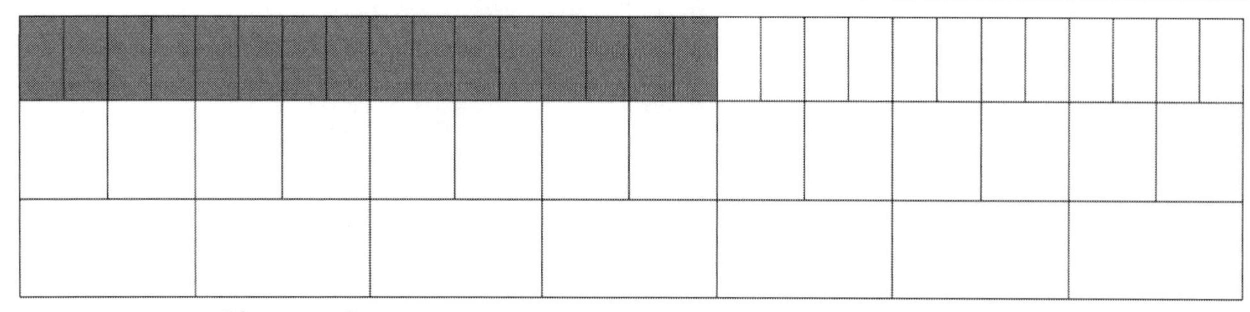

Blue *Green*

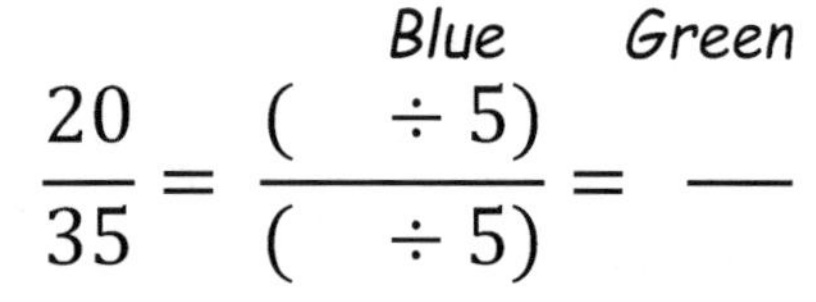

$$\frac{20}{35} = \frac{(\quad \div 5)}{(\quad \div 5)} = \frac{\quad}{\quad}$$

The common factor and the greatest common factor is 5.

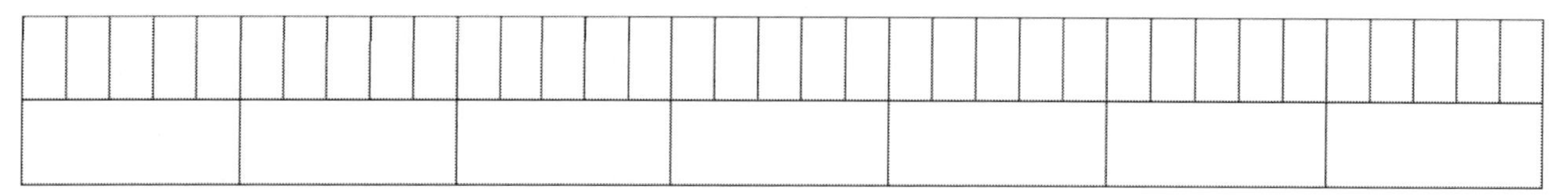

Blue *Green*

$$\frac{14}{21} = \frac{(\quad \div 7)}{(\quad \div 7)} = \frac{\quad}{\quad}$$

The common factor and the greatest common factor is 7.

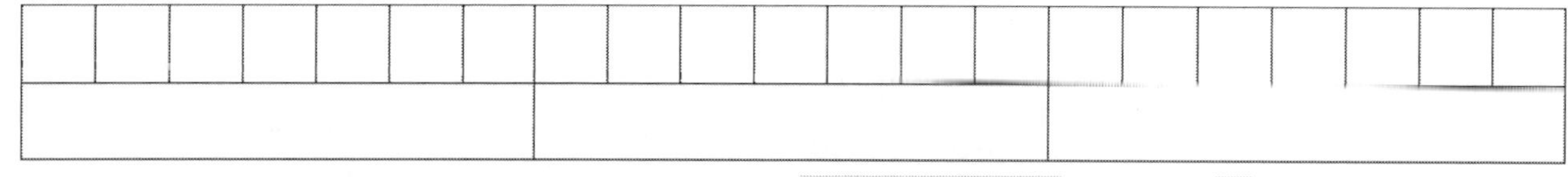

Blue *Green* *Red*

$$\frac{36}{40} = \frac{(\quad \div 2)}{(\quad \div 2)} = \frac{(\quad \div 2)}{(\quad \div 2)} = \frac{\quad}{\quad}$$

Factors of 36: *1, 2, 3, 4, 6, 9, 12, 18, 36.*

Factors of 40: *1, 2, 4, 5, 8, 10, 20, 40.*

*The **greatest common factor (GCF)** is 4.*

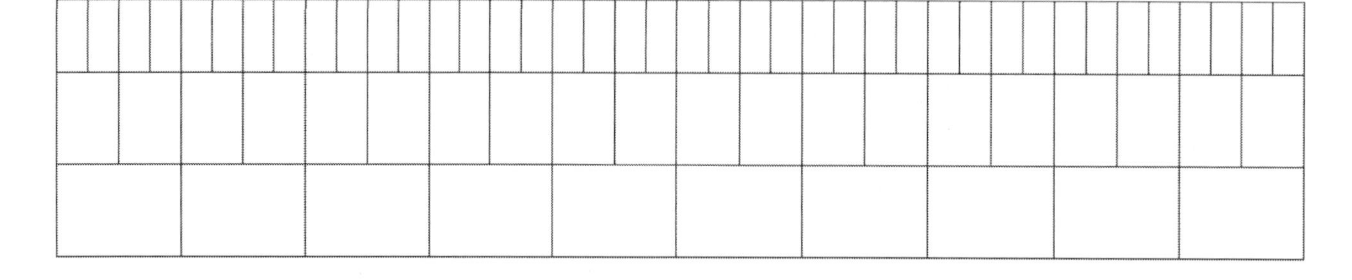

1. Reduce each fraction to its simplest form and color the fraction of each shape. Write the missing numbers.

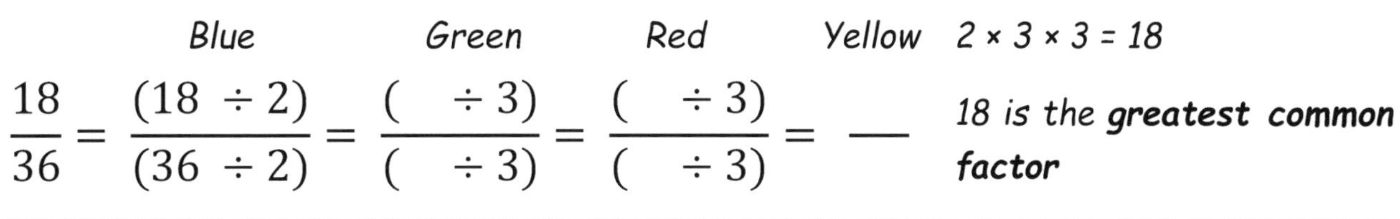

Blue *Green* *Red* *Yellow* *2 × 3 × 3 = 18*

$$\frac{18}{36} = \frac{(18\ \div 2)}{(36\ \div 2)} = \frac{(\quad \div 3)}{(\quad \div 3)} = \frac{(\quad \div 3)}{(\quad \div 3)} = \frac{\quad}{\quad}$$

*18 is the **greatest common factor***

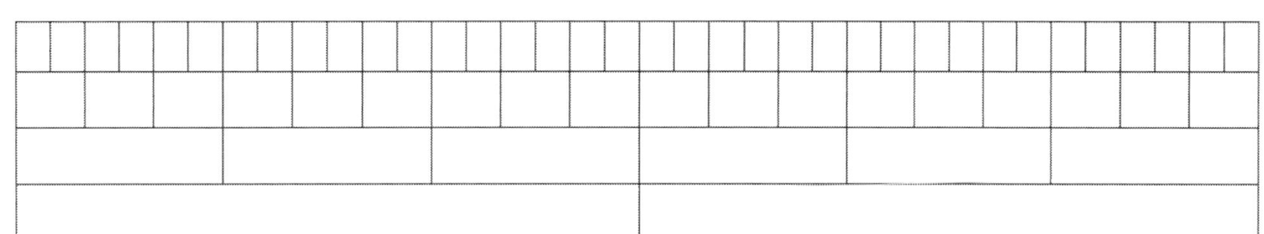

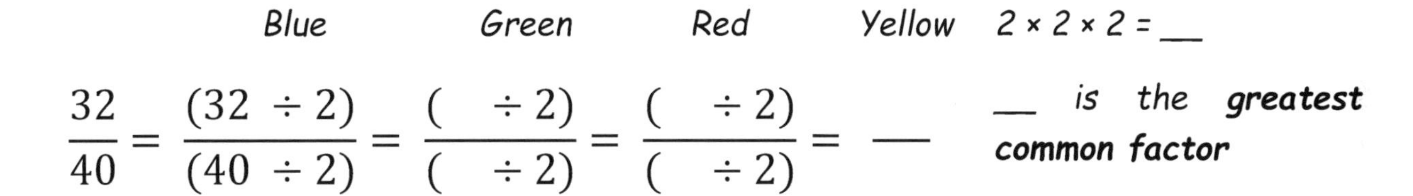

Blue *Green* *Red* *Yellow* *2 × 2 × 2 = __*

$$\frac{32}{40} = \frac{(32\ \div 2)}{(40\ \div 2)} = \frac{(\quad \div 2)}{(\quad \div 2)} = \frac{(\quad \div 2)}{(\quad \div 2)} = \frac{\quad}{\quad}$$

*__ is the **greatest common factor***

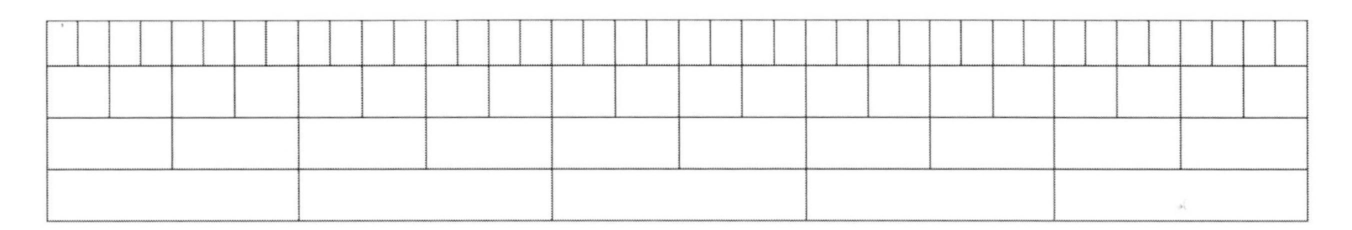

Blue *Green* *Red*

$$\frac{28}{36} = \frac{(28\ \div 2)}{(36\ \div 2)} = \frac{(\quad \div 2)}{(\quad \div 2)} = \frac{\quad}{\quad}$$

2 × 2 = __

*__ is the **greatest common factor***

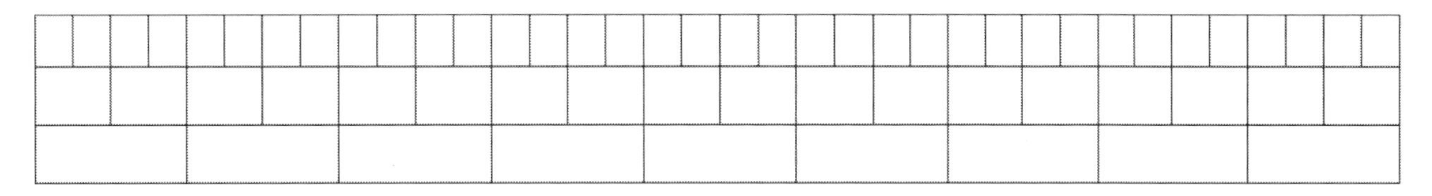

Blue *Yellow*

$$\frac{15}{18} = \frac{(15\ \div\ )}{(18\ \div\ )} = \frac{\quad}{\quad}$$

*__ is the **greatest common factor***

1. Reduce each fraction to its simplest form. Write the missing numbers.

$$\frac{32}{48} = \frac{(32 \div 2)}{(48 \div 2)} = \frac{(\quad \div 2)}{(\quad \div 2)} = \frac{(\quad \div 2)}{(\quad \div 2)} = \frac{(\quad \div 2)}{(\quad \div 2)} = \frac{}{}$$

$$\frac{(32 \div \quad)}{(48 \div \quad)} = \frac{}{}$$

__×__×__×__=______ is the ***greatest common factor***

$$\frac{12}{28} = \frac{(\quad \div 2)}{(\quad \div 2)} = \frac{(\quad \div 2)}{(\quad \div 2)} = \frac{}{}$$

$$\frac{(12 \div \quad)}{(28 \div \quad)} = \frac{}{}$$

________________ is the ***greatest common factor***

$$\frac{36}{48} = \frac{(\quad \div 2)}{(\quad \div 2)} = \frac{(\quad \div 2)}{(\quad \div 2)} = \frac{(\quad \div 3)}{(\quad \div 3)} = \frac{}{}$$

$$\frac{(36 \div \quad)}{(48 \div \quad)} = \frac{}{}$$

________________ is the ***greatest common factor***

$$\frac{50}{75} = \frac{(\quad \div 5)}{(\quad \div 5)} = \frac{(\quad \div 5)}{(\quad \div 5)} = \frac{}{}$$

$$\frac{(50 \div \quad)}{(75 \div \quad)} = \frac{}{}$$

________________ is the ***greatest common factor***

2. Write the fraction name under each picture. Compare the fractions.

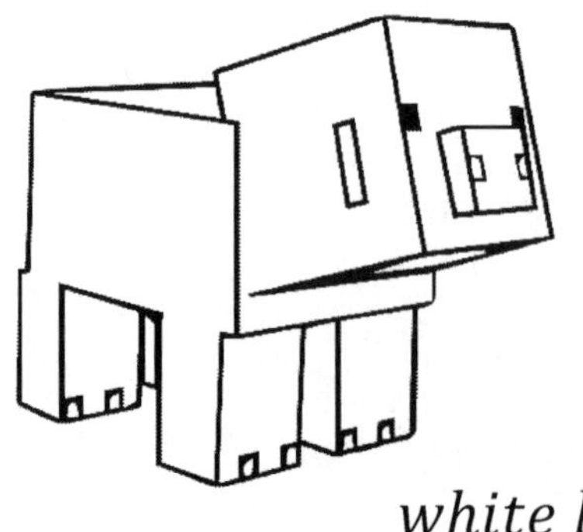

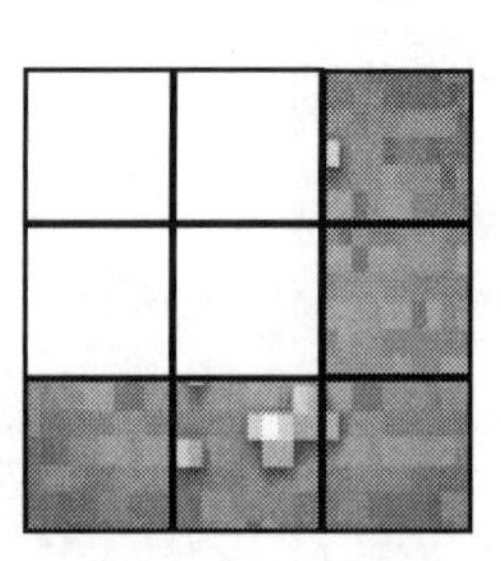

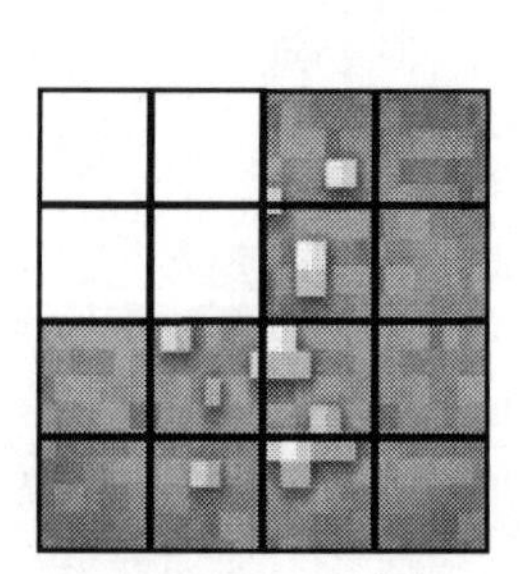

$$\frac{white\ blocks}{blocks\ in\ total} \qquad \frac{\dots}{\dots} \qquad \dots \qquad \frac{\dots}{\dots}$$

1. Reduce each fraction to its simplest form. Write the missing numbers.

$$\frac{56}{72} = \frac{(\quad \div 2)}{(\quad \div \quad)} = \frac{(\quad \div \quad)}{(\quad \div 2)} = \frac{(\quad \div 2)}{(\quad \div \quad)} = \frac{}{} \qquad \frac{(56 \div \quad)}{(72 \div \quad)} = \frac{}{}$$

__________________ *is the **greatest common factor***

$$\frac{36}{81} = \frac{(\quad \div 3)}{(\quad \div \quad)} = \frac{(\quad \div 3)}{(\quad \div \quad)} = \frac{}{} \qquad \frac{(36 \div \quad)}{(81 \div \quad)} = \frac{}{}$$

__________________ *is the **greatest common factor***

$$\frac{60}{96} = \frac{(\quad \div 2)}{(\quad \div \quad)} = \frac{(\quad \div 2)}{(\quad \div \quad)} = \frac{(\quad \div 3)}{(\quad \div \quad)} = \frac{}{} \qquad \frac{(60 \div \quad)}{(96 \div \quad)} = \frac{}{}$$

__________________ *is the **greatest common factor***

$$\frac{48}{96} = \frac{(\quad \div 2)}{(\quad \div \quad)} = \frac{(\quad \div 2)}{(\quad \div \quad)} = \frac{(\quad \div 2)}{(\quad \div \quad)} = \frac{(\quad \div 2)}{(\quad \div \quad)} = \frac{(\quad \div 3)}{(\quad \div \quad)} = \frac{}{}$$

$$\frac{(48 \div \quad)}{(96 \div \quad)} = \frac{}{}$$

__________________ *is the **greatest common factor***

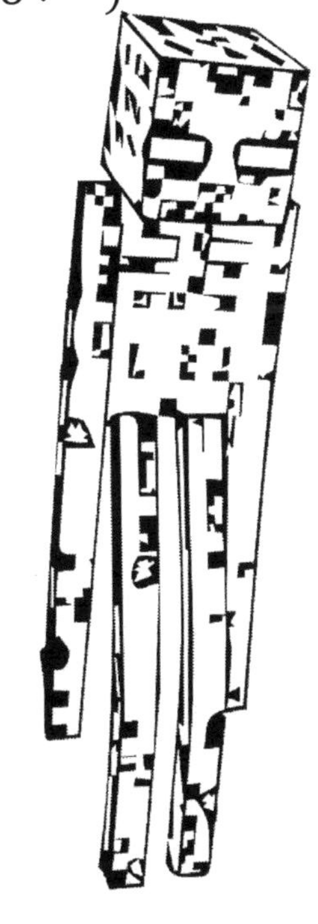

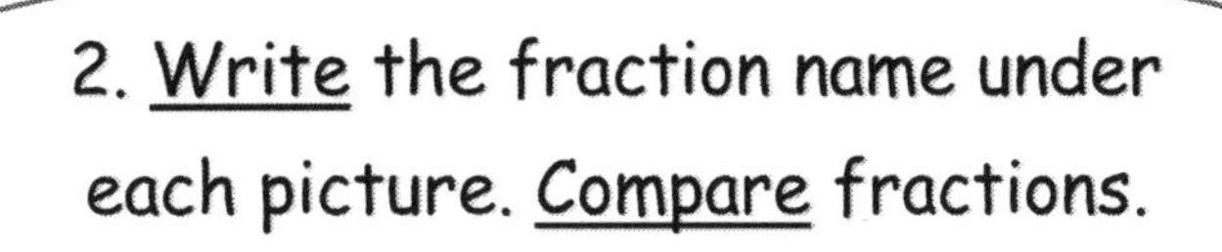
2. Write the fraction name under each picture. Compare fractions.

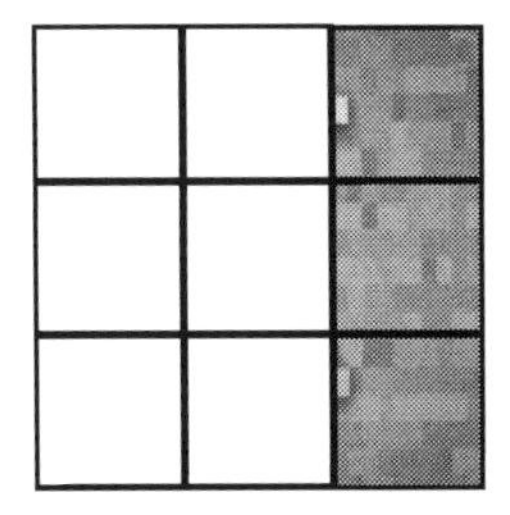

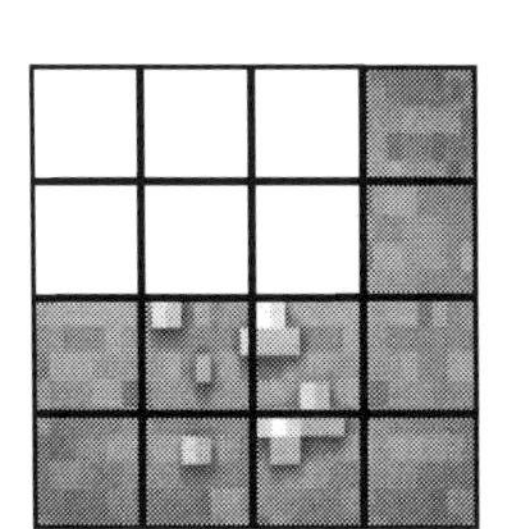

$\frac{}{}$ $\frac{}{}$ $\frac{}{}$

1. Reduce each fraction to its simplest form. Find the greatest common factor (GCF).

$\frac{(20 \div \ )}{(28 \div \ )} = \text{—}$

GCF = ________

$\frac{(18 \div \ )}{(27 \div \ )} = \text{—}$

GCF = ________

$\frac{(12 \div \ )}{(42 \div \ )} = \text{—}$

GCF = ________

$\frac{(6 \div \ )}{(63 \div \ )} = \text{—}$

GCF = ________

$\frac{(15 \div \ )}{(35 \div \ )} = \text{—}$

GCF = ________

$\frac{(49 \div \ )}{(77 \div \ )} = \text{—}$

GCF = ________

$\frac{(24 \div \ )}{(56 \div \ )} = \text{—}$

GCF = ________

$\frac{(16 \div \ )}{(36 \div \ )} = \text{—}$

GCF = ________

$\frac{(20 \div \ )}{(32 \div \ )} = \text{—}$

GCF = ________

$\frac{(10 \div \ )}{(34 \div \ )} = \text{—}$

GCF = ________

$\frac{(21 \div \ )}{(60 \div \ )} = \text{—}$

GCF = ________

$\frac{(9 \div \ )}{(24 \div \ )} = \text{—}$

GCF = ________

$\frac{(25 \div \ )}{(45 \div \ )} = \text{—}$

GCF = ________

$\frac{(40 \div \ )}{(64 \div \ )} = \text{—}$

GCF = ________

$\frac{(35 \div \ )}{(84 \div \ )} = \text{—}$

GCF = ________

2. Color the fraction of each shape. Compare fractions.

$\frac{8}{9}$ ___ $\frac{8}{12}$

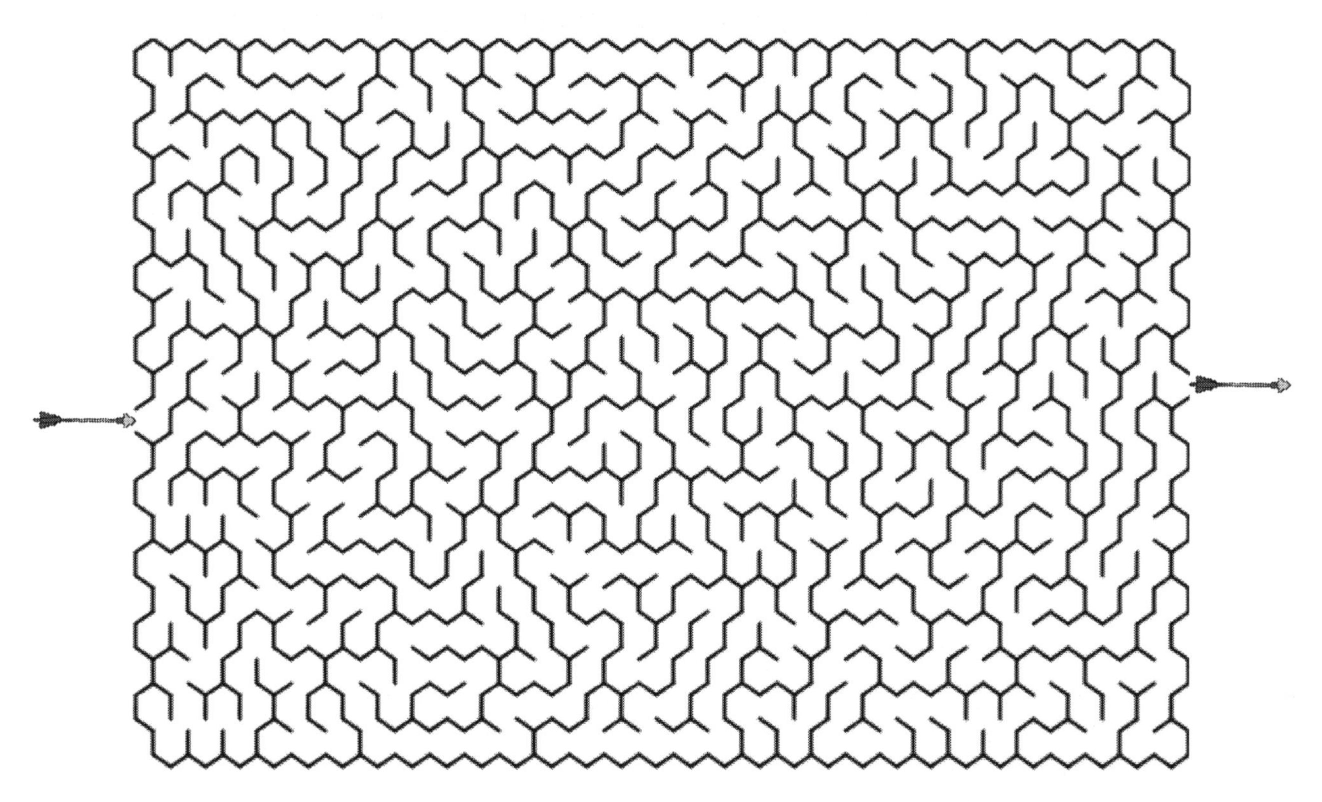

1. Find and circle or cross out the words to find out more about Minecraft.

O	W	S	N	Q	J	Y	N	Z	W	C	W	T
W	U	R	N	A	J	O	G	R	F	O	N	L
S	Q	T	Y	E	I	H	A	N	O	X	I	A
G	N	B	L	S	P	A	X	D	A	J	A	T
Z	H	O	N	I	K	S	S	Z	R	K	T	T
M	M	A	R	D	N	T	M	A	T	W	N	R
Z	M	V	R	D	A	E	I	F	M	T	U	A
Y	T	X	S	I	L	X	S	X	S	U	O	C
E	M	A	R	B	Y	U	U	H	K	E	F	T
T	O	R	C	H	E	S	A	C	A	A	Z	I
B	C	P	Y	H	A	A	I	C	H	P	X	V
Y	V	E	L	I	D	G	O	A	F	U	E	E
E	G	A	T	N	A	V	D	A	S	M	G	S

MANSION

WOOD STAIR

ADVANTAGE

FOUNTAIN

ATTRACTIVE

TORCHES

OUTLINE SHAPE

CAULDRONS

1. <u>Write in</u> the missing numbers on the Minecraft factor tree and <u>cross out</u> the common prime factors.

A prime number is a counting number that has exactly two different factors

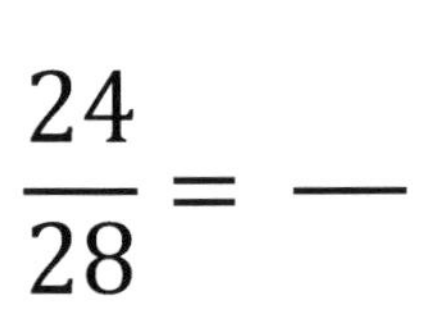

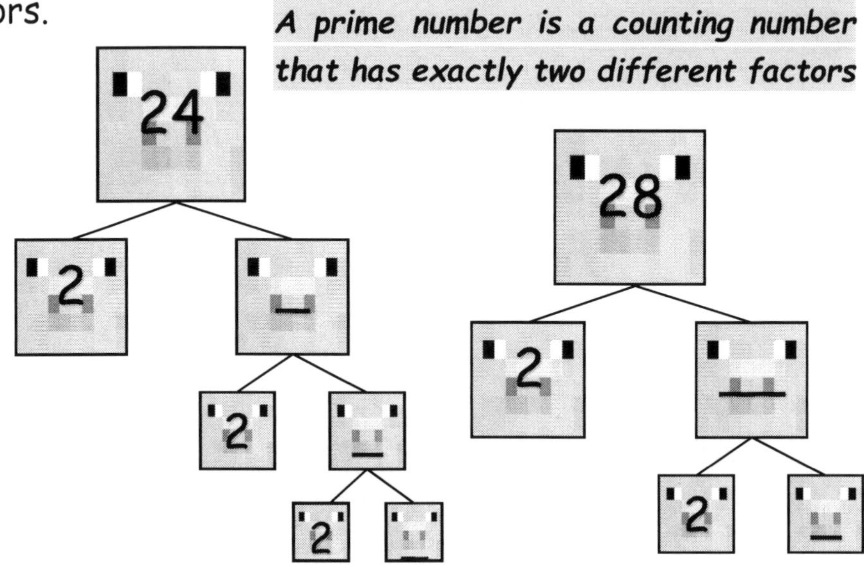

Prime factors for 24:

~~2~~, ~~2~~, 2, ____

Prime factors for 28:

~~2~~, ~~2~~, ____

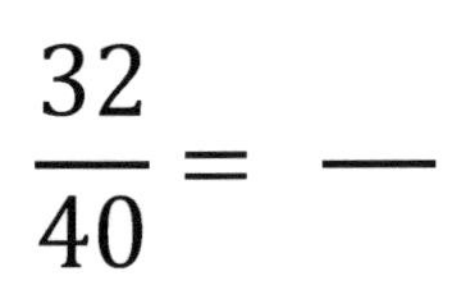

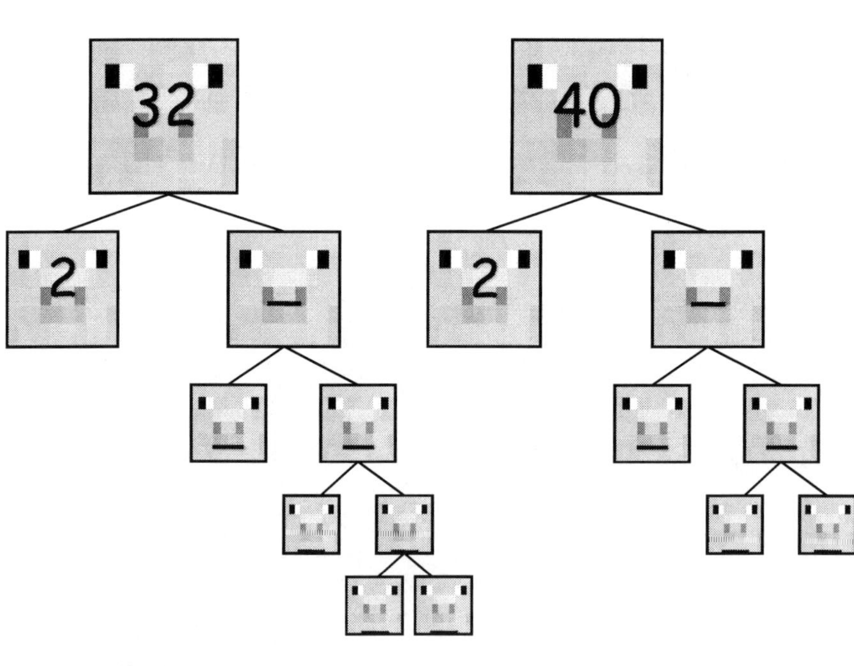

Prime factors for 32:

~~2~~, ____, ____, ____, ____

Prime factors for 40:

~~2~~, ____, ____, ____

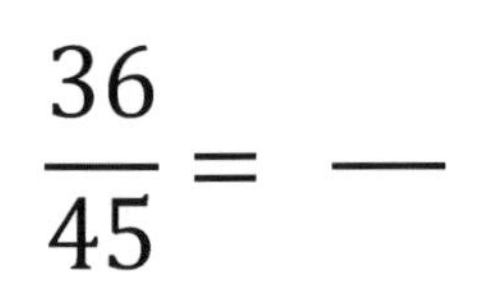

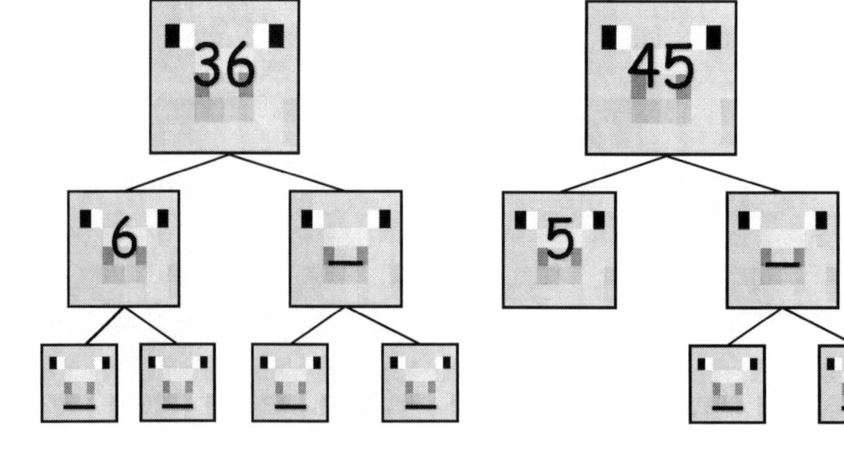

Prime factors for 36:

____, ____, ____, ____

Prime factors for 45:

5, ____, ____

I have another idea! Grumpy, do you like weapons?

Why shouldn't I like them? I love Minecraft!

Great! Imagine, you're fighting with... Greeny, where are you?

Fighting? With Grumpy? Are you mad?

Maybe. Just imagine: you have a sword, a bow, and an axe.

a sword, a bow, an axe
―――――――――――――

← Greeny

Hint: ***To simplify** means to **find both the numerator and the denominator that have no common prime factors** (or weapons).*

I love this idea! So, I have a sword, a bow, an axe, and a trident.

―――――――――――――
a sword, a bow, an axe, and a trident ← Grumpy

***To simplify** means to **divide both the numerator and the denominator by the same number** (or weapon).*

Fine! Now let's compare your weapons and cross out the weapons you BOTH have:

~~A sword~~, ~~a bow~~, and ~~an axe.~~	~~A sword~~, ~~a bow~~, ~~an axe,~~ and a trident.

Hooray! I have a trident! So, guys, cross out the same weapon as your opponent has since you need only a unique weapon to win!

What a disgusting turn of events! Hate weapons! Love fractions! So, I need to find the factors in the numerator and the denominator and then, cross them out.

Factors? Ah, factors are the numbers, when multiplied together, they form a new number. We call that number a product, right?

Aha. For example, what are factors of 15?

15 = 5 × 3. So, 5 and 3 are factors of 15. Now, let's reduce your weapons' fraction to its lowest terms:

$$\frac{\text{a sword, a bow, an axe}}{\text{a sword, a bow, an axe, and a trident}} = \frac{\cancel{\text{a sword}}\text{, a bow, an axe}}{\cancel{\text{a sword}}\text{, a bow, an axe, and a trident}} =$$

$$= \frac{\cancel{\text{a bow}}\text{, an axe}}{\cancel{\text{a bow}}\text{, an axe, and a trident}} = \frac{\cancel{\text{an axe}}}{\cancel{\text{an axe}}\text{, and a trident}} = \frac{1}{\text{a trident}}$$

Can I try?

Hint: A prime number is any number that has only two factors, itself and 1. By the way, the number 1 is neither prime nor composite.

$$\frac{15}{35} = \frac{\cancel{5} \times 3}{\cancel{5} \times 7} = \frac{3}{7}$$

Cancel the 5s in both the numerator and the denominator.

$$\frac{16}{24} = \frac{\cancel{2} \times \cancel{2} \times \cancel{2} \times 2}{\cancel{2} \times \cancel{2} \times \cancel{2} \times 3} = \frac{2}{3}$$

Cancel the 2s in both the numerator and the denominator. So, GCF is 8.

1. Reduce each fraction to its simplest form. Cross out the same factors in the numerator and the denominator.

$\frac{18}{28} = \frac{\not{2} \times 3 \times 3}{\not{2} \times 2 \times 7} = \frac{9}{14}$ ← is in simplest form since 9 and 14 have only 1 as a common factor!

$\frac{20}{30} = \frac{2 \times 2 \times 5}{2 \times 3 \times 5} = \frac{}{}$

$\frac{18}{22} = \frac{2 \times 3 \times 3}{2 \times 11} = \frac{}{}$

$\frac{16}{32} = \frac{2 \times 2 \times 2 \times 2}{2 \times 2 \times 2 \times 2 \times 2} = \frac{}{}$

$\frac{25}{40} = \frac{5 \times 5}{2 \times 2 \times 2 \times 5} = \frac{}{}$

$\frac{24}{48} = \frac{2 \times 2 \times 2 \times 3}{2 \times 2 \times 2 \times 2 \times 3} = \frac{}{}$

$\frac{30}{46} = \frac{2 \times 3 \times 5}{2 \times 23} = \frac{}{}$

$\frac{32}{48} = \frac{2 \times 2 \times 2 \times 2 \times 2}{2 \times 2 \times 2 \times 2 \times 3} = \frac{}{}$

$\frac{48}{64} = \frac{2 \times 2 \times 2 \times 2 \times 3}{2 \times 2 \times 2 \times 2 \times 2 \times 2} = \frac{}{}$

$\frac{44}{55} = \frac{2 \times 2 \times 11}{5 \times 11} = \frac{}{}$

A fraction is written in its simplest form (= lowest term) when both the numerator and the denominator have NO common prime factors!

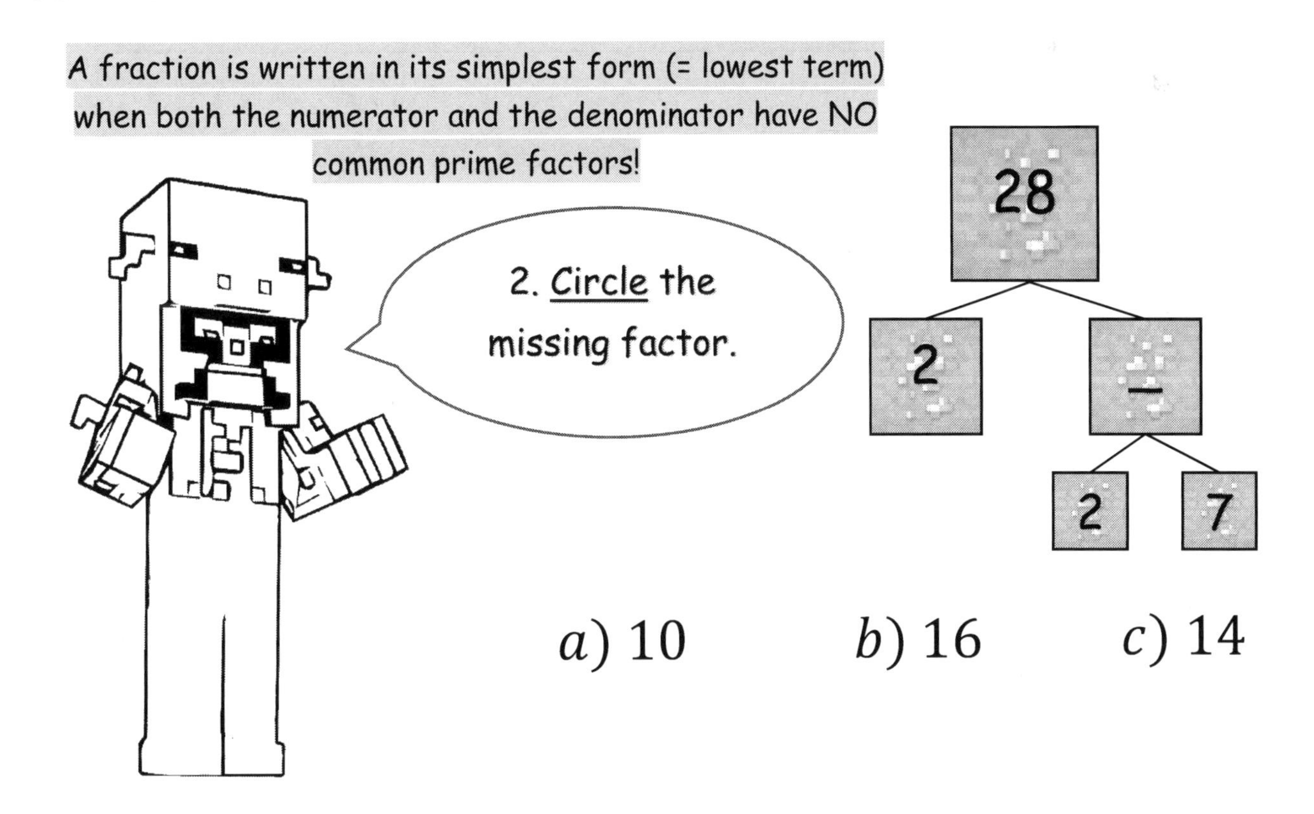

$a)$ 10 $b)$ 16 $c)$ 14

1. Write in the missing numbers on the Minecraft factor tree and cross out the common prime factors.

A composite number is a counting number that has more than two factors

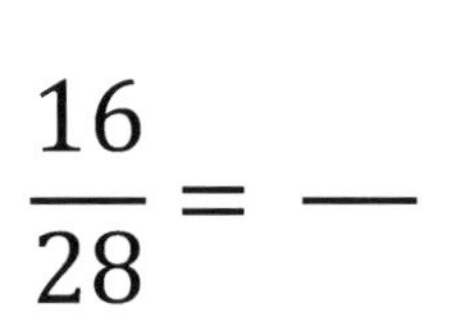

$$\frac{16}{28} = \text{—}$$

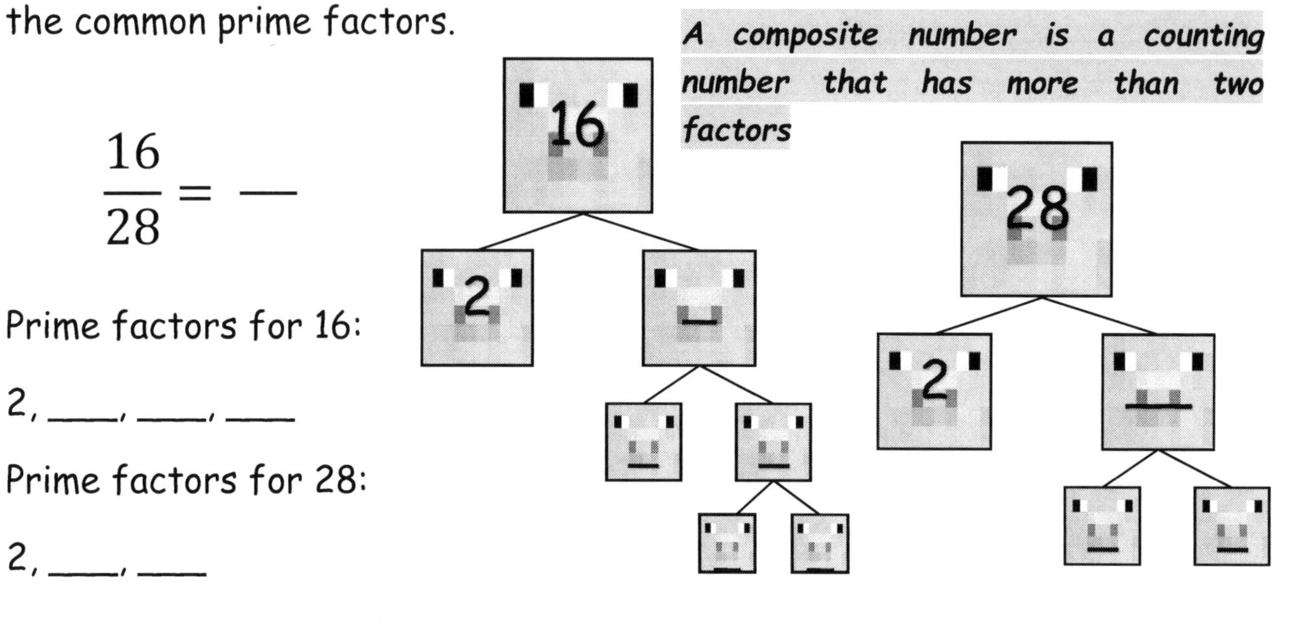

Prime factors for 16:

2, ____, ____, ____

Prime factors for 28:

2, ____, ____

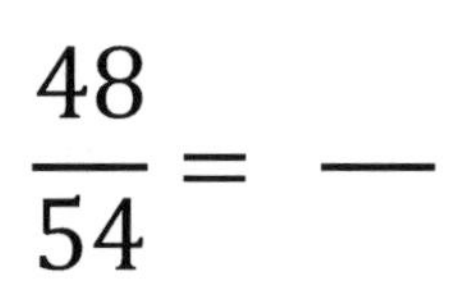

$$\frac{48}{54} = \text{—}$$

48

54

2

Prime factors for 48:

____, ____, ____, ____, ____

Prime factors for 54:

2, ____, ____, ____

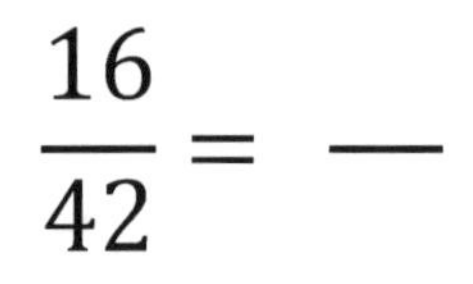

$$\frac{16}{42} = \text{—}$$

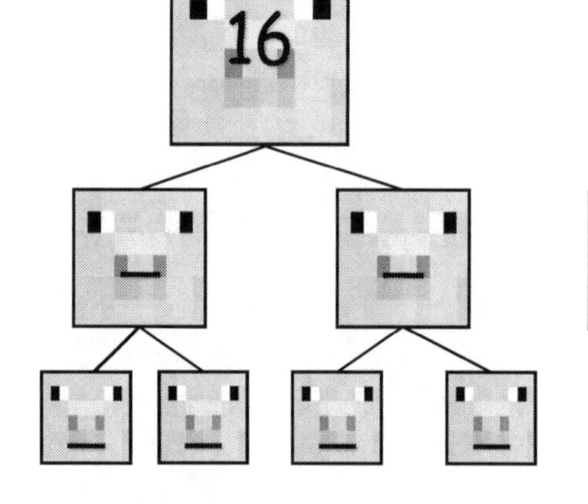

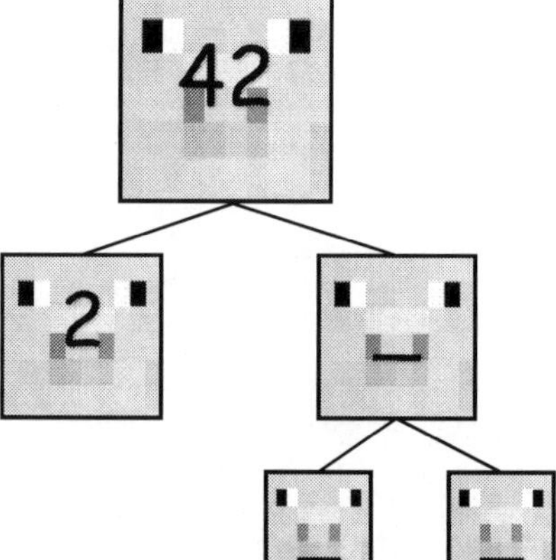

Prime factors for 16:

____, ____, ____, ____

Prime factors for 42:

2, ____, ____

1. Write in the missing numbers on the Minecraft factor tree.

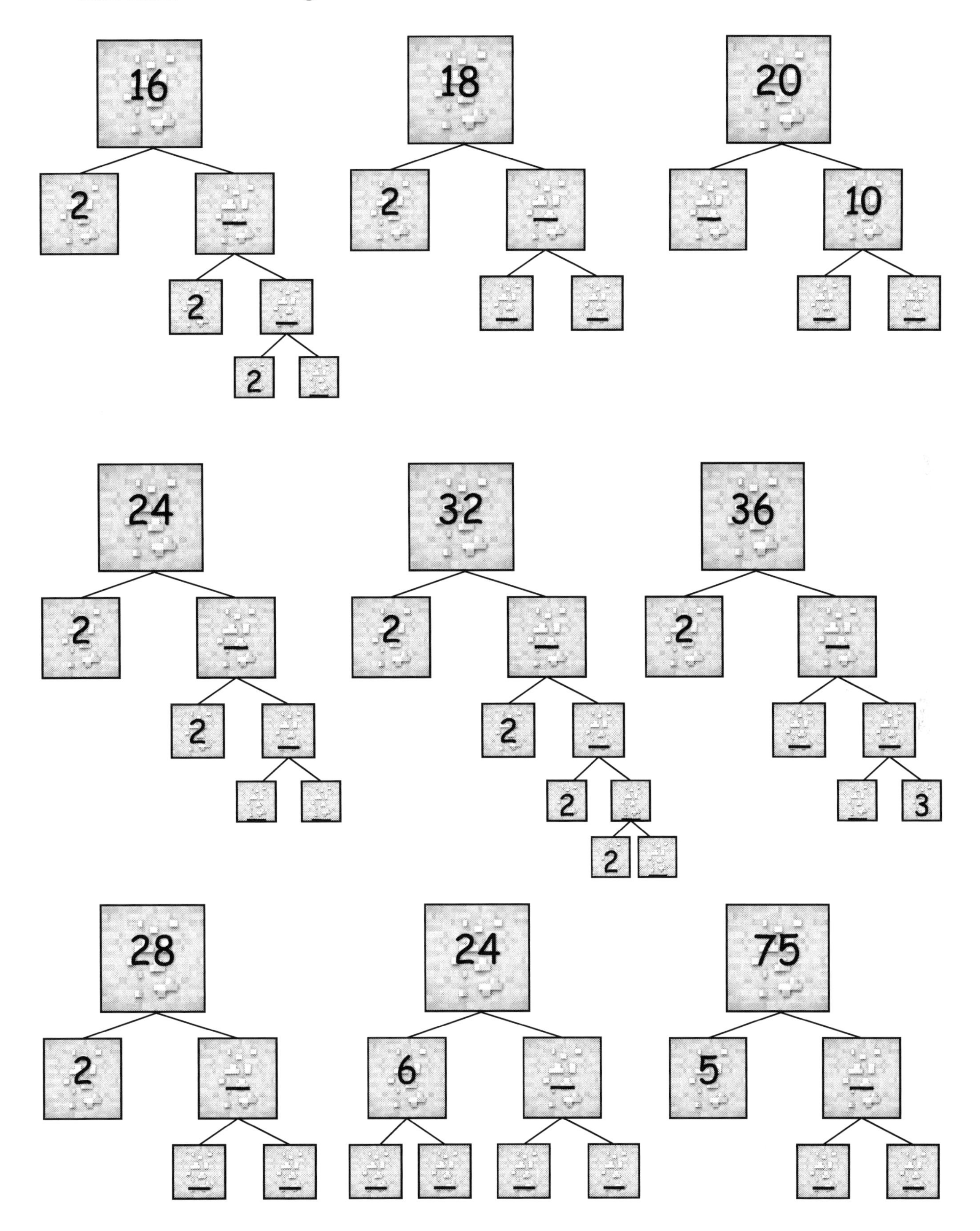

1. Reduce each fraction to simplest form. Cross out the same factors in the numerator and the denominator.

$$\frac{28}{48} = \frac{\quad \times \quad \times \quad}{\times \quad \times \quad \times \quad \times} = \frac{\quad}{\quad} \qquad \frac{36}{54} = \frac{\times \quad \times \quad \times}{\times \quad \times \quad \times} = \frac{\quad}{\quad}$$

$$\frac{52}{64} = \frac{\quad \times \quad \times \quad}{\times \quad \times \quad \times \quad \times \quad \times} = \frac{\quad}{\quad} \qquad \frac{68}{75} = \frac{\times \quad \times \quad \times}{\times \quad \times \quad \times} = \frac{\quad}{\quad}$$

$$\frac{50}{70} = \frac{\times \quad \times \quad \times}{\times \quad \times \quad \times} = \frac{\quad}{\quad} \qquad \frac{72}{80} = \frac{\times \quad \times \quad \times \quad \times}{\times \quad \times \quad \times \quad \times} = \frac{\quad}{\quad}$$

$$\frac{76}{84} = \frac{\times \quad \times \quad \times}{\times \quad \times \quad \times} = \frac{\quad}{\quad} \qquad \frac{48}{88} = \frac{\times \quad \times \quad \times \quad \times}{\times \quad \times \quad \times} = \frac{\quad}{\quad}$$

$$\frac{80}{96} = \frac{\times \quad \times \quad \times \quad \times}{\times \quad \times \quad \times \quad \times \quad \times} = \frac{\quad}{\quad} \qquad \frac{84}{100} = \frac{\times \quad \times \quad \times}{\times \quad \times \quad \times} = \frac{\quad}{\quad}$$

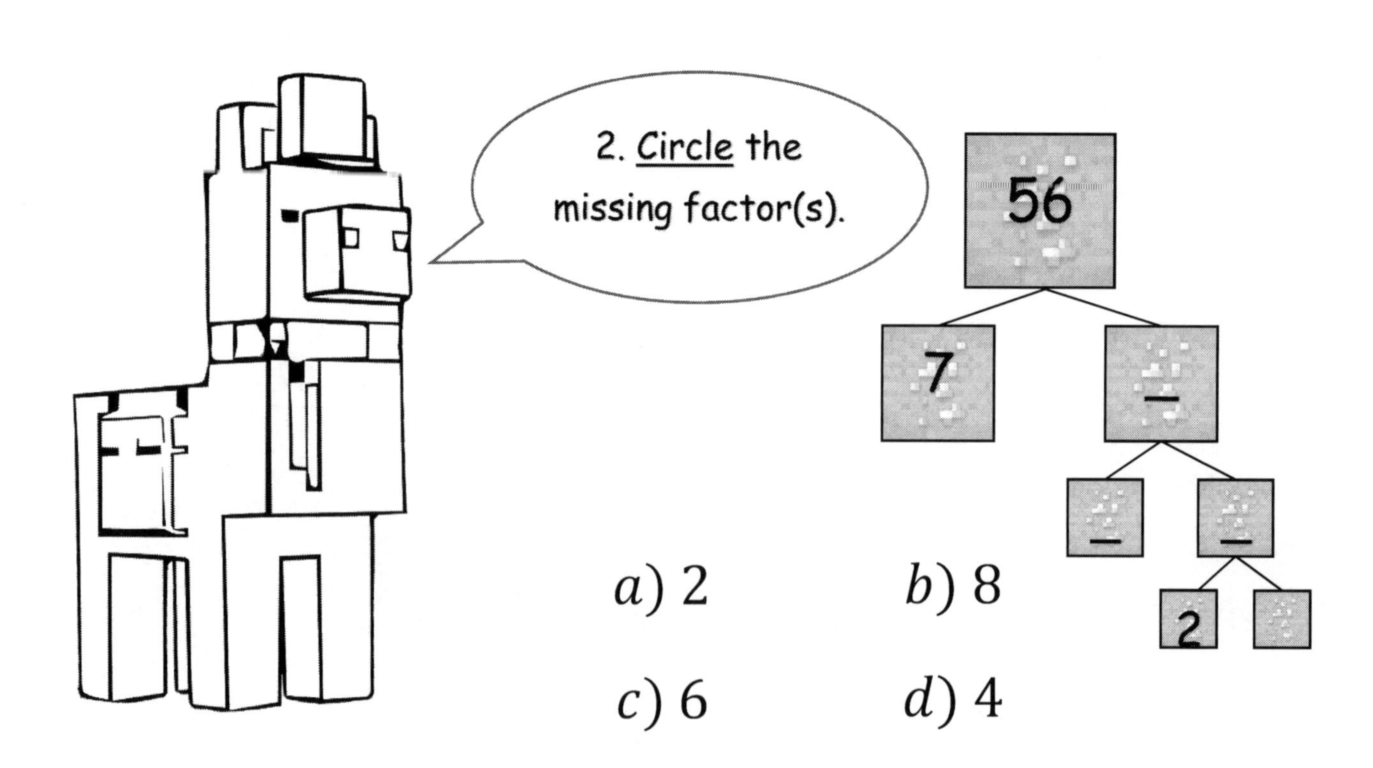

$a)$ 2 $\qquad b)$ 8

$c)$ 6 $\qquad d)$ 4

1. Write in the missing numbers on the Minecraft factor tree.

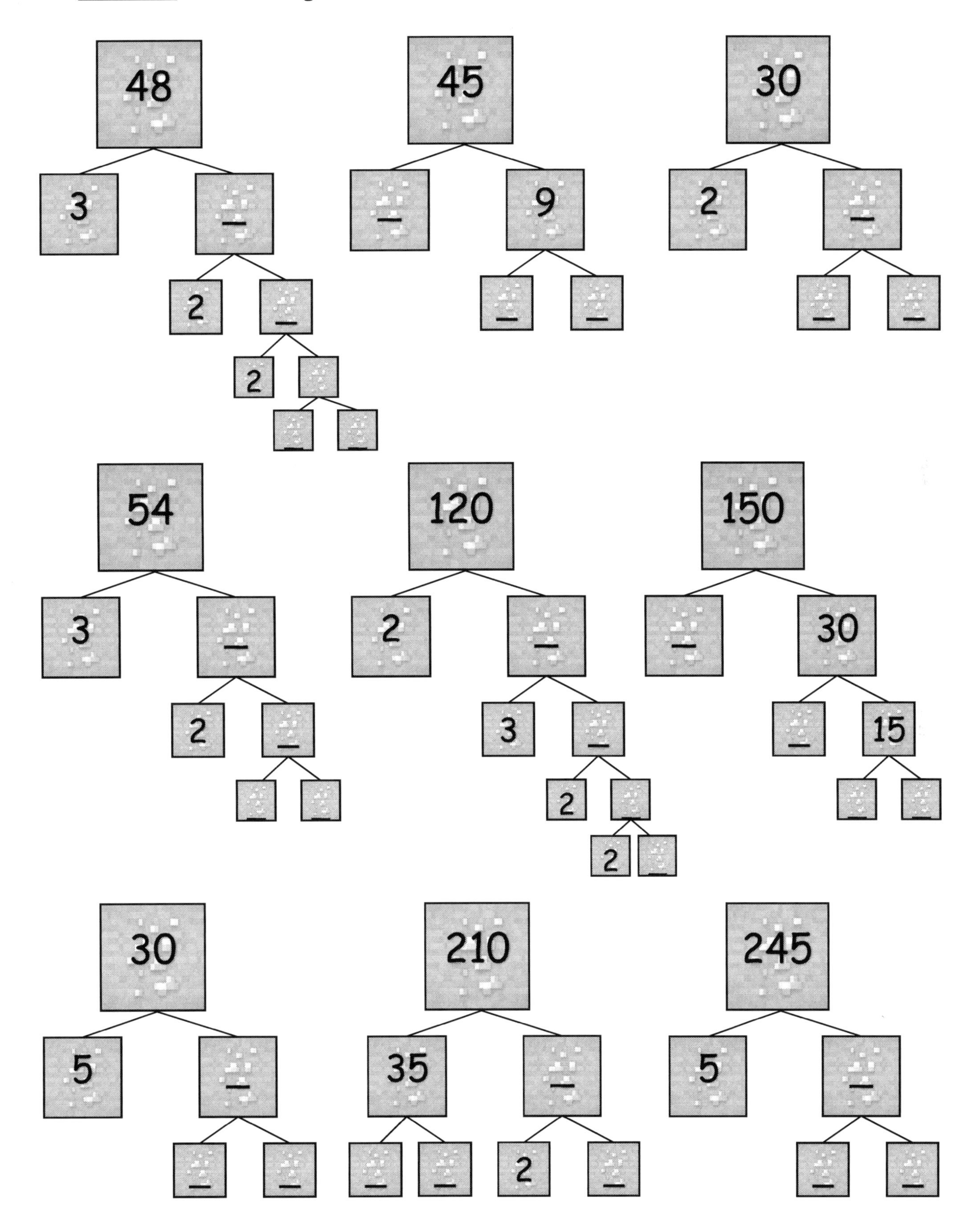

1. Find and circle or cross out the words to find out more about Minecraft.

S	K	F	N	E	Z	J	H	X	S	Q	I	J
X	E	P	L	T	E	M	O	S	E	W	A	U
A	Z	I	Q	A	E	J	G	G	K	G	S	W
T	J	G	T	N	M	I	W	J	U	E	L	E
Y	G	L	E	I	I	M	C	I	R	P	C	O
E	D	O	A	M	R	T	A	N	K	N	D	C
N	I	W	V	U	E	A	A	B	E	J	X	H
Q	S	S	X	L	F	M	L	D	L	H	Q	Q
V	H	T	I	L	E	S	I	I	H	E	A	F
I	E	O	K	I	Z	F	M	M	M	L	G	T
N	Y	N	D	E	N	J	O	F	H	I	H	C
S	Q	E	V	O	Y	C	I	R	V	N	S	A
Z	Y	A	C	O	M	P	L	E	M	E	N	T

FLAMMABLE

SIMILARITIES

GLOWSTONE

ILLUMINATE

AWESOME

CONFIDENCE

USERNAME

COMPLEMENT

1. Compare fractions using "<," ">," or "=."

Help! SOS! Help!!!

Quick, Smarty, help him! I found a chest with diamonds! I need to craft a diamond sword!

Congratulations, Grumpy! Purple-y, look, to compare fractions means to find out which fraction and smaller or larger.

When fractions have the same denominator, the fraction with the greatest (largest) numerator has the greatest value.

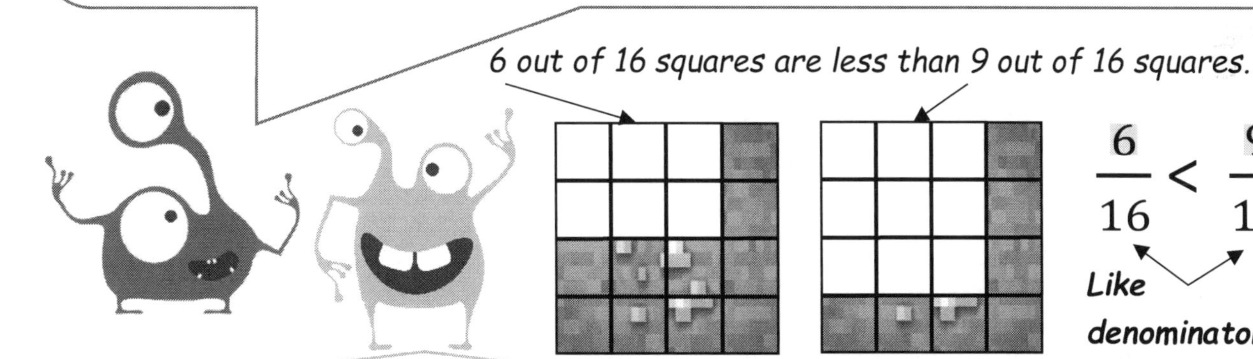

$$\frac{6}{16} < \frac{9}{16}$$

I compare the unlike fractions (= fractions with the different numerators or denominators). Let's start with fractions with like numerators but with unlike denominators:

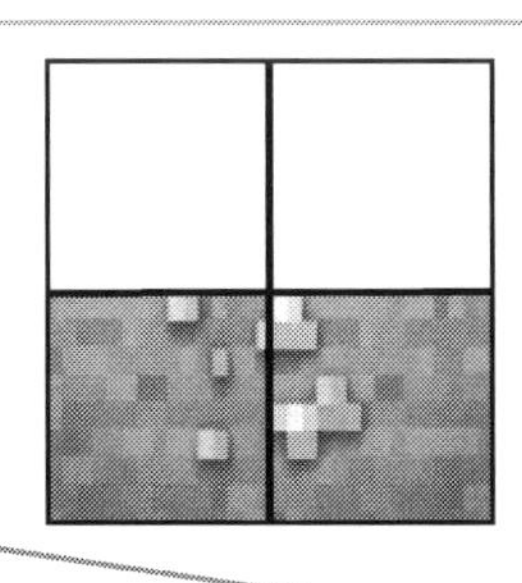

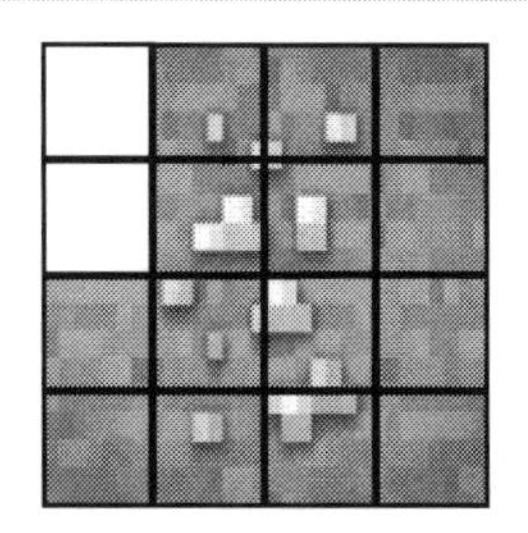

$$\frac{2}{4} \_\_\_ \frac{2}{16}$$

Unlike denominators

I divided my block into 4 huge equal parts and colored 2 parts white. Then, I divided another block into 16 tiny equal parts and colored 2 equal parts white. They look different!

Hm... If the fractions have like numerators, then, the fraction divided into a smaller number of equal parts is bigger.

1 part out of 2 equal parts is bigger than

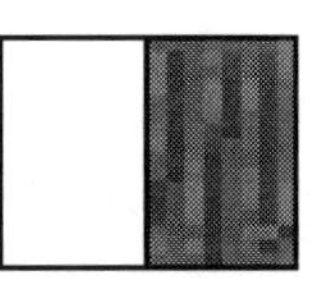

1 part out of 4 equal parts.

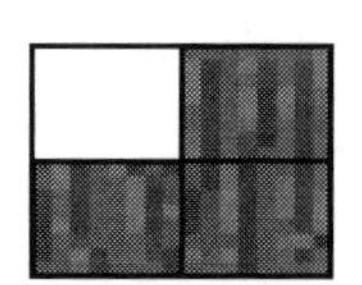

Look, I divided a candy into 4 equal parts and I ate 1 part out of the four equal parts. Or I divided a candy into 2 equal parts and I ate 1 part out of the two equal parts.

When did I eat more? Of course, when the candy was divided into 2 equal parts.

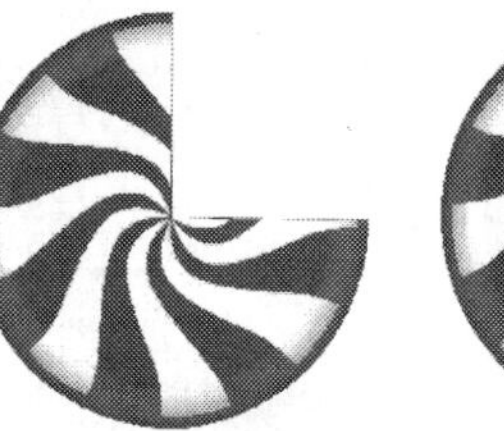

$$\frac{1}{4} < \frac{1}{2}$$

1 out of 4 equal parts is smaller than 1 out of 2 equal parts.

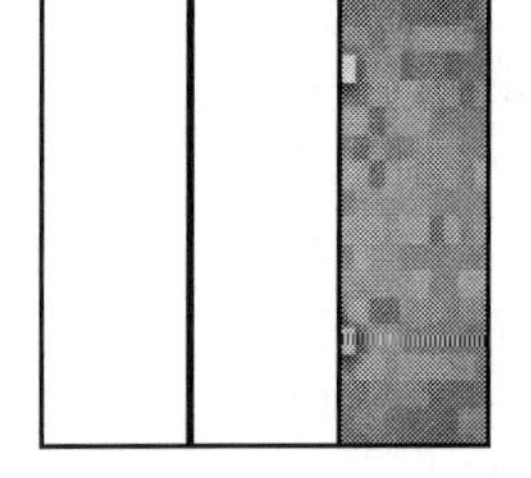

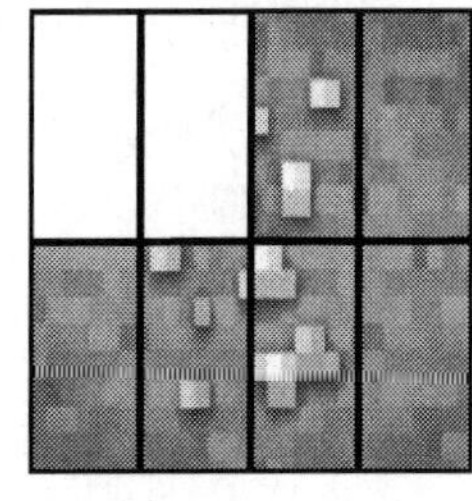

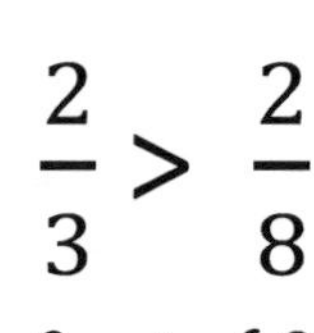

$$\frac{2}{3} > \frac{2}{8}$$

2 out of 3 equal parts are bigger than 2 out of 8 equal parts.

I divided my block into 3 huge equal parts and colored 2 parts white.

Then, I divided another block into 8 small equal parts and colored 2 equal parts white.

Which fraction is largest?

The larger fraction is the one with the smaller denominator (that means the whole is divided into a smaller number of equal parts).

1. <u>Compare</u> fractions using "<," ">," or "=." <u>Color</u> the fraction of each shape. <u>Cross out</u> the wrong word. The first one is done for you.

$\frac{2}{6} < \frac{2}{4}$

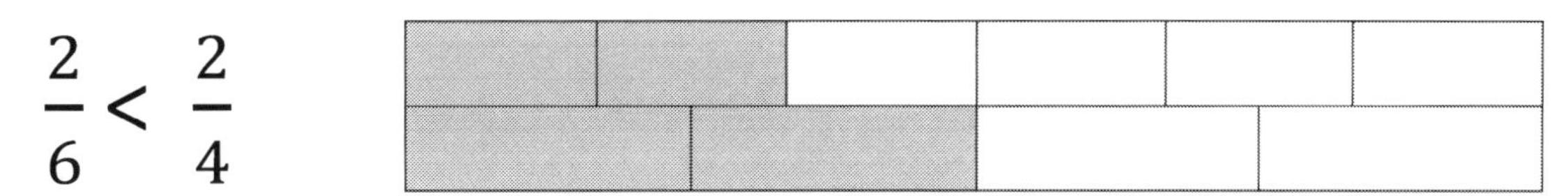

2 out of 6 equal parts are ~~bigger~~/smaller than 2 out of 4 equal parts.

$\frac{5}{8}$ ___ $\frac{5}{11}$

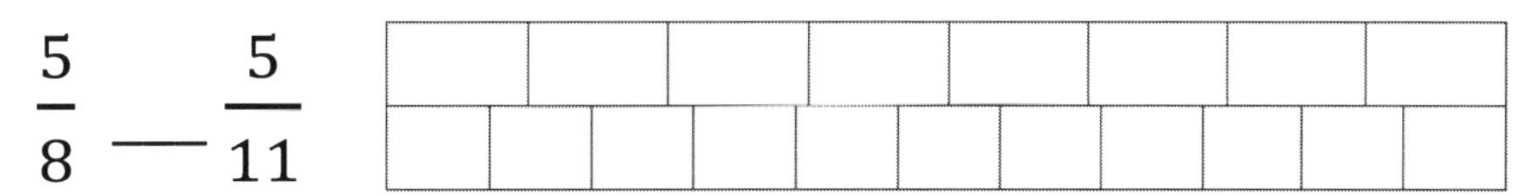

5 out of 8 equal parts are bigger/smaller than 5 out of 11 equal parts.

$\frac{3}{5}$ ___ $\frac{3}{3}$

3 out of 5 equal parts are bigger/smaller than 3 out of 3 equal parts.

$\frac{4}{8}$ ___ $\frac{4}{12}$

4 out of 8 equal parts are bigger/smaller than 4 out of 12 equal parts.

2. <u>How many fourths</u> make a half? <u>How many sixths</u> make a half? <u>Write</u> the missing numbers. <u>Draw</u> and <u>color</u> the fraction of each shape.

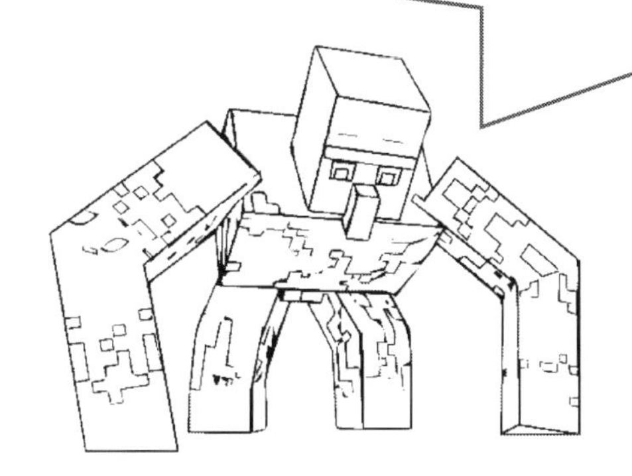

$\frac{1}{2} = \frac{\_\_}{4}$

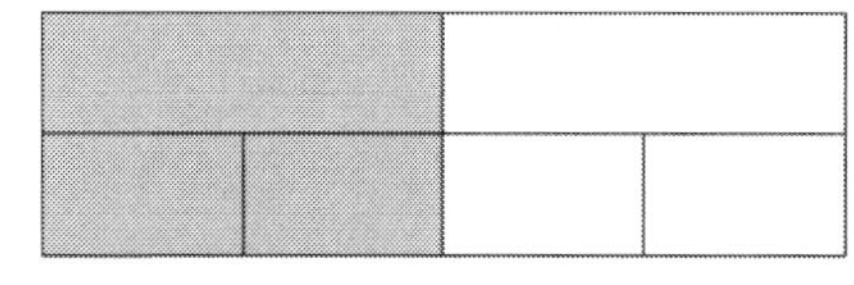

$\frac{1}{2} = \frac{\_\_}{\_\_}$

1. Compare fractions using "<," ">," or "=." Color the fraction of each shape. Cross out the wrong word.

$\frac{5}{7}$ ___ $\frac{5}{14}$

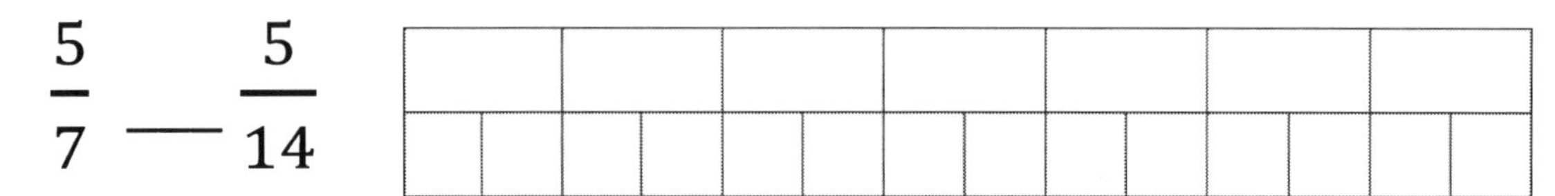

5 out of 7 equal parts are bigger/smaller than 5 out of 14 equal parts.

$\frac{5}{16}$ ___ $\frac{5}{10}$

5 out of 16 equal parts are bigger/smaller than 5 out of 10 equal parts.

$\frac{6}{6}$ ___ $\frac{6}{10}$

6 out of 6 equal parts are bigger/smaller than 6 out of 10 equal parts.

$\frac{8}{20}$ ___ $\frac{8}{14}$

8 out of 20 equal parts are bigger/smaller than 8 out of 14 equal parts.

2. How many tenths make a half? How many twelfths make a quarter? Write the missing numbers. Draw and color the fraction of each shape.

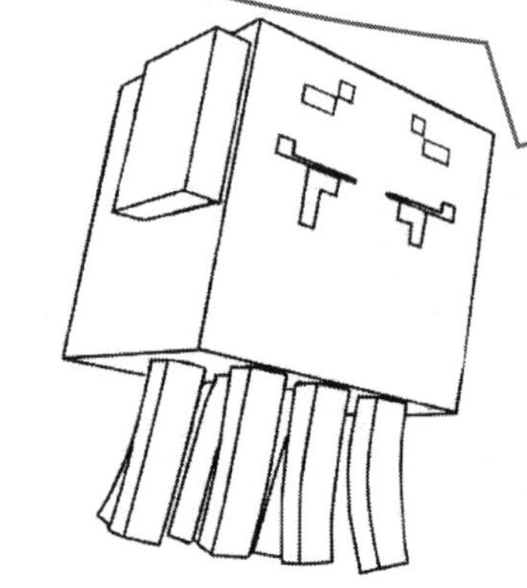

$\frac{1}{2} = \frac{}{}$

$\frac{1}{4} = \frac{}{}$

Steve meets $\frac{1}{3}$ of as many zombies as Alex. If Alex meets 12 zombies, <u>how many zombies</u> does Steve meet?

You need to find parts of a number. Remember that the word OF in fractions means MULTIPLY. So, Steve meets $\frac{1}{3}$ of 12 zombies in Math means $\frac{1}{3} \times 12$.

I have an idea! I need to divide my rectangle into 12 equal parts and color $\frac{1}{3}$ of them brown.

Step 1: I find an equivalent fraction with the denominator 12: $\frac{1}{3} = \frac{4}{12}$ (remember, equivalent fractions are fractions that represent the same value).

Step 2: The numerator shows how many parts I need to color brown. I need to color 4 parts brown since 4 parts out of 12 equal parts is the same as 1 part out of 3 equal parts.

$$\frac{1}{3} = \frac{4}{12}$$

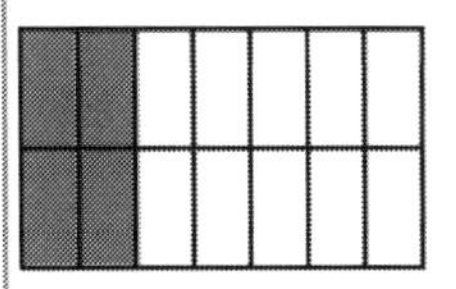

Another algorithm:

Step 1: OF means multiply: $\frac{1}{3} \times 12$

Step 2: write 12 as $\frac{12}{1}$: $\frac{1}{3} \times \frac{12}{1}$

Step 3: Multiply numerators and denominators, cancel the 3s:

$$\frac{1}{3} \times 12 = \frac{1}{3} \times \frac{12}{1} = \frac{1 \times 12}{3 \times 1} = \frac{1 \times \cancel{12}^{4}}{\cancel{3} \times 1} = \frac{1 \times 4}{1 \times 1} = 4$$

Aha... I have another idea! All the zombies are divided equally into 3 equal groups. And I meet only $\frac{1}{3}$ of them.

The question is: If there are 12 zombies altogether, how many zombies are in one of the 3 equal parts?

Step 1: find how many zombies are in one of the three equal parts: $12 \div 3 = 4$

(4 zombies are in each of the three equal groups)

Step 2: find how many zombies are in one group: $1 \times 4 = 4$

(4 zombies in 1 group).

I wonder, if I meet $\boxed{\frac{2}{3}}$ of $\boxed{12}$ zombies, <u>how many zombies</u> do I meet?

That's the question I like. My favorite algorithm:

$$\frac{2}{3} \times 12 = \frac{2}{3} \times \frac{12}{1} = \frac{2 \times 12}{3 \times 1} = \frac{2 \times \cancel{12}}{\cancel{3} \times 1} = \frac{2 \times 4}{1 \times 1} = 8 \text{ (zombies)}$$

I like to solve it slower: $\frac{2}{3} \times 12$

Step 1: $12 \div 3 = 4$ (find $\frac{1}{3}$ out of 12)

(4 zombies in each of the three equal groups)

Step 2: $2 \times 4 = 8$ (find how many zombies are in 2 groups)

(8 zombies in 2 groups).

1. <u>Multiply</u> and <u>simplify</u> if possible. <u>Color</u> the correct number of images.

Find $\frac{1}{5}$ of 15 pigs

$\frac{1}{5} \times 15 = \frac{\_\_}{\_\_} \times \frac{\_\_}{\_\_} = \frac{\_\_ \times \_\_}{\_\_ \times \_\_} = \_\_\_$

Find $\frac{3}{4}$ of 16 pigs

Step 1: ____ ÷ ____ = ____ (find $\frac{1}{4}$ out of 16)

Step 2: ____ × ____ = ____

Find $\frac{5}{7}$ of 21 pigs

$\frac{\_\_}{\_\_} \times \_\_\_ = \frac{\_\_}{\_\_} \times \frac{\_\_}{\_\_} = \frac{\_\_ \times \_\_}{\_\_ \times \_\_} = \_\_\_$

Find $\frac{5}{8}$ of 24 pigs

Step 1: ____ ÷ ____ = ____ (find $\frac{1}{8}$ out of 24)

Step 2: ____ × ____ = ____

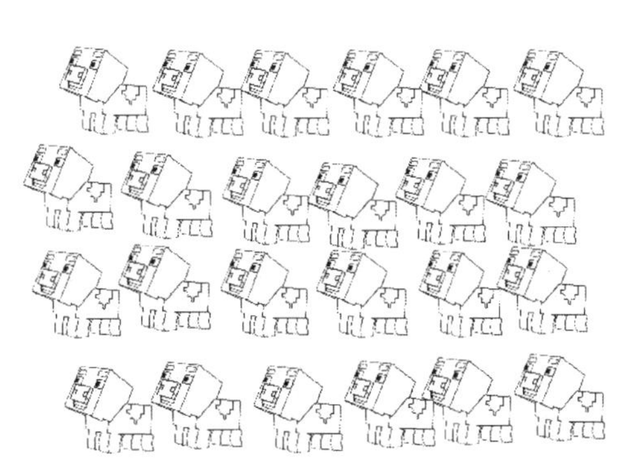

1. Reduce each fraction to simplest form, write the missing numbers, and compare fractions using "<," ">," or "=." Color the fraction of each shape. Cross out the wrong word.

$\frac{14}{16} = -$ ___ $\frac{14}{20} = -$

__ *out of* __ *equal parts are bigger/smaller than* __ *out of* __ *equal parts.*

$\frac{16}{18} = -$ ___ $\frac{16}{22} = -$

__ *out of* __ *equal parts are bigger/smaller than* __ *out of* __ *equal parts.*

$\frac{20}{24} = -$ ___ $\frac{10}{14} = -$

__ *out of* __ *equal parts are bigger/smaller than* __ *out of* __ *equal parts.*

2. Find and circle the area of a field to plant potatoes that was drawn by a villager. Hint: Figures are not drawn to scale.

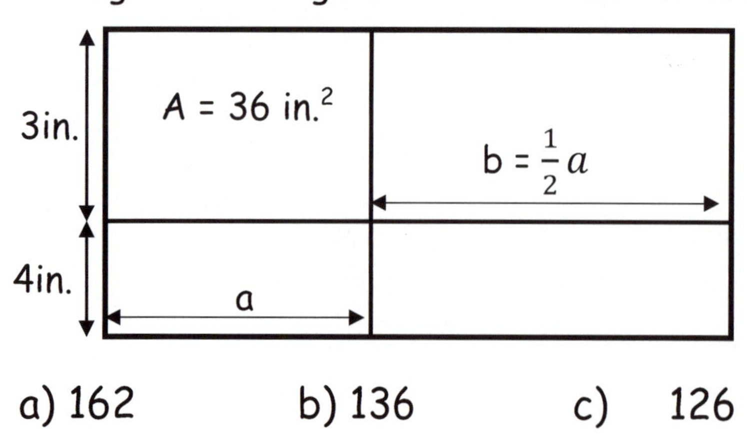

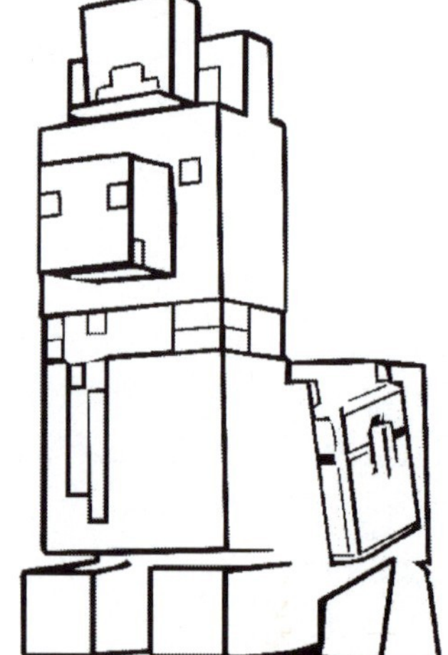

a) 162 b) 136 c) 126

1. Write in the missing numbers on the Minecraft factor tree and cross out the common prime factors.

A prime number is a counting number that has exactly two different factors

$\frac{56}{66} = \frac{\quad}{\quad}$

56

2

66

2

Prime factors for 56:

~~2~~, ____, ____, ____,

Prime factors for 66:

~~2~~, ____, ____

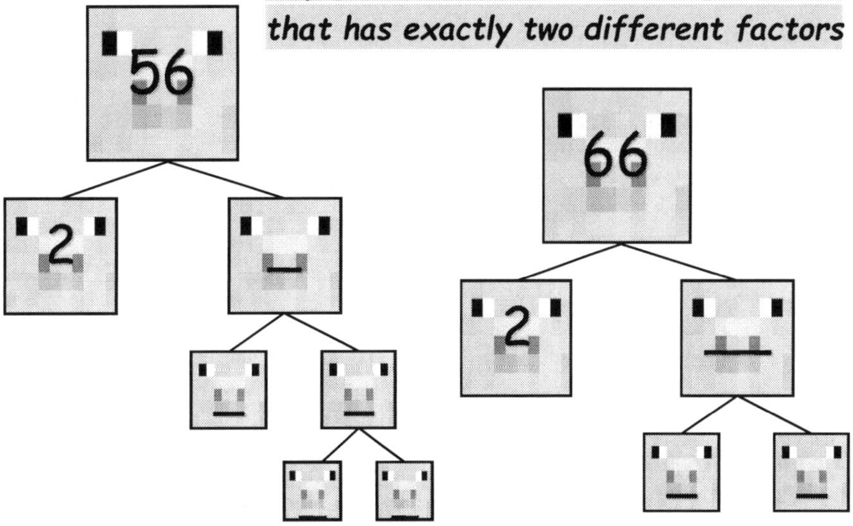

$\frac{64}{68} = \frac{\quad}{\quad}$

64

68

Prime factors for 64:

____, ____, ____, ____, ____, ____

Prime factors for 68:

____, ____, ____

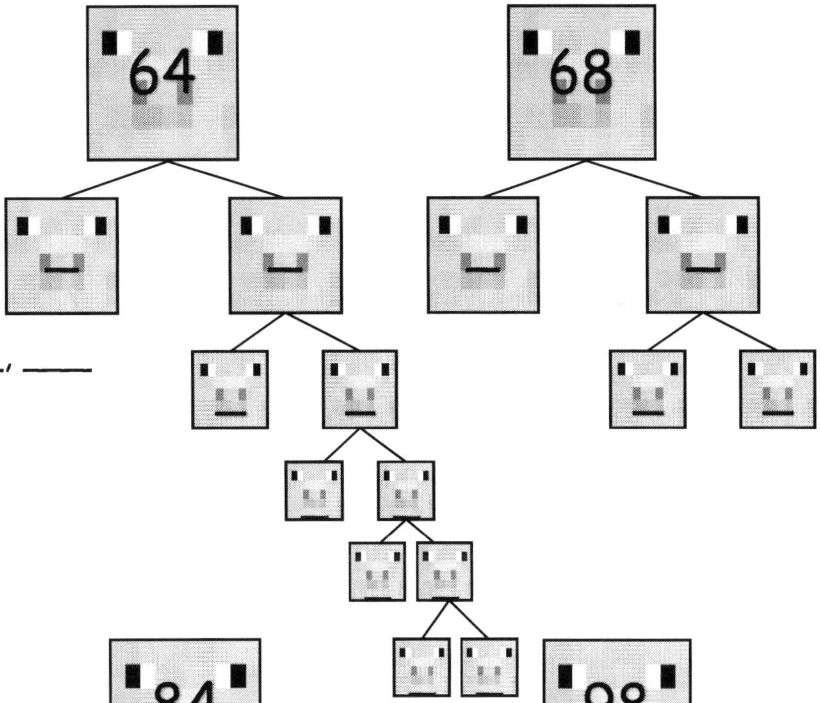

$\frac{84}{98} = \frac{\quad}{\quad}$

84

98

Prime factors for 84:

____, ____, ____, ____

Prime factors for 98:

____, ____, ____

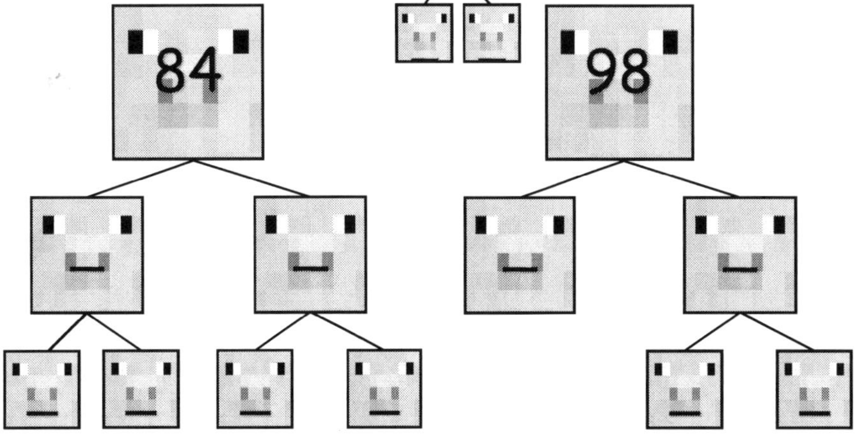

1. Find and circle or cross out the words to find out more about Minecraft.

S	B	Q	A	S	A	F	P	S	Y	Z	N	Q	A
S	E	G	H	V	A	A	I	R	S	O	D	R	T
O	O	I	Z	F	T	N	E	E	I	Z	C	Z	S
V	G	R	B	H	O	W	D	T	U	M	D	P	M
R	W	Y	W	M	D	D	A	S	P	C	F	O	Y
O	Q	A	J	V	O	N	N	N	T	O	P	T	S
N	Y	S	Z	P	I	Z	V	O	J	O	N	H	T
D	N	D	L	B	W	Y	E	M	I	X	N	W	I
D	I	J	M	H	S	E	J	L	U	X	W	E	C
P	S	O	P	U	E	Z	B	I	I	U	K	W	A
O	C	U	W	E	X	L	L	V	N	T	C	H	L
Y	A	W	K	L	A	W	T	E	D	Q	S	U	D
W	O	O	D	P	L	A	N	K	S	Q	R	O	G
F	G	S	B	J	B	H	G	F	X	H	E	Z	H

WOOD PLANKS

COMBINATION

HOSTILE ZOMBIES

MYSTICAL

SANDSTONE

PATHWAY

EVIL MONSTERS

WALKWAY

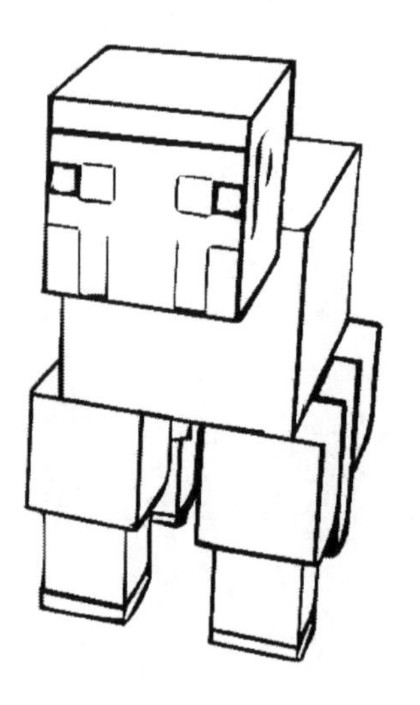

1. Reduce each fraction to simplest form, write the missing numbers, and compare fractions using "<," ">," or "=." Color the fraction of each shape. Cross out the wrong word.

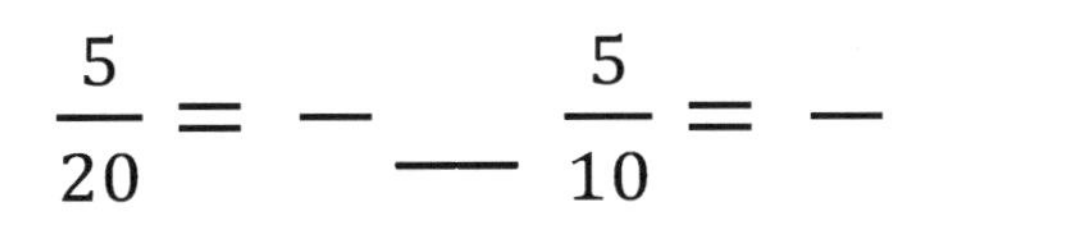

$\frac{5}{20} = -$ ___ $\frac{5}{10} = -$

__ *out of* __ *equal parts are bigger/smaller than* __ *out of* __ *equal parts.*

$\frac{6}{8} = -$ ___ $\frac{6}{10} = -$

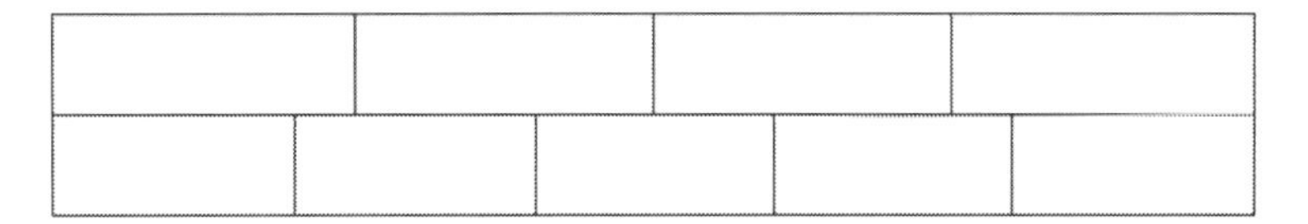

__ *out of* __ *equal parts are bigger/smaller than* __ *out of* __ *equal parts.*

$\frac{4}{12} = -$ ___ $\frac{4}{8} = -$

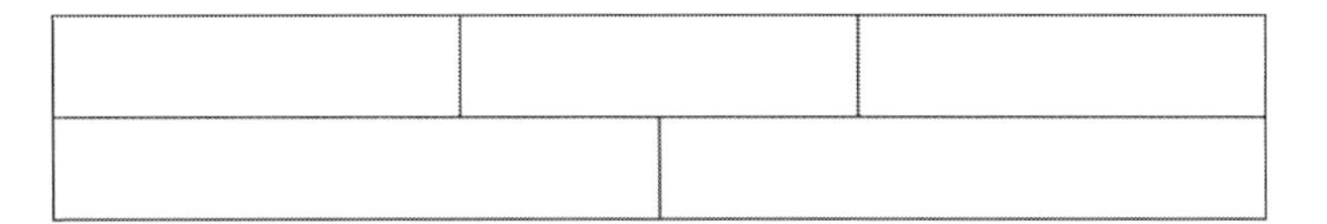

__ *out of* __ *equal parts are bigger/smaller than* __ *out of* __ *equal parts.*

$\frac{2}{8} = -$ ___ $\frac{2}{12} = -$

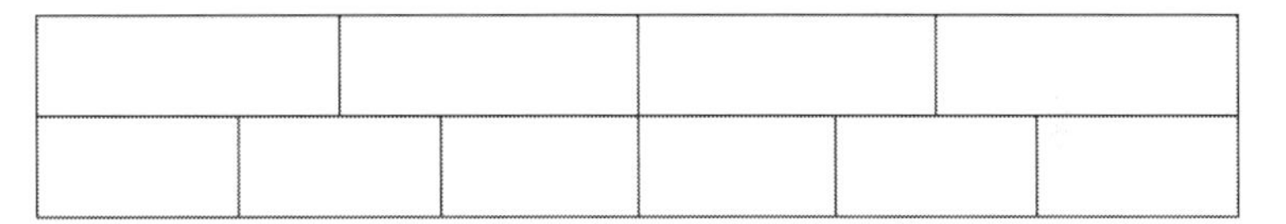

__ *out of* __ *equal parts are bigger/smaller than* __ *out of* __ *equal parts.*

2. Write 4 fractions that you cannot reduce to their simplest forms.

— ; — ; — ; — ; —

1. Write in the missing numbers on the Minecraft factor tree.

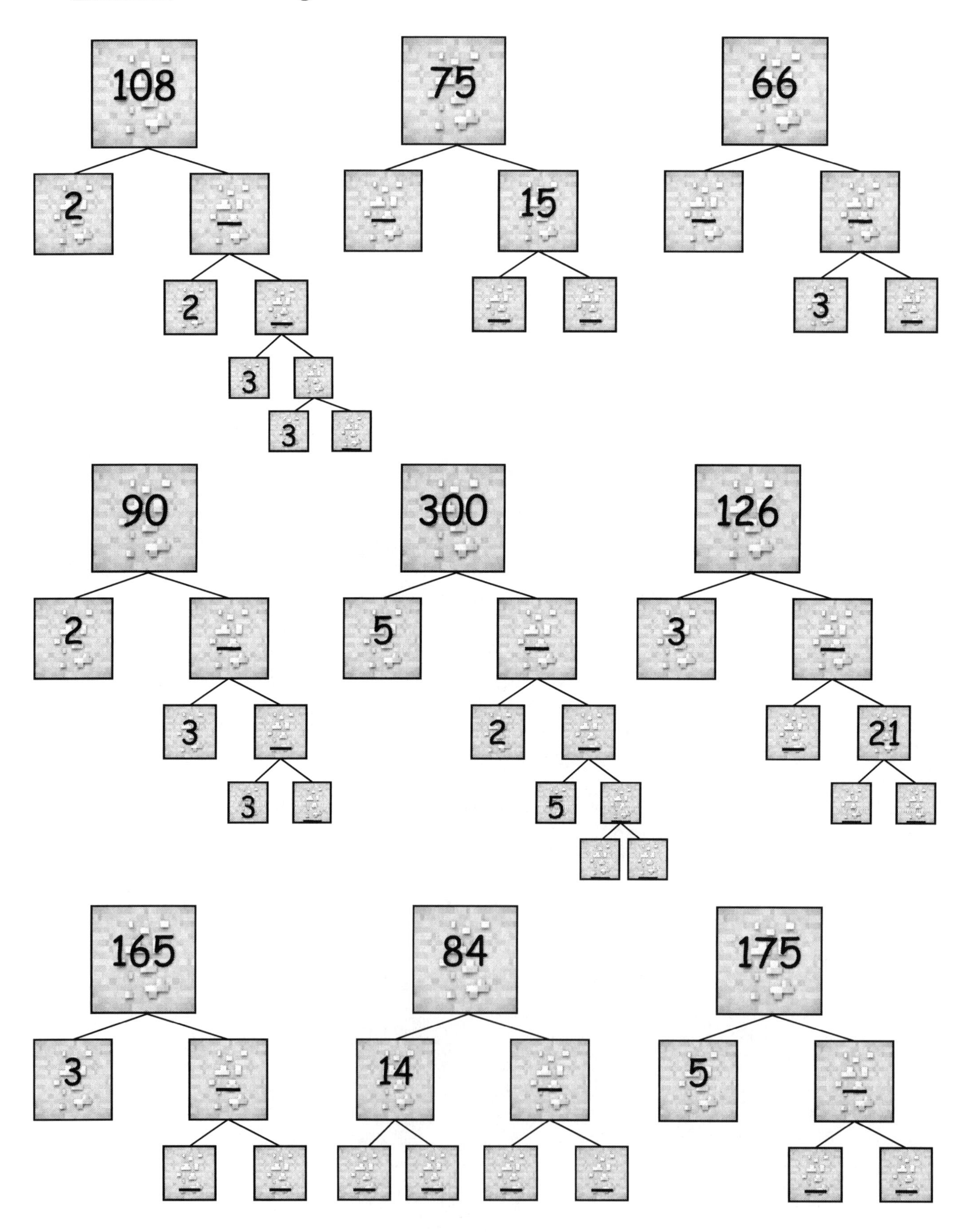

1. Write equivalent fractions. Fill in the missing numbers.

$\frac{2 \times}{3 \times} = \frac{}{6}$

Common factor = 6 ÷ 3 = 2

$\frac{1 \times}{2 \times} = \frac{}{8}$

Common factor = 8 ÷ 2 = 4

$\frac{3 \times}{4 \times} = \frac{}{12}$

Common factor = ______________

$\frac{1 \times}{3 \times} = \frac{}{18}$

Common factor = ______________

$\frac{7 \times}{8 \times} = \frac{}{24}$

Common factor = ______________

$\frac{2 \times}{3 \times} = \frac{}{12}$

Common factor = ______________

$\frac{5 \times}{6 \times} = \frac{}{36}$

Common factor = ______________

$\frac{5 \times}{7 \times} = \frac{}{28}$

Common factor = ______________

$\frac{4 \times}{5 \times} = \frac{}{35}$

Common factor = ______________

$\frac{2 \times}{7 \times} = \frac{}{42}$

Common factor = ______________

$\frac{3 \times}{5 \times} = \frac{}{60}$

Common factor = ______________

$\frac{4 \times}{9 \times} = \frac{}{45}$

Common factor = ______________

$\frac{2 \times}{5 \times} = \frac{}{30}$

Common factor = ______________

$\frac{1 \times}{8 \times} = \frac{}{48}$

Common factor = ______________

$\frac{2 \times}{3 \times} = \frac{}{60}$

Common factor = ______________

$\frac{1 \times}{4 \times} = \frac{}{40}$

Common factor = ______________

2. I need 4 pastels to make a block. What fraction are 3 pastels of 2 blocks?

a) $\frac{3}{7}$ b) $\frac{6}{8}$ c) $\frac{3}{8}$

1. Multiply and simplify if possible.

Find and color $\frac{2}{3}$ of 21 blocks.

$\frac{2}{3} \times 21 = \frac{}{} \times \frac{}{} = \frac{\times}{\times} = \_\_\_$

Find and color $\frac{6}{11}$ of 33 blocks.

Step 1: ___ ÷ ___ = ___ (find $\frac{1}{11}$ out of 33)

Step 2: ___ × ___ = ___

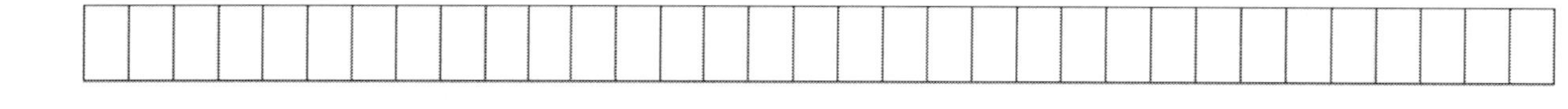

Find and color $\frac{7}{9}$ of 27 blocks.

$\frac{7}{9} \times 27 = \frac{}{} \times \frac{}{} = \frac{\times}{\times} = \_\_\_$

Find and color $\frac{4}{7}$ of 28 blocks.

Step 1: ___ ÷ ___ = ___ (find $\frac{1}{7}$ out of 28)

Step 2: ___ × ___ = ___

1. <u>Reduce</u> each fraction to simplest form, <u>write</u> the missing numbers, and <u>compare</u> fractions using "<," ">," or "=." <u>Color</u> the fraction of each shape. <u>Cross out</u> the wrong word.

$\frac{36}{48} = -$ ___ $\frac{36}{60} = -$

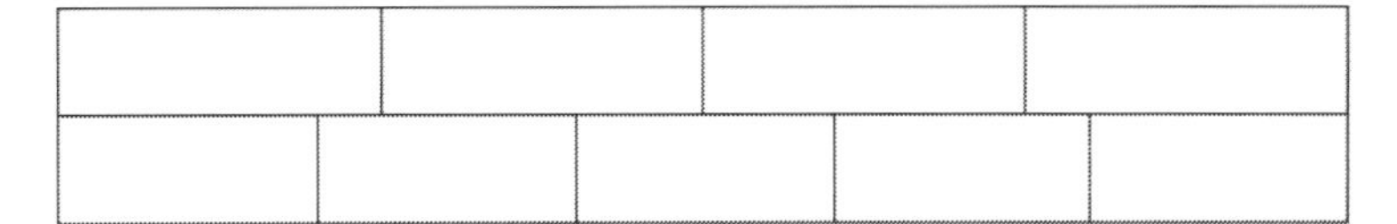

___ *out of* ___ *equal parts are bigger/smaller than* ___ *out of* ___ *equal parts.*

$\frac{45}{70} = -$ ___ $\frac{45}{80} = -$

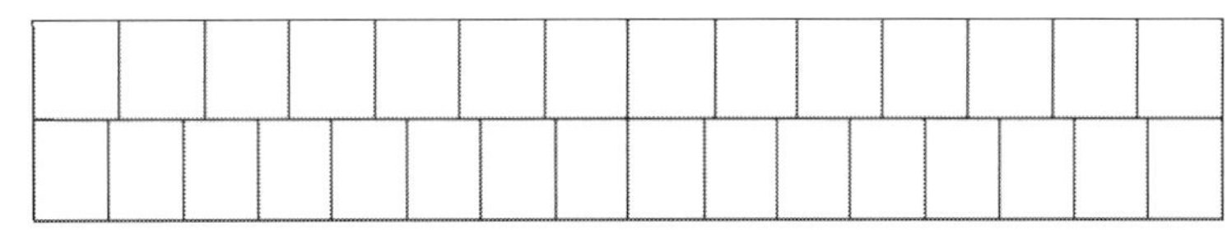

___ *out of* ___ *equal parts are bigger/smaller than* ___ *out of* ___ *equal parts.*

$\frac{40}{64} = -$ ___ $\frac{60}{108} = -$

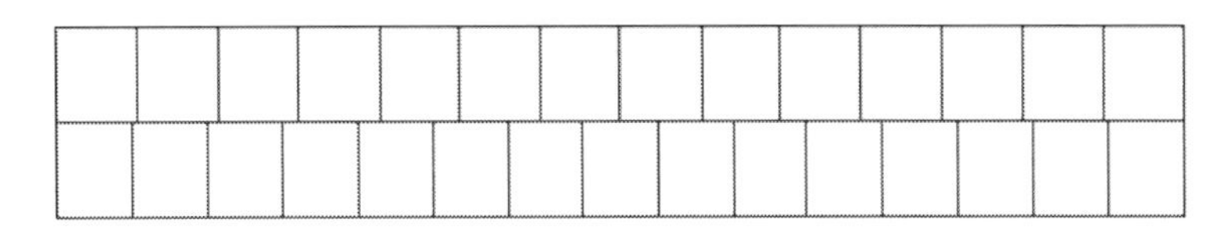

___ *out of* ___ *equal parts are bigger/smaller than* ___ *out of* ___ *equal parts.*

$\frac{28}{49} = -$ ___ $\frac{28}{63} = -$

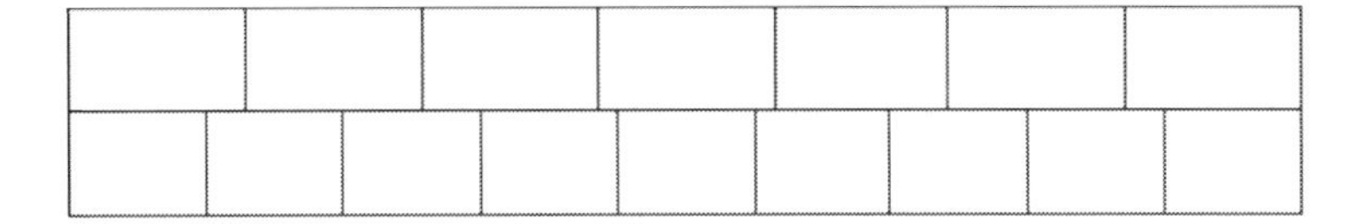

___ *out of* ___ *equal parts are bigger/smaller than* ___ *out of* ___ *equal parts.*

2. <u>Find d and circle</u> the correct answer. Hint: Figures are not drawn to scale.

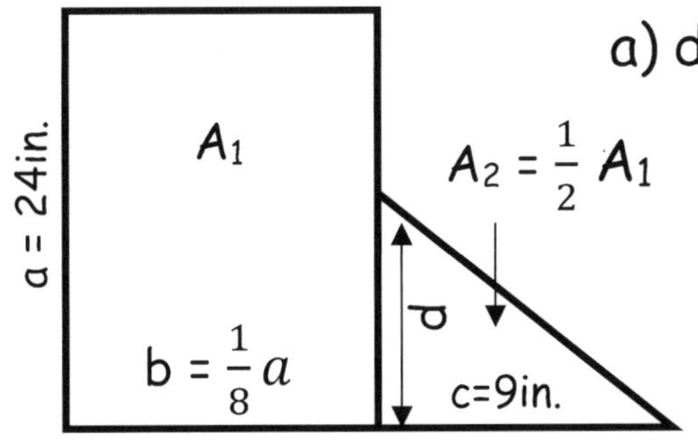

a) d = 9 b) d = 8 c) d = 7

1. Multiply and simplify if possible. Circle the right answer (letter).

Find and color $\frac{6}{11}$ of 22 blocks.

a) 10
b) 12
c) 15

$\frac{6}{11} \times 22 = \frac{\quad}{\quad} \times \frac{\quad}{\quad} = \frac{\quad \times \quad}{\quad \times \quad} =$ ____

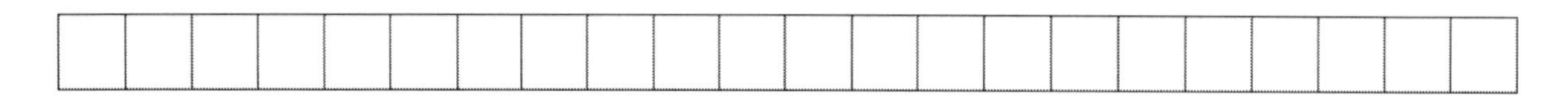

Find and color $\frac{7}{10}$ of 30 blocks.

a) 21
b) 35
c) 20

Step 1: ____ ÷ ____ = ____ (find $\frac{1}{10}$ out of 30)

Step 2: ____ × ____ = ____

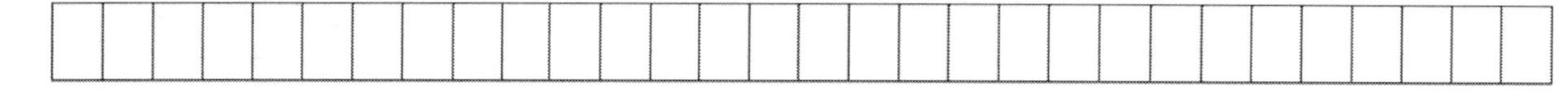

Find and color $\frac{3}{8}$ of 32 blocks.

a) 16
b) 10
c) 12

$\frac{3}{8} \times 32 = \frac{\quad}{\quad} \times \frac{\quad}{\quad} = \frac{\quad \times \quad}{\quad \times \quad} =$ ____

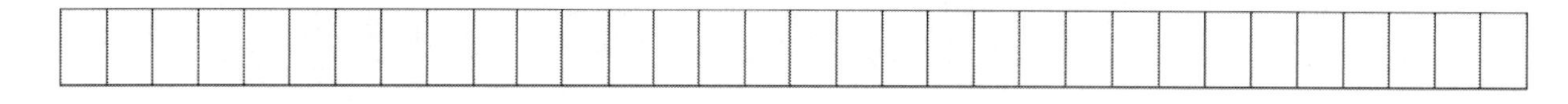

Find and color $\frac{5}{7}$ of 35 blocks.

a) 20
b) 35
c) 25

Step 1: ____ ÷ ____ = ____ (find $\frac{1}{7}$ out of 35)

Step 2: ____ × ____ = ____

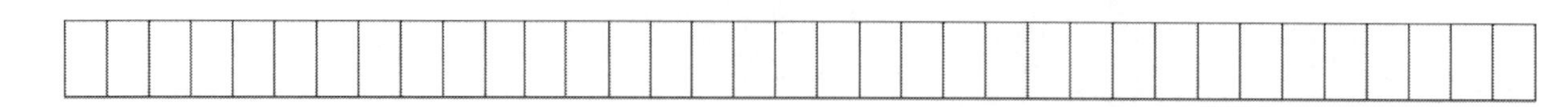

1. Reduce each fraction to simplest form, write the missing numbers, and compare fractions using "<," ">," or "=." Color the fraction of each shape. Cross out the wrong word.

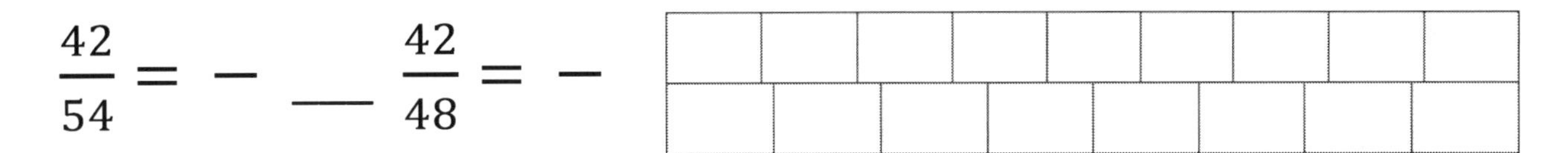

$\frac{42}{54} = -$ ___ $\frac{42}{48} = -$

__ out of __ equal parts are bigger/smaller than __ out of __ equal parts.

$\frac{24}{40} = -$ ___ $\frac{24}{64} = -$

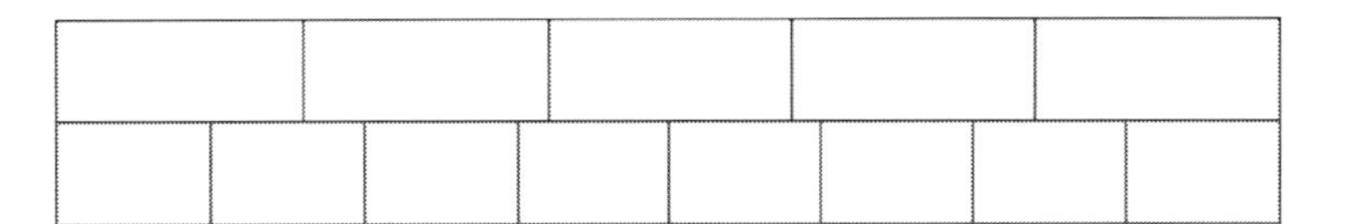

__ out of __ equal parts are bigger/smaller than __ out of __ equal parts.

$\frac{20}{24} = -$ ___ $\frac{40}{64} = -$

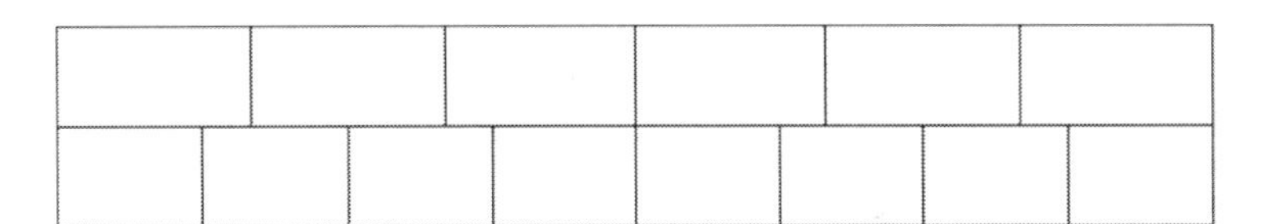

__ out of __ equal parts are bigger/smaller than __ out of __ equal parts.

2. Find and circle the area of a field to plant potatoes that was drawn by a villager. Hint: Figures are not drawn to scale.

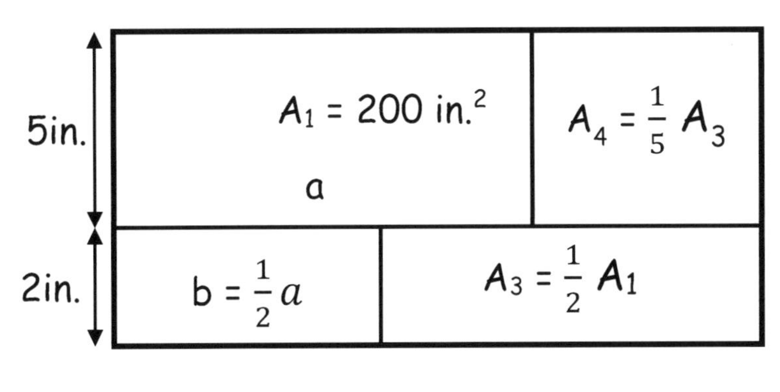

a) 390 b) 260 c) 360

1. Find and circle or cross out the words to find out more about Minecraft.

H R G E G S F D E Y I K M R
E O N C F N L A V G L S Q R
C L I P J O O O E X A H Y A
H L Y D E I A R N Y B I R G
V E F G O T T L T I V P Y V
A R I P J C I E U N Y S U Y
V C R O U E N V A T G H Z Q
Z O R I M S G A L G G U L T
W A E D T R I R L Z T L T E
V S T O D E S G Y H Z L V N
L T P R C P L R R F Y H H Y
G E C Y C P A O V Q U L I S
Z R Y Z P U N G Q T A Y S C
M I Y T P E D S J T E P B T

UPPER SECTIONS

GRAVEL ROAD

EVENTUALLY

ROLLER COASTER

THRONE

FLOATING ISLAND

SHIP'S HULL

TERRIFYING

1. <u>Compare</u> fractions with unlike numerators and denominators.

Well, I cannot compare fractions until they have been changed into equivalent fractions, right?

Yeah, you need fractions with the same denominators!

Okay, first, fractions with different denominators are called unlike fractions.

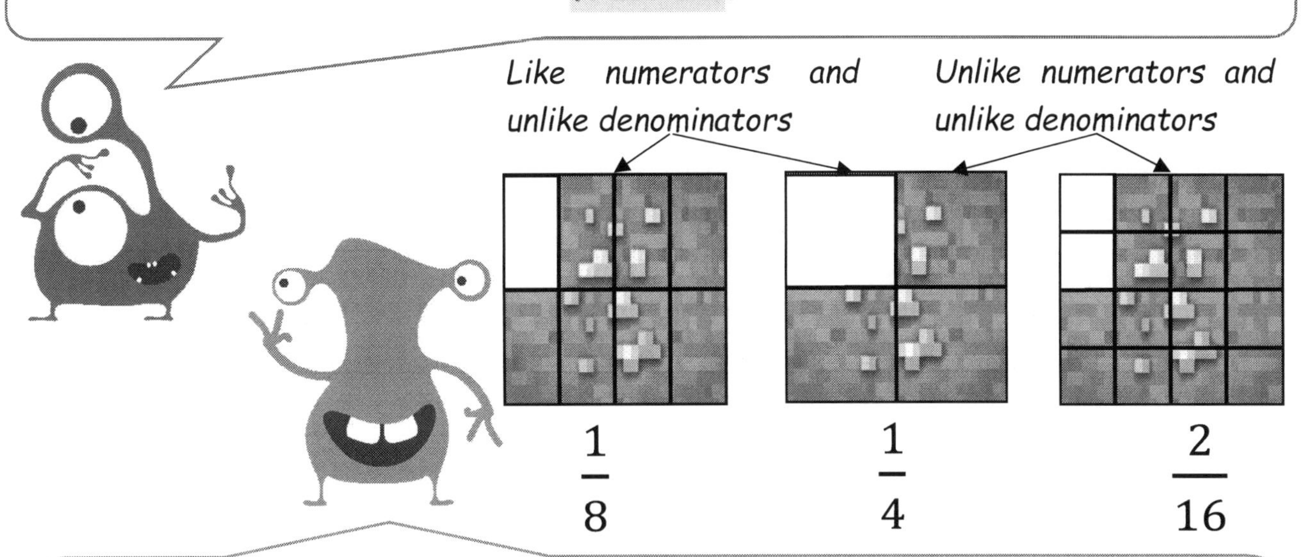

Hints: If a whole is divided into more equal parts (12 > 8), each equal part will be smaller ($\frac{1}{8} > \frac{1}{12}$.)

The larger the number shown by the denominator (5 > 2), the smaller is the size of this denominator ($\frac{1}{5} < \frac{1}{2}$.)

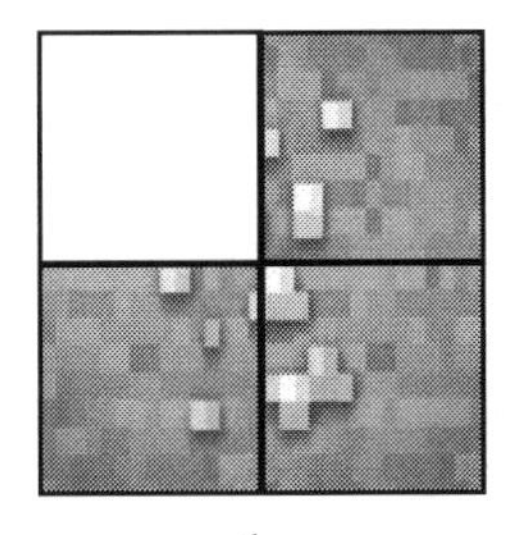

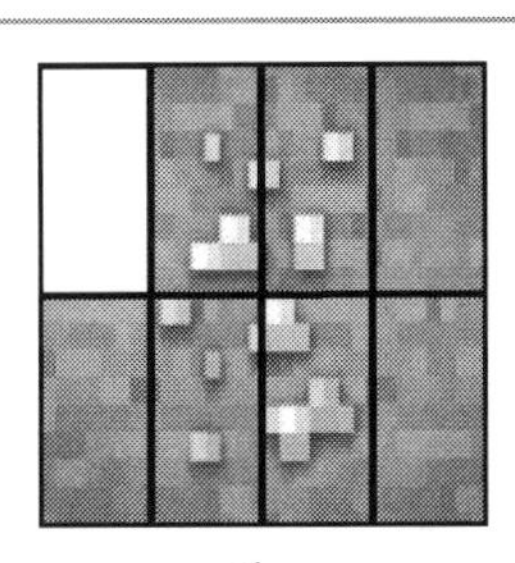

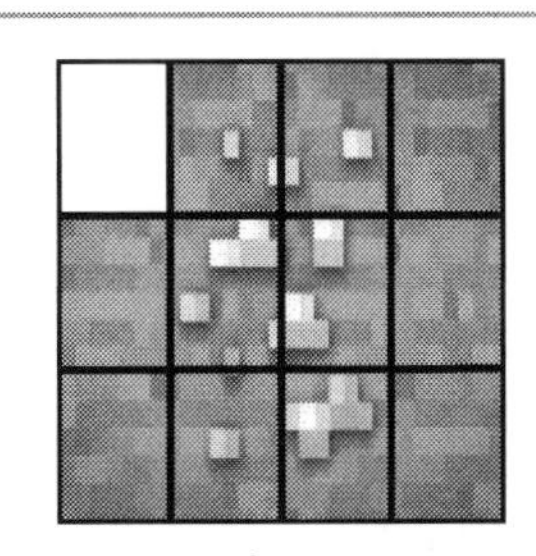

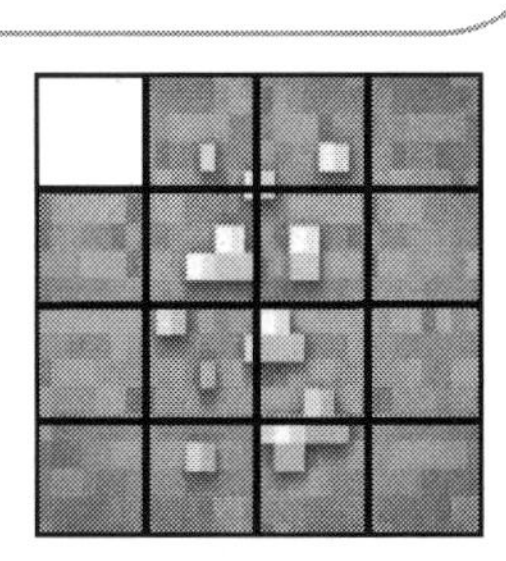

$$\frac{1}{4} > \frac{1}{8} > \frac{1}{12} > \frac{1}{16}$$

The question is: How can I compare unlike fractions with unlike denominators and unlike numerators?

☺ *Hint: Find **equivalent fractions with the same denominator**.*

Look, I divided a candy into 4 equal parts and I ate 3 parts out of the four equal parts. Or I divided a candy into 2 equal parts and I ate 1 part out of the two equal parts.

When did I eat more?

***Change the original denominators** by **multiplying both the numerator and the denominator by the same number** to get equivalent fractions.*

Which fraction is bigger?

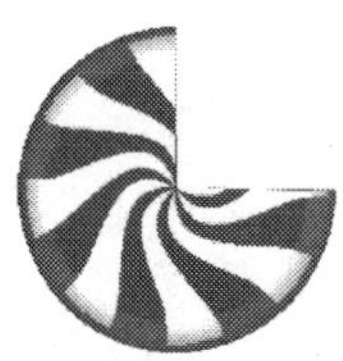

$\frac{3}{4} \; ? \; \frac{1}{2}$

To compare unlike fractions - for example, $\frac{3}{4}$ and $\frac{5}{8}$, I need to change $\frac{3}{4}$ to an equivalent fraction with the denominator 8. Thus, I can compare the two fractions. To do that, I need to find an equivalent fraction to $\frac{3}{4}$:

$\frac{3}{4} = \frac{6}{8}$

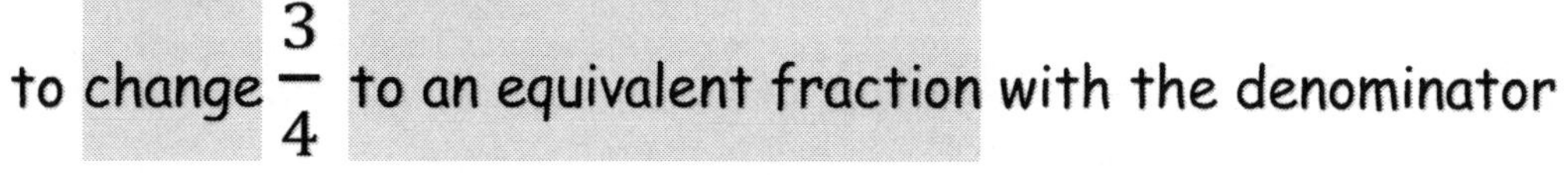

Multiply both the numerator and the denominator by the common factor 2: 3 × 2 = 6; 4 × 2 = 8

$\frac{6}{8} > \frac{5}{8}$

These fractions are equivalent in size since they represent the whole divided into 8 equal parts

The same denominator for both fractions is called a **common denominator**. Now, I can easily compare the fractions:

$\frac{3}{4} > \frac{5}{8}$

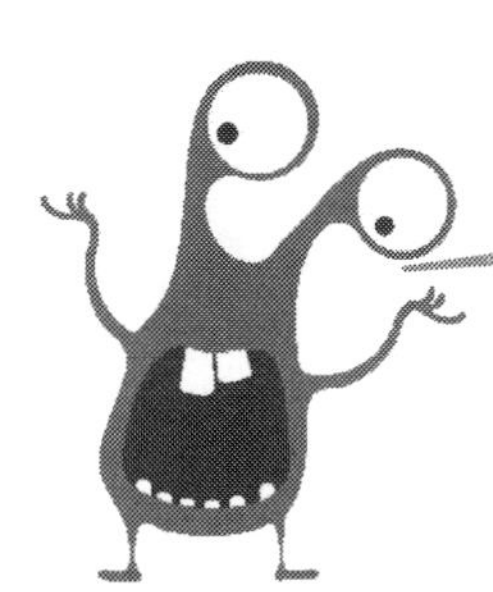

Help me compare these unlike fractions:

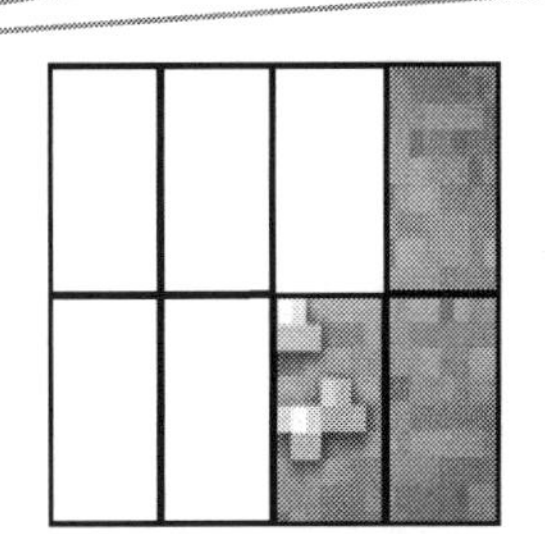

$$\frac{5}{8} \quad ? \quad \frac{6}{16}$$

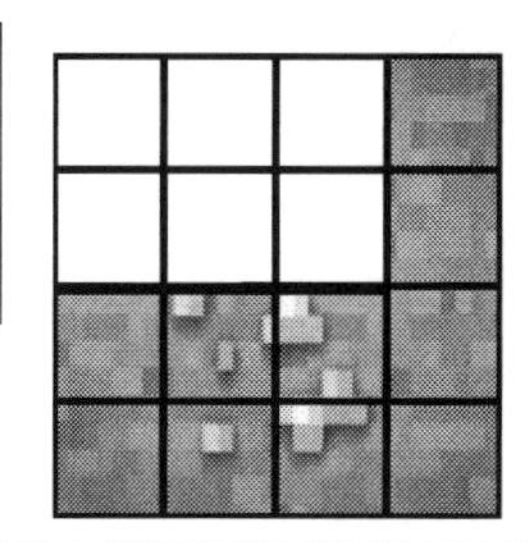

Step 1: Find the Least Common Denominator (LCD). Change one or both fractions into equivalent fractions.

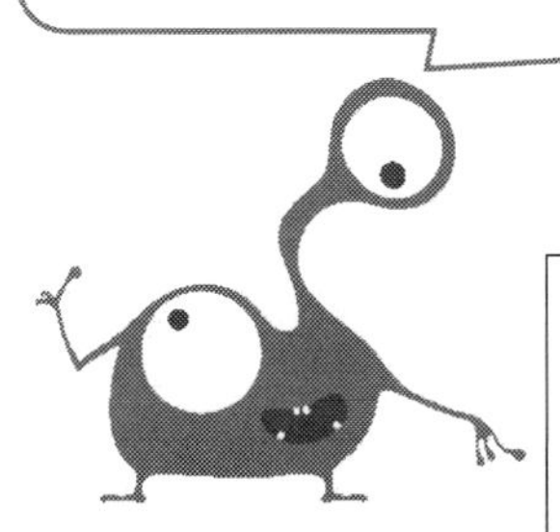

Multiples of 8: *8, **16**, 24, **32**, 40, **48**, 56, **64**, ...*

Multiples of 16: ***16, 32, 48, 64**, 80, 96, ...*

*The **Least Common Denominator** is 16*

$$\frac{5}{8} \quad \frac{7}{16} \rightarrow \frac{5\times 2}{8\times 2} \quad \frac{7}{16} \rightarrow \frac{10}{16} \quad \frac{7}{16}$$

*The **Least Common denominator** is the **least whole number that is a multiple of both numbers**.*

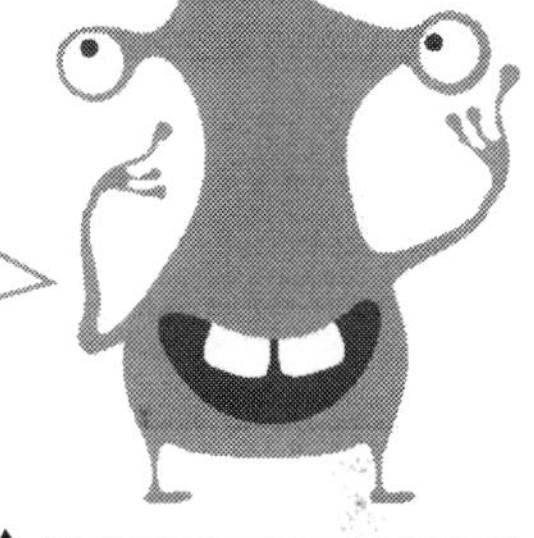

Step 2: Compare like fractions: the larger fraction is the one with the larger numerator:

$$\frac{10}{16} > \frac{6}{16}$$

What about this problem?

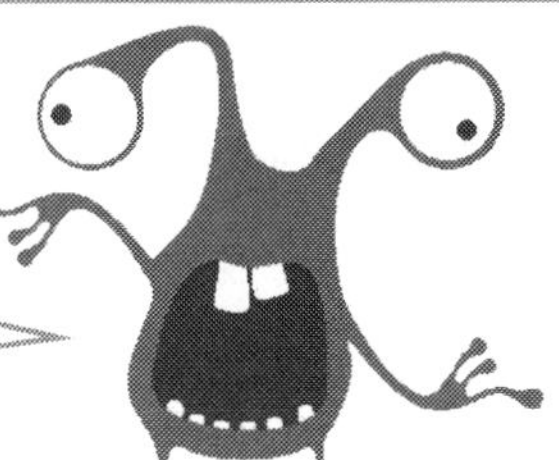

$$\frac{3}{12} \; ? \; \frac{27}{30}$$

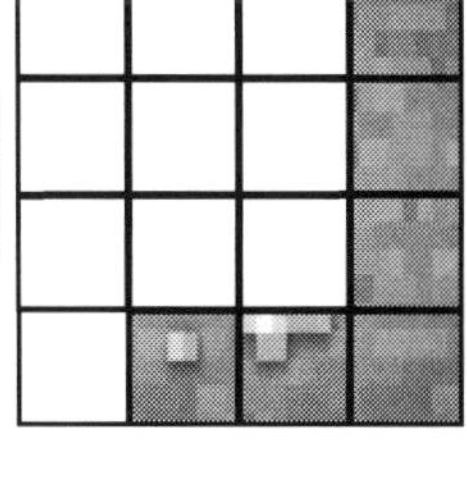

Let me think. Step1: <u>Reduce</u> each fraction to lowest terms:

Divide both the numerator and the denominator by their greatest common factor (GCF).

$$\frac{(3\div 3)}{(12\div 3)} \; ? \; \frac{(27\div 3)}{(30\div 3)} \rightarrow \frac{1}{4} \; ? \; \frac{9}{10}$$

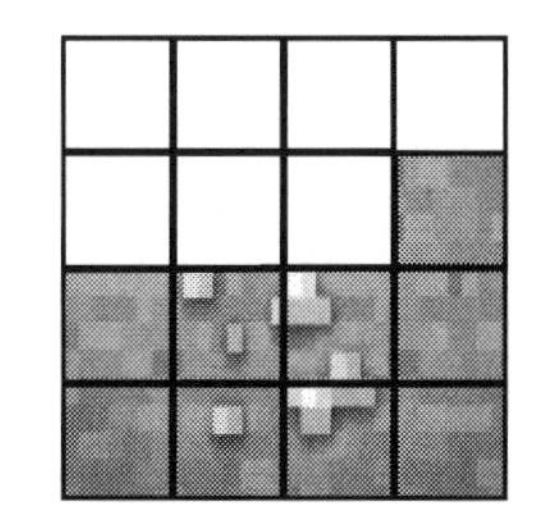

Step 2: Find a Common Denominator (change both fractions into equivalent fractions so that they both have the same denominator).

Multiples of 4: *4, 8, 12, 16, 20, 24, 28, 32, 36, 40, ...*

Multiples of 10: *10, 20, 30, 40, 50, 60, 70, 80, 90, ...*

*The **Least Common Denominator** is 20.*

What is the number that both the 4 and 10 can be divided by?

20 and 40 can be divided by 4 and 10. Here, the **least** common denominator is 20.

Step 3: 20 divided by 4 is 5, so, I need to multiply both the numerator and the denominator of $\frac{1}{4}$ by 5.

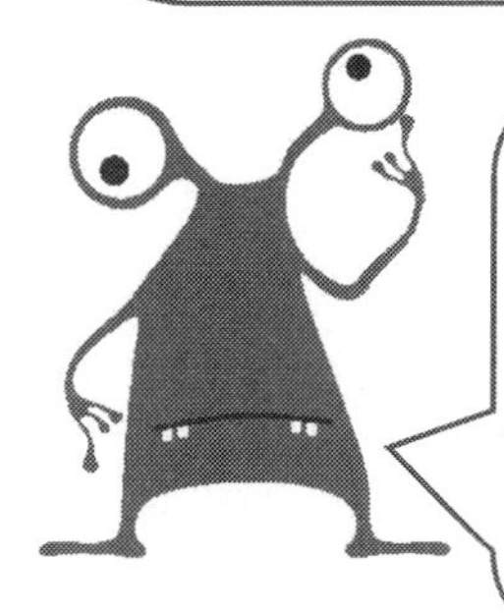

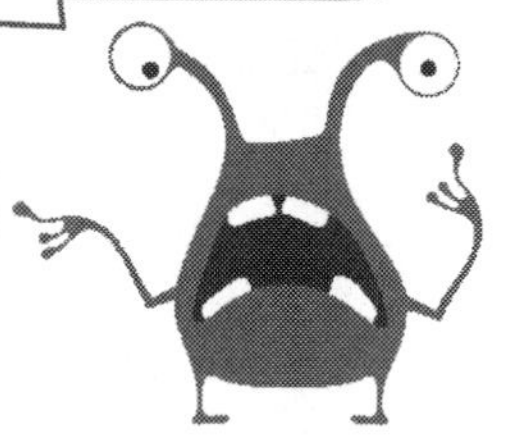

Step 4: 20 divided by 10 is 2, so, I need to multiply both the numerator and the denominator of $\frac{9}{10}$ by 2.

$$\frac{1}{4} \;\_\_\; \frac{9}{10} \longrightarrow \frac{1 \times 5}{4 \times 5} \;\_\_\; \frac{9 \times 2}{10 \times 2} \longrightarrow \frac{5}{20} < \frac{18}{20}$$

They are like fractions now. Compare the numerators!

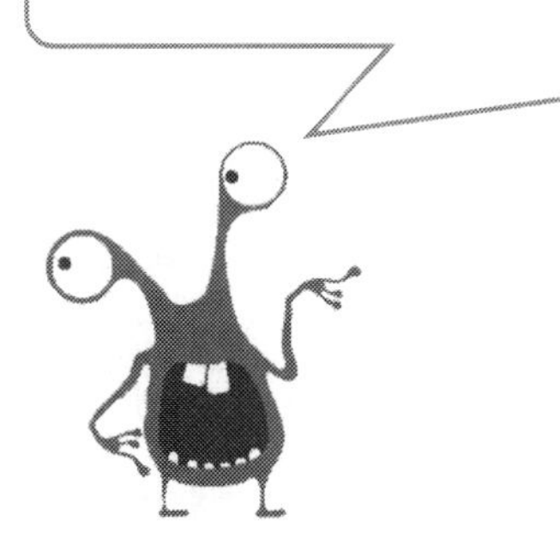

$$\frac{5}{20} < \frac{18}{20}$$

since 5 is less than 18 (or you can say that 18 is greater than 5): any fractions can be compared by getting a common denominator

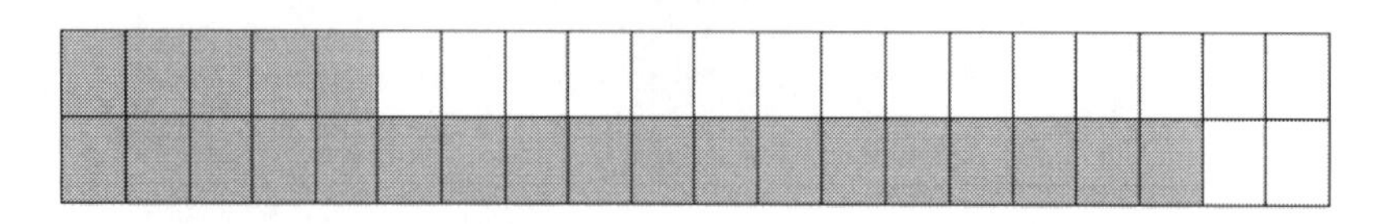

1. Color the correct fraction of each shape. Write the missing numbers. Compare unlike fractions using ">," "<," or "=."

$\frac{1}{6}$ ___ $\frac{2}{4}$ The Least Common Denominator is 12: $\frac{1 \times 2}{6 \times 2}$ ___ $\frac{2 \times 3}{4 \times 3}$ → $\frac{2}{12}$ ___ $\frac{6}{12}$

Multiples of 6: 6, 12, 18, 24, ...

Multiples of 4: 4, 8, 12, 16, 20

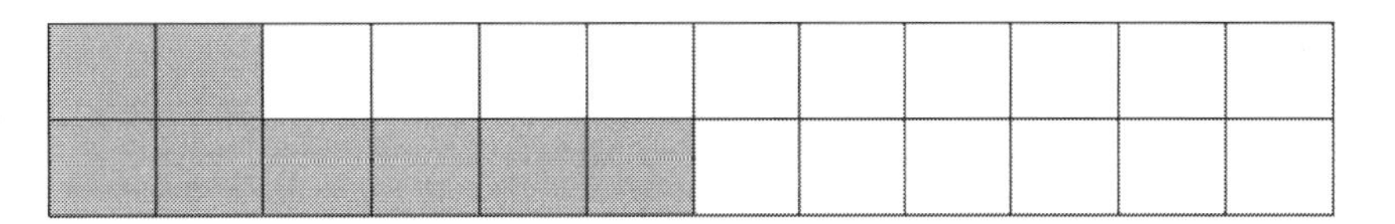

$\frac{3}{5}$ ___ $\frac{2}{3}$ The Least Common Denominator is ________: $\frac{3 \times 3}{5 \times 3}$ ___ $\frac{2 \times 5}{3 \times 5}$ → $\frac{9}{15}$ ___ $\frac{10}{15}$

Multiples of 5: ____________

Multiples of 3: ____________

$\frac{4}{8}$ ___ $\frac{5}{6}$ The Least Common Denominator is ________: $\frac{4 \times}{8 \times}$ ___ $\frac{5 \times}{6 \times}$ → $\frac{\ }{24}$ ___ $\frac{\ }{\ }$

Multiples of 8: ____________

Multiples of 6: ____________

$\frac{3}{4}$ ___ $\frac{5}{6}$ The Least Common Denominator is ________: $\frac{\times}{\times}$ ___ $\frac{\times}{\times}$ → $\frac{\ }{\ }$ ___ $\frac{\ }{\ }$

Multiples of 4: ____________

Multiples of 6: ____________

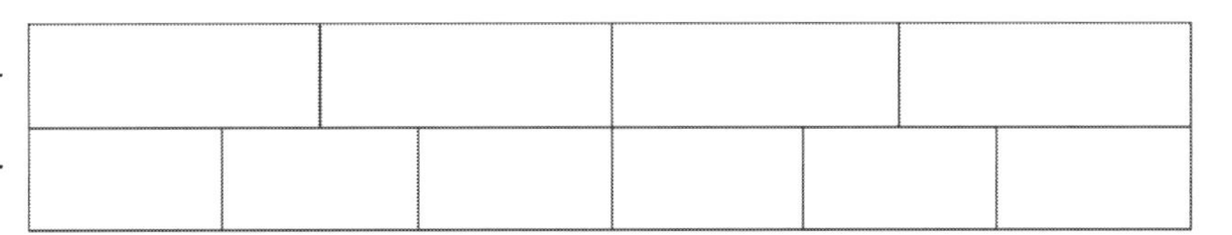

1. Color the correct fraction of each shape. Write the missing numbers. Compare unlike fractions using ">," "<," or "=."

$\frac{5}{6}$ ___ $\frac{7}{9}$ The Least Common Denominator is ________: $\frac{\quad\times\quad}{\quad\times\quad}$ ___ $\frac{\quad\times\quad}{\quad\times\quad}$ → $\frac{\quad}{18}$ ___ $\frac{\quad}{18}$

Multiples of 6: ____________

Multiples of 9: ____________

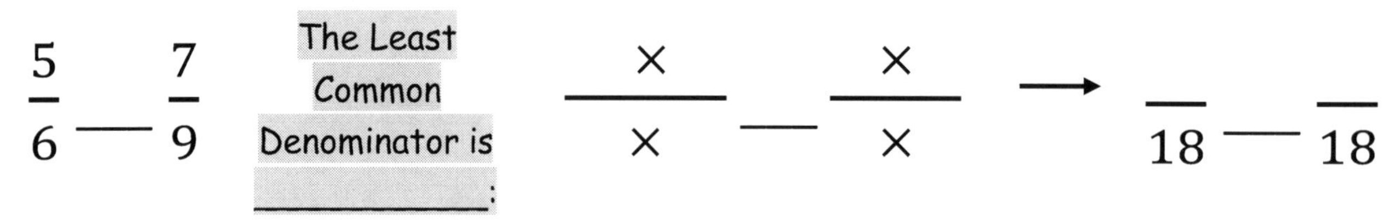

$\frac{2}{3}$ ___ $\frac{9}{10}$ The Least Common Denominator is ________: $\frac{2\times}{3\times}$ ___ $\frac{9\times}{10\times}$ → $\frac{\quad}{\quad}$ ___ $\frac{\quad}{\quad}$

Multiples of 3: ____________

Multiples of 10: ____________

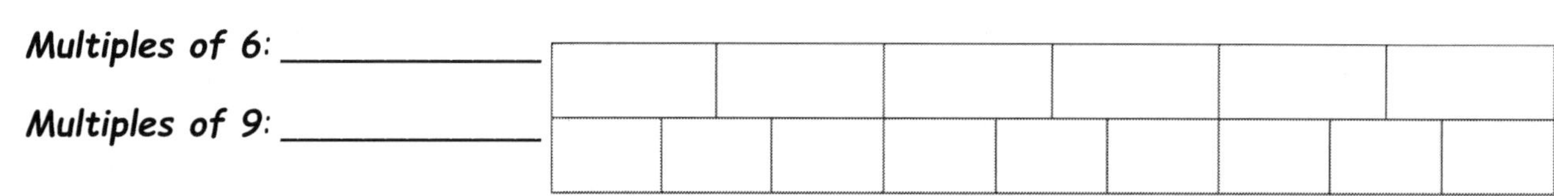

2. Multiply and simplify if possible. Circle the right answer (letter).

Find and color red $\frac{4}{5}$ of 20 blocks. Find and cross out $\frac{3}{4}$ of these red blocks.

$\frac{4}{5} \times 20 = \frac{\quad}{\quad} \times \frac{\quad}{\quad} = \frac{\quad\times\quad}{\quad\times\quad} =$ ____

$\frac{3}{4} \times$ ____ $= \frac{\quad}{\quad} \times \frac{\quad}{\quad} = \frac{\quad\times\quad}{\quad\times\quad} =$ ____

a) 12

b) 16

c) 10

1. Color the correct fraction of each shape. Write the missing numbers.

Compare unlike fractions using ">," "<," or "=."

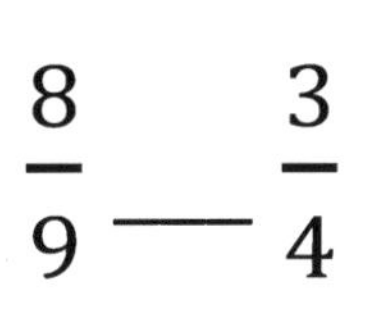

Common Denominator is ________:

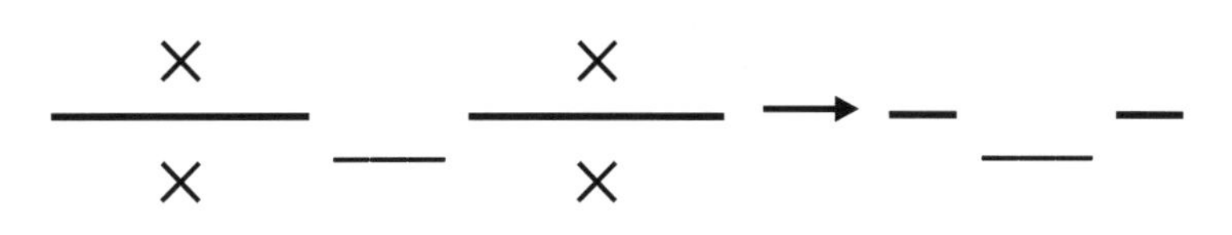

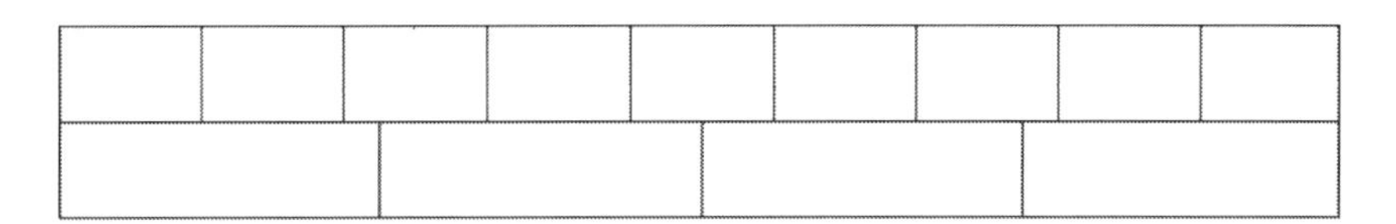

$\frac{5}{8}$ ___ $\frac{7}{11}$ Common Denominator is ________: $\frac{\times}{\times}$ ___ $\frac{\times}{\times}$ → — ___ —

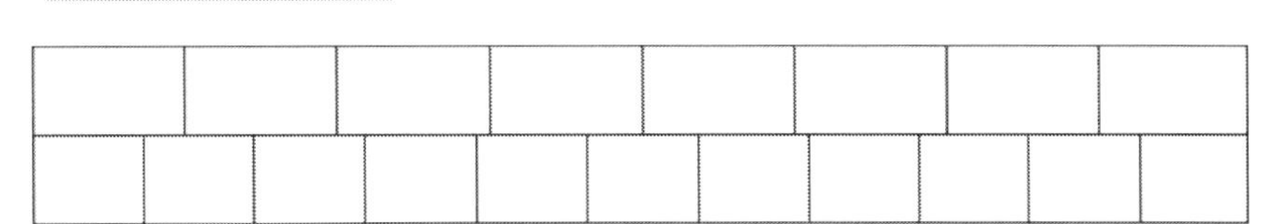

2. Multiply and simplify if possible. Circle the right answer (letter).

Find and color red $\frac{4}{7}$ of 21 blocks. Find and cross out $\frac{1}{2}$ of these red blocks.

$\frac{4}{7} \times 21 = — \times — = \frac{\times}{\times} =$ ___

$\frac{1}{2} \times$ ___ $= — \times — = \frac{\times}{\times} =$ ___

a) 9

b) 10

c) 6

1. <u>Color</u> the correct fraction of each shape. <u>Write</u> the missing numbers.

<u>Compare</u> unlike fractions using “>,” “<,” or “=.”

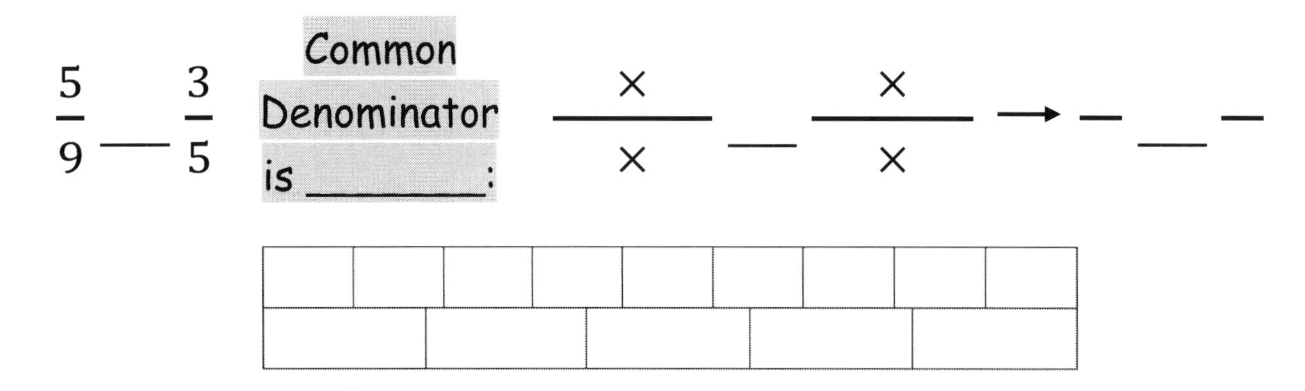

$\frac{5}{9}$ ___ $\frac{3}{5}$ Common Denominator is ________: $\frac{\quad \times \quad}{\quad \times \quad}$ ___ $\frac{\quad \times \quad}{\quad \times \quad}$ → $\frac{\ }{\ }$ ___ $\frac{\ }{\ }$

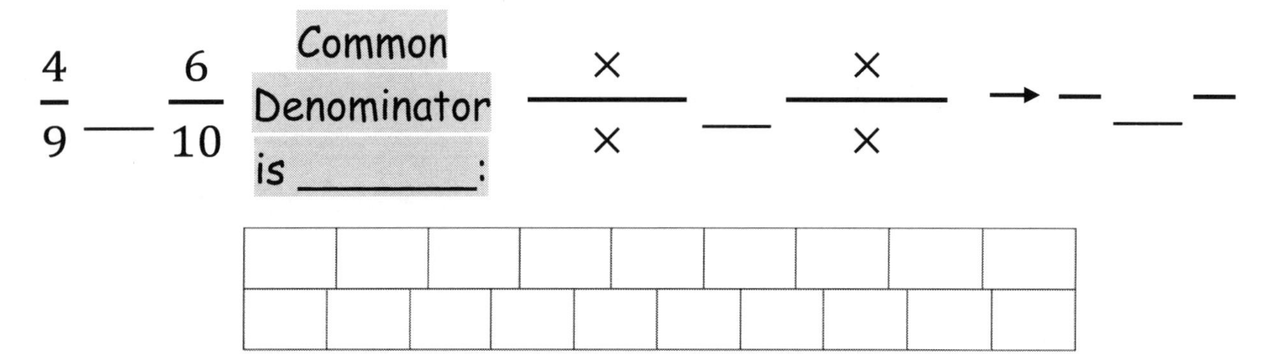

$\frac{4}{9}$ ___ $\frac{6}{10}$ Common Denominator is ________: $\frac{\quad \times \quad}{\quad \times \quad}$ ___ $\frac{\quad \times \quad}{\quad \times \quad}$ → $\frac{\ }{\ }$ ___ $\frac{\ }{\ }$

2. I created an enchanted table placing 1 book, 2 diamonds, and 4 obsidian. <u>What fraction(s)</u> are the diamonds?

a) $\frac{2}{7}$ b) $\frac{2}{9}$ c) $\frac{2}{3}$

3. What fraction is made of zombies?

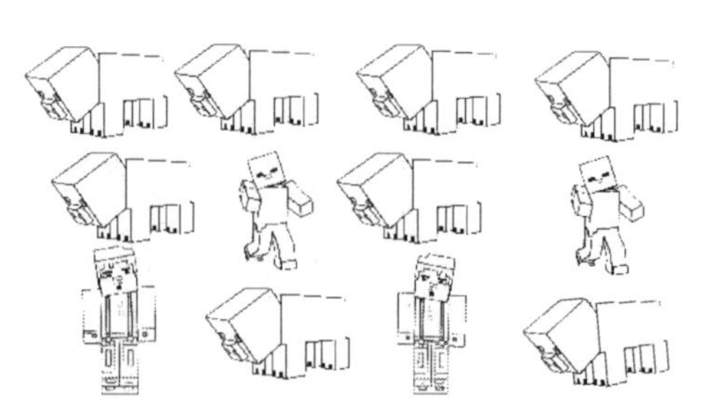

1. Find and circle or cross out the words to find out more about Minecraft.

B	H	S	Q	V	V	J	E	X	S	C	E	E	B
U	A	N	E	E	F	N	Y	L	K	N	Z	Q	R
B	F	S	P	G	O	Z	R	Y	C	D	Q	G	U
N	J	F	I	T	N	A	E	H	A	R	V	N	N
Q	F	Q	S	C	M	E	A	C	Q	D	X	O	D
L	Z	D	F	P	C	N	L	S	L	F	L	I	E
T	E	V	Z	U	T	I	E	L	L	L	O	T	R
R	R	T	M	M	Y	I	R	T	A	Z	B	A	G
I	C	G	E	Y	B	A	J	C	X	H	D	V	R
Y	I	N	B	Z	Z	W	X	L	U	V	C	I	O
W	T	D	O	O	R	W	A	Y	O	I	U	T	U
N	O	I	T	A	N	I	T	S	E	D	T	C	N
Z	Y	X	Z	A	J	E	X	J	Z	O	J	A	D
D	N	N	U	F	L	S	Q	R	L	U	Q	N	S

REDSTONE

CHALLENGES

UNDERGROUND

ENCHANTMENT

BASIC CIRCUIT

DOORWAY

DESTINATION

ACTIVATION

1. Color the correct fraction of each shape. Write the missing numbers. Compare unlike fractions using ">," "<," or "=."

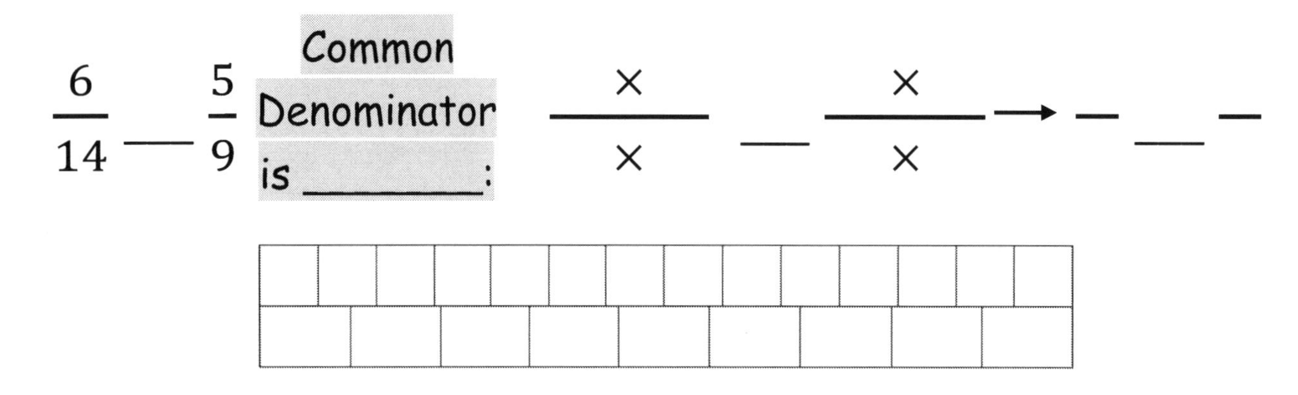

$\frac{6}{14}$ ___ $\frac{5}{9}$ Common Denominator is ________: $\frac{\quad\times\quad}{\quad\times\quad}$ ___ $\frac{\quad\times\quad}{\quad\times\quad}$ → — ___ —

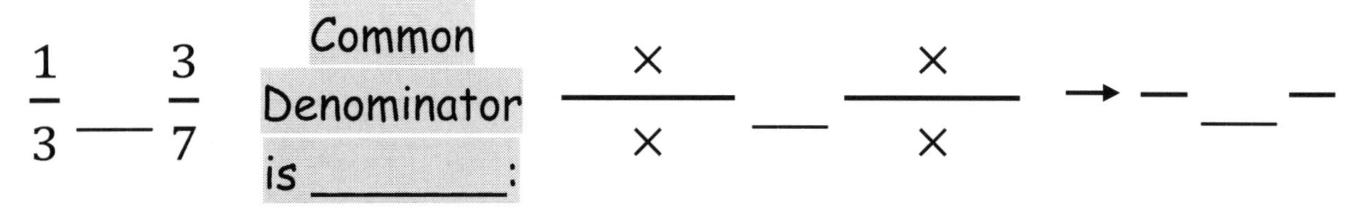

$\frac{1}{3}$ ___ $\frac{3}{7}$ Common Denominator is ________: $\frac{\quad\times\quad}{\quad\times\quad}$ ___ $\frac{\quad\times\quad}{\quad\times\quad}$ → — ___ —

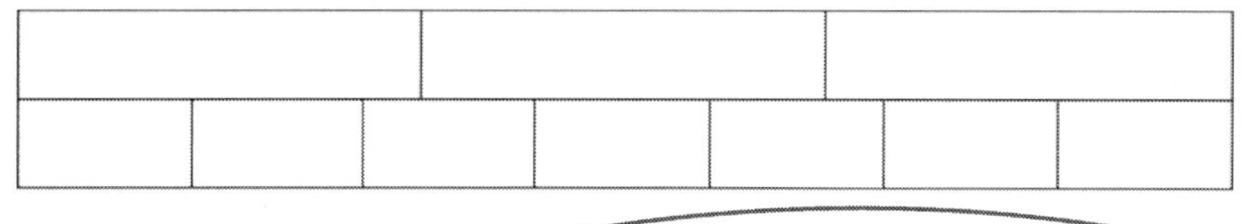

2. I crafted a dropper using 1 redstone and 7 cobblestones. What fraction(s) are 4 cobblestones?

a) $\frac{4}{9}$ b) $\frac{4}{8}$ c) $\frac{1}{2}$

3. What fraction is made of pigs?

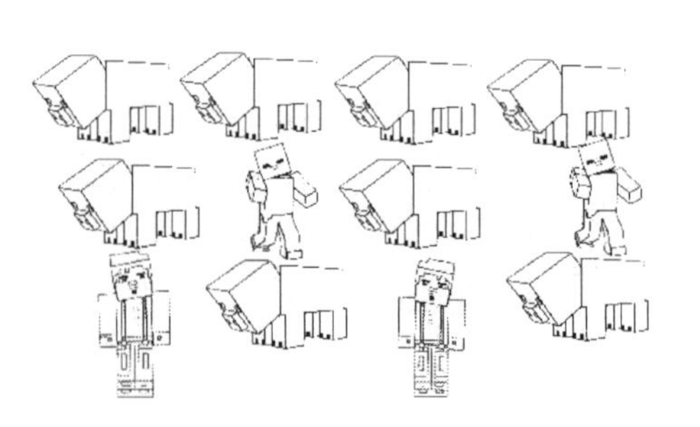

1. <u>Color</u> the numerator. <u>Find</u> the Common Denominator. <u>Compare</u> unlike fractions.

I usually avoid all the tricks, but now I want to help you with the tricks since they really work.

I compare one fourth (or a quarter) and five sixths.

$$\frac{1}{4} \ ? \ \frac{5}{6}$$

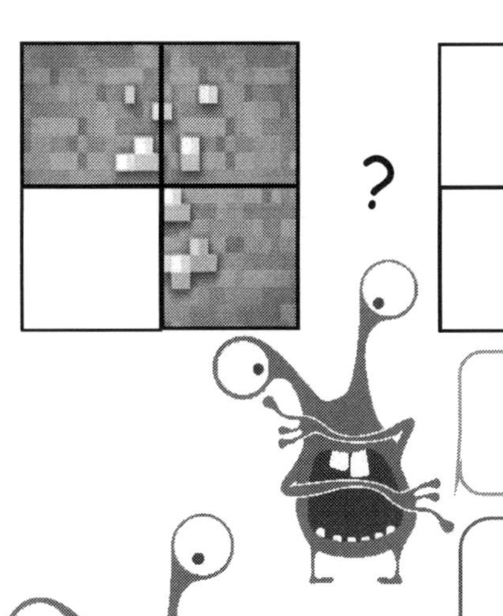

?

Compare the halves!

I don't see any halves!

Of course, there are no halves. But is one fourth more or less than a half?

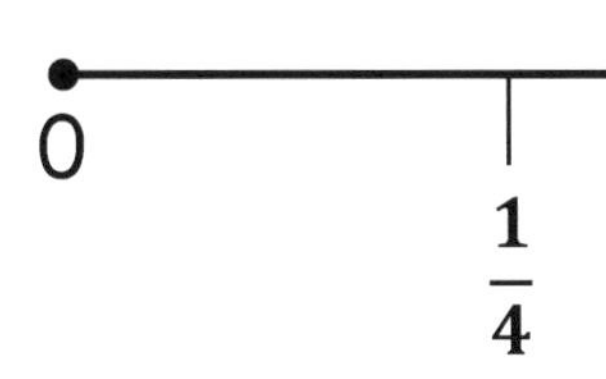

One fourth is less than a half!

0 $\frac{1}{4}$ $\frac{1}{2} = \frac{2}{4}$ $\frac{3}{4}$ 1

$$\frac{1}{4} < \frac{1}{2}$$

Because a half with the denominator 4 is two fourths: $\frac{2}{4}$.

And $\frac{1}{4} < \frac{2}{4}$.

Great job! Now, are five sixths more or less than a half?

Five sixths are more than a half.

$\frac{5}{6} > \frac{1}{2}$

since a half with the denominator 6 is three sixths: $\frac{3}{6}$. And $\frac{5}{6} > \frac{3}{6}$.

I compare one fourth (=a quarter) and five sixths through halves.

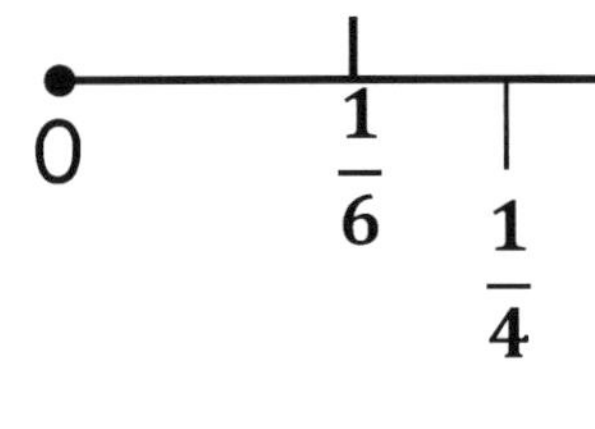

0 $\frac{1}{6}$ $\frac{1}{4}$ $\frac{1}{2} = \frac{2}{4} = \frac{3}{6}$ $\frac{3}{4}$ $\frac{3}{6}$ 1

$\frac{1}{4} < \frac{1}{2} < \frac{5}{6}$

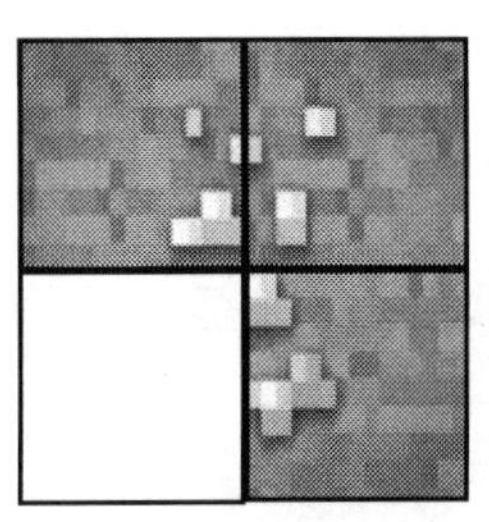

<

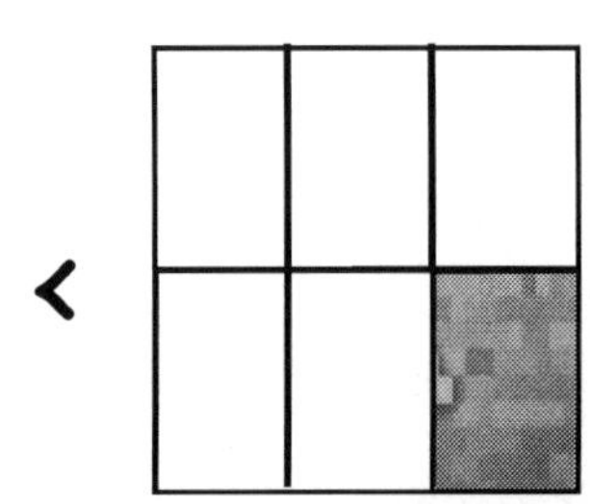

Good job, me! You're better than all the Brainers!

Back to math! How to find a half of a fraction? Look at the denominator and divide it by 2!

1. Compare fractions using my trick and write ">" or "<." Color the correct fraction of each shape. I'll compare $\frac{5}{6}$ and $\frac{1}{4}$.

$\frac{any\ number\ (x)}{6}$ → a half with the denominator 6 is three sixths ($\frac{3}{6}$) → compare $\frac{3}{6}$ and $\frac{5}{6}$ → $\frac{5}{6} > \frac{3}{6}$

$\frac{any\ number\ (x)}{4}$ → a half with the denominator 4 is two fourths ($\frac{2}{4}$) → compare $\frac{2}{4}$ and $\frac{1}{4}$ → $\frac{1}{4} < \frac{2}{4}$

$\frac{2}{6}$ ___ $\frac{7}{12}$

$\frac{2}{6}$ ___ $\frac{1}{2}$ ___ $\frac{7}{12}$

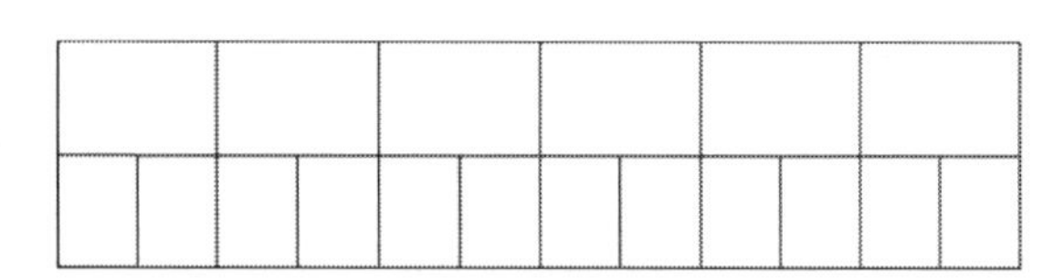

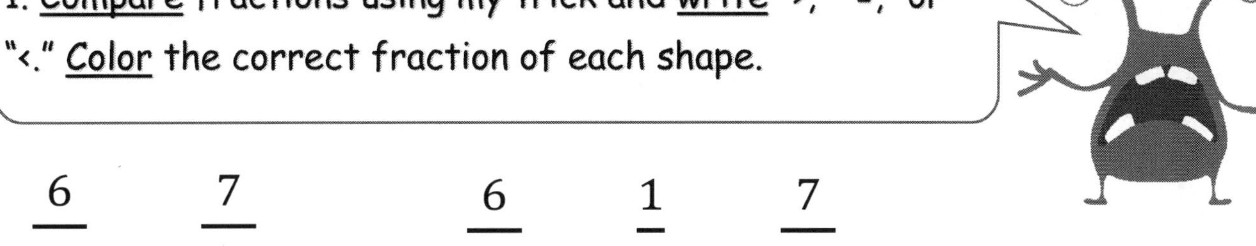

$\frac{6}{10}$ ___ $\frac{7}{16}$ $\qquad$ $\frac{6}{10}$ ___ $\frac{1}{2}$ ___ $\frac{7}{16}$

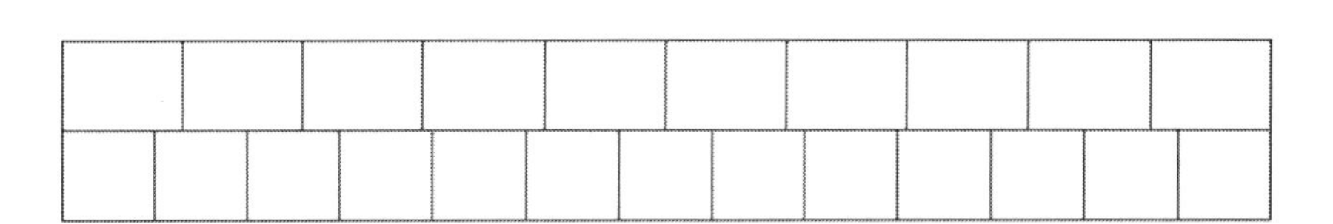

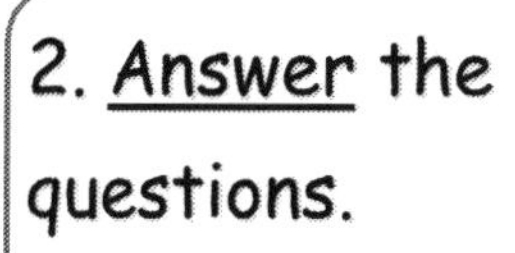

$\frac{3}{8}$ ___ $\frac{8}{11}$ $\qquad$ $\frac{3}{8}$ ___ $\frac{1}{2}$ ___ $\frac{8}{11}$

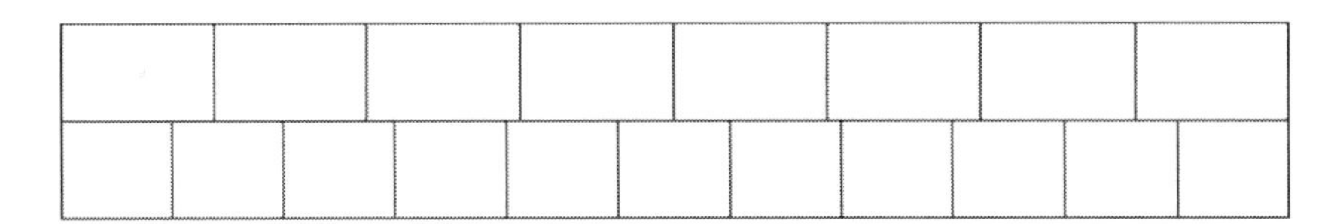

Find $\frac{1}{5}$ of zombies.

$\frac{4}{7}$ ___ $\frac{5}{12}$ $\qquad$ $\frac{4}{7}$ ___ $\frac{1}{2}$ ___ $\frac{5}{12}$

Find $\frac{3}{5}$ of zombies.

$\frac{7}{9}$ ___ $\frac{6}{13}$ $\qquad$ $\frac{7}{9}$ ___ $\frac{1}{2}$ ___ $\frac{6}{13}$

Find $\frac{2}{5}$ of zombies

$\frac{3}{10}$ ___ $\frac{11}{18}$ $\qquad$ $\frac{3}{10}$ ___ $\frac{1}{2}$ ___ $\frac{11}{18}$

$\frac{4}{12}$ ___ $\frac{6}{9}$ $\qquad$ $\frac{4}{12}$ ___ $\frac{1}{2}$ ___ $\frac{6}{9}$

$\frac{4}{16}$ ___ $\frac{10}{15}$ $\qquad$ $\frac{4}{16}$ ___ $\frac{1}{2}$ ___ $\frac{10}{15}$

1. Find and circle or cross out the words to find out more about Minecraft.

T	S	B	N	Z	S	R	S	B	A	R	H	T	J
I	O	J	M	L	K	E	L	W	R	Z	H	J	O
U	C	V	I	W	C	P	I	Q	R	V	T	U	Q
C	U	G	M	K	O	E	D	K	O	B	E	Z	X
R	V	F	C	W	L	A	I	F	W	N	B	Z	I
I	X	X	M	X	B	T	N	F	D	E	L	B	C
C	N	L	E	Q	D	E	G	S	I	T	Y	R	Y
K	G	W	T	K	I	R	F	C	S	W	E	X	J
C	E	L	A	W	L	S	L	K	P	O	G	P	D
O	D	R	Y	F	O	J	O	P	E	R	B	H	P
L	M	M	O	T	S	R	O	E	N	Y	H	F	L
C	E	T	A	L	P	E	R	U	S	S	E	R	P
T	E	Q	Z	L	O	C	K	M	E	M	O	R	Y
N	O	I	T	A	E	R	C	Z	R	H	Z	A	Y

CREATION

SOLID BLOCKS

CLOCK CIRCUIT

REPEATERS

PRESSURE PLATE

LOCK MEMORY

SLIDING FLOOR

ARROW DISPENSER

What's the difference between a proper fraction and an improper fraction?

That's easy! An improper fraction always misbehaves😉! Just like me!

A proper fraction is less than a whole: that means the numerator (top number) is less than the denominator (bottom number):

a whole > $\frac{1}{2}$ (a half);

a whole > $\frac{3}{4}$ (three fourths).

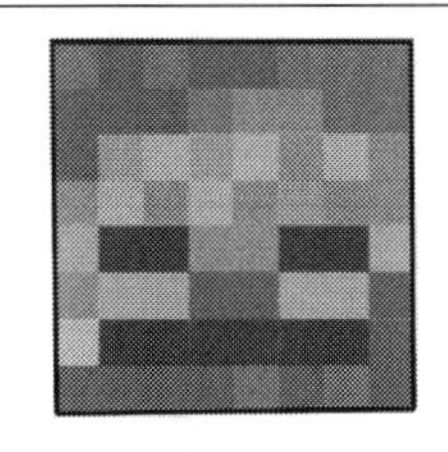 > 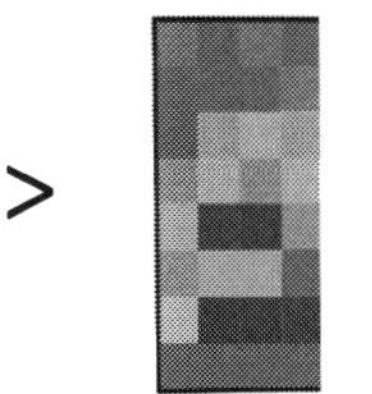

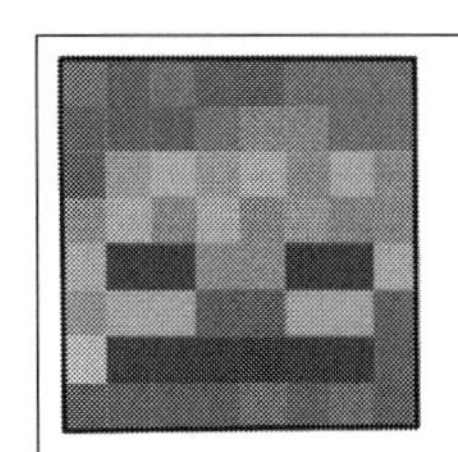 > 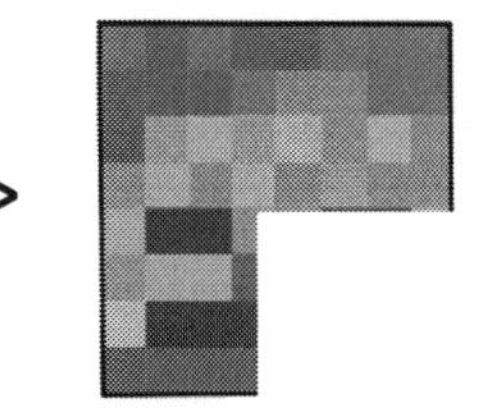

Here, an improper fraction is equal to or more than a whole: that means the numerator (top number) is equal to or greater than the denominator (bottom number):

$\frac{4}{4}$ (four fourths) = a whole

$\frac{3}{2}$ (three halves) > a whole.

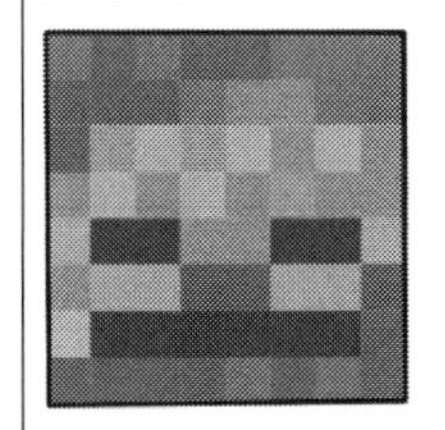 = 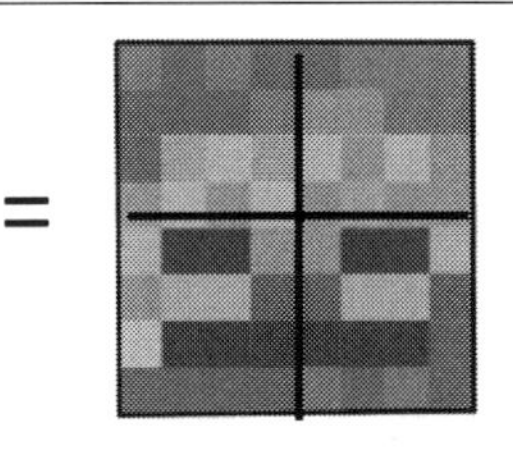

An improper fraction has a value equal to or greater than 1.

 > 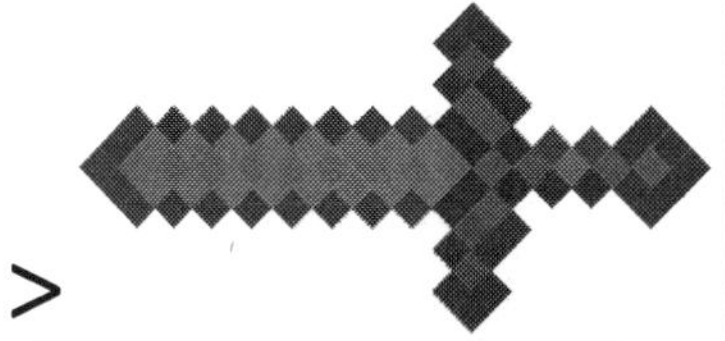

3 units of a half are greater than 2 units of a half: $\frac{3}{2} > \frac{2}{2}$.

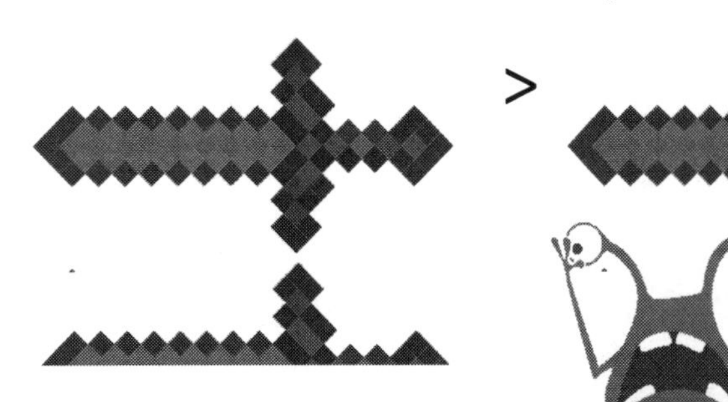

I would say that we compare 1 whole plus $\frac{1}{2}$ and 1 whole.

Numbers like 1, 2, 3, 4, etc. are called whole numbers

You mentioned a mixed fractions. A combination of a whole number with a fraction is called a mixed number.

$1\frac{1}{2}$ or $3\frac{2}{5}$ are called mixed numbers.

Hint: A mixed number = the sum of a whole number and a fraction: $1\frac{1}{2} = 1 + \frac{1}{2}$

1. Cross out proper fractions and circle improper fractions. Simplify if possible. Be quick! I need to build a house!

$\frac{8}{3} = -$ $\frac{5}{9} = -$ $\frac{4}{8} = -$ $\frac{6}{2} = -$

$\frac{1}{12} = -$ $\frac{4}{3} = -$ $\frac{15}{16} = -$ $\frac{20}{11} = -$

$\frac{7}{10} = -$ $\frac{3}{8} = -$ $\frac{21}{16} = -$ $\frac{18}{9} = -$

$\frac{3}{9} = -$ $\frac{14}{3} = -$ $\frac{1}{5} = -$ $\frac{3}{6} = -$

$\frac{7}{5} = -$ $\frac{6}{8} = -$ $\frac{3}{24} = -$ $\frac{36}{12} = -$

1. Draw the swords to match each fraction. Circle the proper fractions.

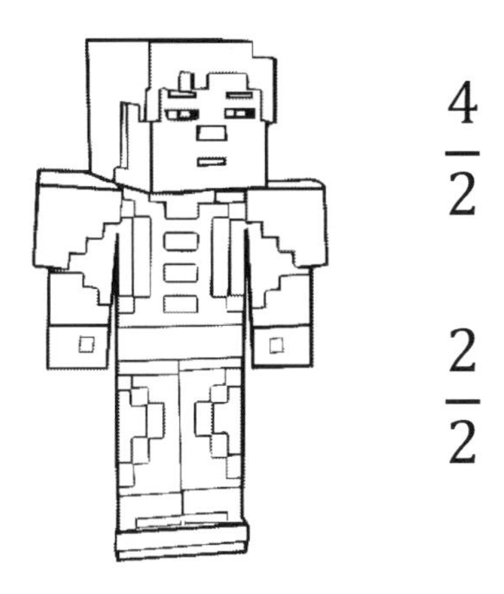

$\frac{4}{2}$

$\frac{2}{2}$

2. Color the blocks black to match each fraction. Circle the improper fraction(s). Cross out the block(s) you do not need.

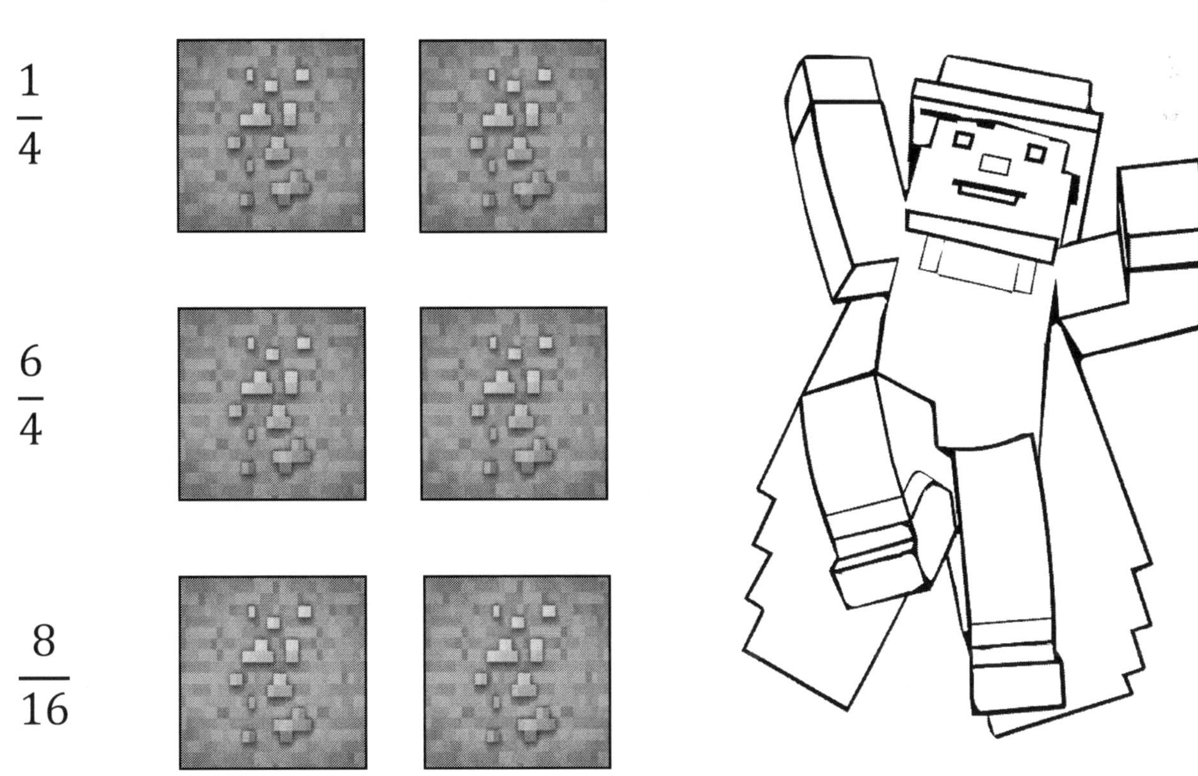

$\frac{1}{4}$

$\frac{6}{4}$

$\frac{8}{16}$

3. Four-sevenths of 42 villagers are harvesting crops, and 7 villagers are planting seeds. How many villagers are picking up carrots?

Answer: ____

How can I convert an improper fraction to a mixed number? For example, $\frac{3}{2}$ equals What?

Hint: The improper fraction equals the mixed number: $\frac{3}{2} = 1\frac{1}{2}$

Aha... The mixed number ($1\frac{1}{2}$) is the sum of a whole number and a fraction:

$1\frac{1}{2} = 1 + \frac{1}{2}$ (a mixed number consists of a whole number part and a fraction part.)

Step 1: Ask a question: The improper fraction equals what mixed number? For example, the improper fraction $\frac{3}{2}$ equals what mixed number? $\frac{2}{2}$ is a whole. Then, $\frac{3}{2} - \frac{2}{2} = \frac{1}{2}$.

So, $\frac{3}{2} = \frac{2}{2} + \frac{1}{2} = 1 + \frac{1}{2} = 1\frac{1}{2}$

Another algorithm:

$\frac{5}{2}$ = ? mixed number

To convert an improper fraction to a mixed number, you need to divide the numerator by the denominator to get the whole number 2 and the remainder 1.

The whole number part → 2

The denominator of a fraction part → 2 | 5

-4

1 ← The numerator of a fraction part

The whole number in the quotient is the whole part of your mixed number.

The remainder is the numerator of a fraction part of your mixed number.

$\frac{11}{3}$ = ? mixed number

3 = the whole number part of a mixed number

denominator → 3 | 11 ← = numerator of an improper fraction

9

2 ← = remainder = the new numerator of a fraction part of a mixed number

$\frac{11}{3} = 3\frac{2}{3}$

1. Convert an improper fraction to a mixed number. Color the blocks to match each fraction.

$\frac{11}{5} = \_\,\frac{\_}{\_}$ $\frac{19}{8} = \_\,\frac{\_}{\_}$ $\frac{13}{4} = \_\,\frac{\_}{\_}$ $\frac{15}{2} = \_\,\frac{\_}{\_}$

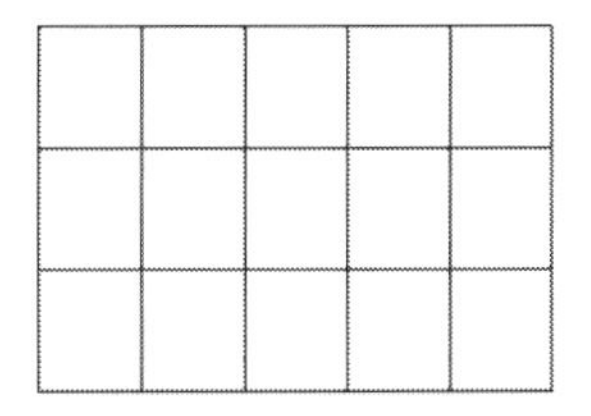
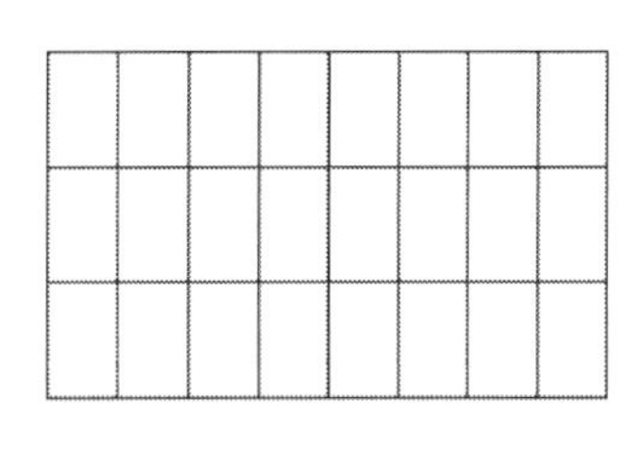
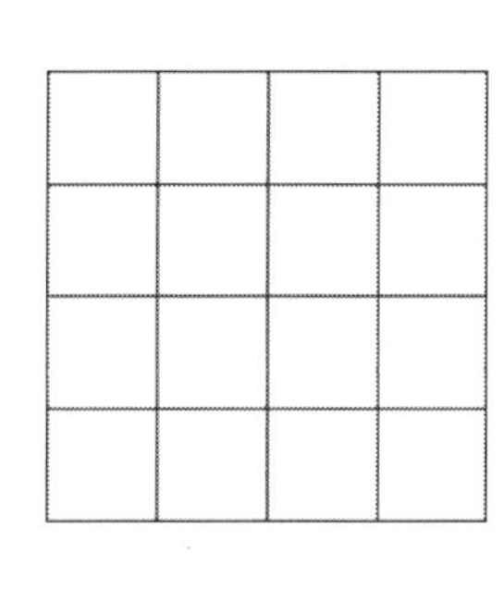
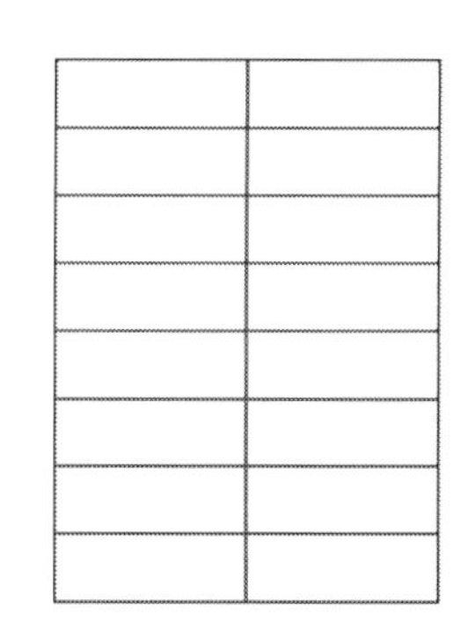

2. Two-eighths of 72 villagers are fishermen, and four-ninths of the rest of the villagers are shepherds. How many villagers are shepherds?

Answer: ____

3. Convert an improper fraction to a mixed number.

$\frac{8}{3} = \_\,\frac{\_}{\_}$ $\frac{9}{5} = \_\,\frac{\_}{\_}$ $\frac{8}{4} = \_\,\frac{\_}{\_}$ $\frac{6}{2} = \_\,\frac{\_}{\_}$

$\frac{12}{5} = \_\,\frac{\_}{\_}$ $\frac{10}{3} = \_\,\frac{\_}{\_}$ $\frac{16}{8} = \_\,\frac{\_}{\_}$ $\frac{20}{11} = \frac{\_}{\_}$

$\frac{18}{10} = \_\,\frac{\_}{\_}$ $\frac{24}{7} = \_\,\frac{\_}{\_}$ $\frac{20}{16} = \_\,\frac{\_}{\_}$ $\frac{18}{9} = \_\,\frac{\_}{\_}$

1. Convert an improper fraction to a mixed number.

$\frac{9}{3} = \_\frac{\_}{\_}$ $\frac{14}{3} = \_\frac{\_}{\_}$ $\frac{5}{2} = \_\frac{\_}{\_}$ $\frac{7}{6} = \_\frac{\_}{\_}$

$\frac{7}{5} = \_\frac{\_}{\_}$ $\frac{8}{5} = \_\frac{\_}{\_}$ $\frac{24}{3} = \_\frac{\_}{\_}$ $\frac{36}{12} = \_\frac{\_}{\_}$

$\frac{22}{3} = \_\frac{\_}{\_}$ $\frac{34}{6} = \_\frac{\_}{\_}$ $\frac{11}{2} = \_\frac{\_}{\_}$ $\frac{28}{5} = \_\frac{\_}{\_}$

$\frac{42}{9} = \_\frac{\_}{\_}$ $\frac{25}{4} = \_\frac{\_}{\_}$ $\frac{63}{8} = \_\frac{\_}{\_}$ $\frac{36}{10} = \_\frac{\_}{\_}$

$\frac{31}{7} = \_\frac{\_}{\_}$ $\frac{46}{5} = \_\frac{\_}{\_}$ $\frac{25}{12} = \_\frac{\_}{\_}$ $\frac{41}{15} = \_\frac{\_}{\_}$

2. Two-sevenths of 35 spiders are cave spiders. How many more or less cave spiders than regular spiders are there?

Answer: ____

3. I have run away from five-eighths of the 32 zombies I had met in my game. My brother run away from four-ninths of the 27 zombies he had met in his game. Who was luckier and run away from the greatest number of zombies?

Answer: ____

1. Find and circle or cross out the words to find out more about Minecraft.

E	O	L	M	H	W	G	K	P	H	S	C
C	Z	U	E	P	O	J	O	C	C	V	N
E	N	V	D	G	T	R	R	U	G	Q	O
I	U	F	I	Q	P	O	S	P	D	C	I
P	V	N	E	V	P	W	G	E	H	J	T
R	V	B	V	R	B	S	A	W	S	M	C
E	F	L	A	B	Y	R	I	N	T	H	E
T	O	E	L	G	N	I	N	W	A	P	S
N	R	I	F	E	O	U	P	U	P	M	S
E	F	D	O	M	A	O	V	D	W	D	K
C	X	F	P	I	L	L	A	R	S	G	T
B	R	F	V	A	J	X	X	U	K	W	Z

HORSES

REAR PORCH

LABYRINTH

SECTION

CENTERPIECE

MEDIEVAL

PILLARS

SPAWNING

How can I convert a mixed number to an improper fraction?

For example, $2\frac{2}{3}$ equals What?

Hint: The mixed number equals the improper fraction:

$2\frac{1}{2}$ (two swords and a half) = $\frac{5}{2}$ (five halves of swords)

Correct. Let's think!

Step 1: The whole number part equals what fraction? The whole number part of 1 equals: $1 = \frac{2}{2}$. The whole number part of 2 equals: $2 = \frac{4}{2}$.

Step 2: Add the two fractions: $\frac{4}{2} + \frac{1}{2} = \frac{4+1}{2} = \frac{5}{2}$

Another algorithm: $2\frac{1}{2} = ?$

Step 1: Multiply the whole number part by the denominator → $2 \times 2 = 4$ (the product is the numerator of the whole number part)

(in our problem we multiply 2 by 2).

Step 2: Add the product to the numerator of a fraction part → $4 + 1 = 5$ (we added the numerator of the whole number part and the numerator of the fraction part that equals the numerator of the improper fraction)

(in our problem it's 1).

$$2\frac{1}{2} = \frac{5}{2}$$

1. Convert a mixed number to an improper fraction.

$1\frac{3}{5} = -$ $2\frac{5}{8} = -$ $3\frac{1}{4} = -$

1) 1 × 5 = ____ 1) __ × __ = ____ 1) __ × __ = ____

2) __ + 3 = ____ 2) __ + __ = ____ 2) __ + __ = ____

2. Three-tenths of 50 blooming flowers were blue orchid. If sunflower were one-tenths more than blue orchid, and the rest of the flowers were divided by peonies and lilacs, and roses respectively, how many lilacs were blooming?

Answer: ______________________________

3. Convert a mixed number to an improper fraction. Write the missing numbers.

$4\frac{2}{5} = \frac{(4\times5)+2}{5} = -$ $7\frac{1}{2} = \frac{(\_\times\_)+\_}{\_} = -$

$2\frac{1}{4} = \frac{(\_\times\_)+\_}{\_} = -$ $6\frac{2}{3} = \frac{(\_\times\_)+\_}{\_} = -$

$3\frac{1}{2} = \frac{(\_\times\_)+\_}{\_} = -$ $3\frac{5}{8} = \frac{(\_\times\_)+\_}{\_} = -$

1. Convert a mixed number to an improper fraction. Write the missing numbers.

$7\frac{3}{4} = \frac{(\_\times\_)+\_}{\_} = \frac{}{}$ $\quad$ $9\frac{4}{5} = \frac{(\_\times\_)+\_}{\_} = \frac{}{}$

$3\frac{3}{8} = \frac{(\_\times\_)+\_}{\_} = \frac{}{}$ $\quad$ $5\frac{5}{8} = \frac{(\_\times\_)+\_}{\_} = \frac{}{}$

$6\frac{1}{10} = \frac{(\_\times\_)+\_}{\_} = \frac{}{}$ $\quad$ $3\frac{2}{9} = \frac{(\_\times\_)+\_}{\_} = \frac{}{}$

$10\frac{3}{5} = \frac{}{} = \frac{}{}$ $\quad$ $7\frac{9}{11} = \frac{}{} = \frac{}{}$

2. I met some horses of two colors. Three-sixths of the horses were white. 24 horses were black. How many horses did I meet?

Answer: ____

3. The villager planted some melon and pumpkin seeds. Two-fifths of them were melon seeds. If he planted 45 pumpkin seeds, how many melon and pumpkin seeds did he have at first?

Answer: ____

1. Convert an improper fraction to a mixed number. Circle the right answer.

$\frac{9}{2} = \_\ \frac{\_}{\_}$ $\frac{25}{8} = \_\ \frac{\_}{\_}$ $\frac{31}{4} = \_\ \frac{\_}{\_}$ $\frac{37}{5} = \_\ \frac{\_}{\_}$

a) $5\frac{2}{2}$ a) $2\frac{5}{8}$ a) $5\frac{3}{4}$ a) $7\frac{2}{5}$

b) $1\frac{3}{2}$ b) $3\frac{1}{8}$ b) $8\frac{1}{4}$ b) $6\frac{7}{5}$

c) $4\frac{1}{2}$ c) $5\frac{1}{8}$ c) $4\frac{3}{4}$ c) $5\frac{21}{5}$

d) $4\frac{2}{2}$ d) $3\frac{2}{8}$ d) $7\frac{3}{4}$ d) $4\frac{4}{5}$

2. I am an improper fraction. The sum of my numerator and denominator is 14, their difference is 2^3 (2 × 2 × 2). What fraction am I?

a) $\frac{8}{6}$ b) $\frac{15}{1}$ c) $\frac{6}{8}$ d) $\frac{11}{3}$

3. True or false? Explain. Circle the right answer.

To convert an improper fraction to a mixed number, you need to divide the denominator by the numerator.

____

____

a) TRUE b) FALSE

1. Convert a mixed number to an improper fraction. Circle the right answer.

$2\frac{1}{4} = \frac{\_}{\_}$ $7\frac{3}{8} = \frac{\_}{\_}$ $5\frac{7}{12} = \frac{\_}{\_}$ $3\frac{2}{5} = \frac{\_}{\_}$

a) $\frac{8}{4}$	a) $\frac{60}{8}$	a) $\frac{66}{12}$	a) $\frac{17}{5}$
b) $\frac{3}{4}$	b) $\frac{56}{8}$	b) $\frac{70}{12}$	b) $\frac{10}{5}$
c) $\frac{9}{4}$	c) $\frac{58}{8}$	c) $\frac{67}{12}$	c) $\frac{21}{5}$
d) $\frac{10}{4}$	d) $\frac{59}{8}$	d) $\frac{77}{12}$	d) $\frac{16}{5}$

2. I am an improper fraction. The product of my numerator and denominator is 27, their quotient is 3 What fraction am I?

a) $\frac{27}{9}$ b) $\frac{15}{14}$ c) $\frac{12}{4}$ d) $\frac{9}{3}$

3. True or false? Explain. Circle the right answer.

The smaller the denominator the greater the fraction.

____

____

b) TRUE b) FALSE

1. Find and circle or cross out the words to find out more about Minecraft.

L	C	O	O	X	D	I	E	L	O	E	K	H	E	J
Y	I	C	U	U	X	D	G	G	D	C	C	O	D	U
S	Q	A	C	T	N	L	B	F	A	F	P	R	I	O
K	Q	A	S	T	S	A	M	R	G	E	F	I	R	Q
V	K	J	T	R	D	T	T	A	T	Y	U	Z	G	B
N	G	T	M	O	A	G	A	R	L	L	S	O	N	U
X	S	G	L	I	N	L	I	N	P	R	H	N	I	S
A	A	V	W	I	P	F	U	F	D	S	B	T	L	S
Y	O	V	K	W	Y	N	U	G	C	I	F	A	L	J
Z	R	A	J	I	I	B	W	P	N	M	N	L	I	I
H	N	A	N	F	C	C	I	M	E	A	I	G	R	J
S	H	G	D	P	N	L	C	V	V	P	I	P	H	R
F	A	N	T	A	S	Y	S	T	Y	L	E	R	T	Q
O	X	P	L	Q	L	T	X	R	K	U	F	V	T	W
X	S	Z	U	K	R	C	R	E	I	G	C	T	J	H

SNAKING TRACK

OUTSTANDING

HORIZONTAL

PETRIFYING

MASTS

FANTASY-STYLE

TRIANGULAR SAIL

THRILLING RIDE

I need to add fractions:

$$\frac{1}{4} + \frac{5}{6} = \ ?$$

That's impossible!

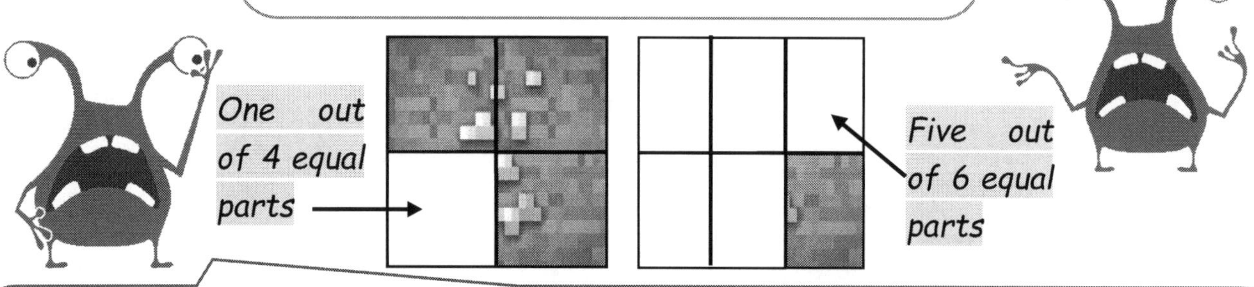

Just imagine, I ate $\frac{1}{4}$ of a Strawberry Chocolate bar and $\frac{3}{8}$ of a Hazelnut Chocolate bar. How much did I eat?

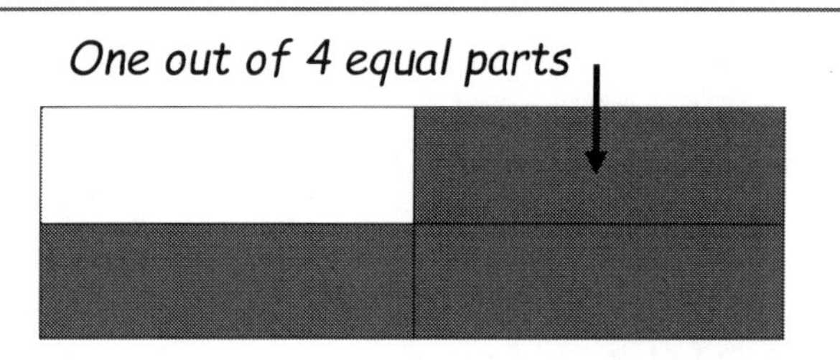

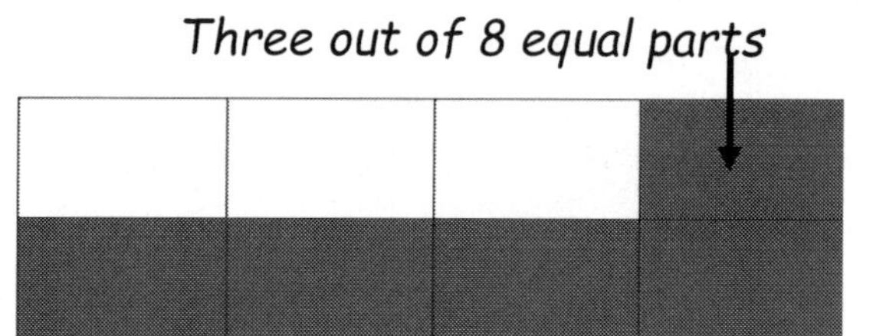

Step 1: Find out equivalent fractions to get the same common denominator (the common denominator is LCD)

Step 2: Add the numerators only

Step 3: Simplify where possible

Yeah, first, we need to find out equivalent fractions!

$$\frac{1}{4} = \frac{2}{8} = \frac{3}{12} = \frac{4}{16} = \frac{5}{20} = \frac{6}{24} = \frac{7}{28} = \frac{8}{32}$$

$$\frac{3}{8} = \frac{6}{16} = \frac{9}{24} = \frac{12}{32}$$

Equivalent fractions are found out by multiplying both the numerator and denominator by the same number

I have several like fractions with 8, 16, 24, 32 in the denominator. I colored the fractions strips grey.

$$\frac{1}{4} = \frac{2}{8} = \frac{3}{12} = \frac{4}{16} = \frac{5}{20} = \frac{6}{24} = \frac{7}{28} = \frac{8}{32}$$

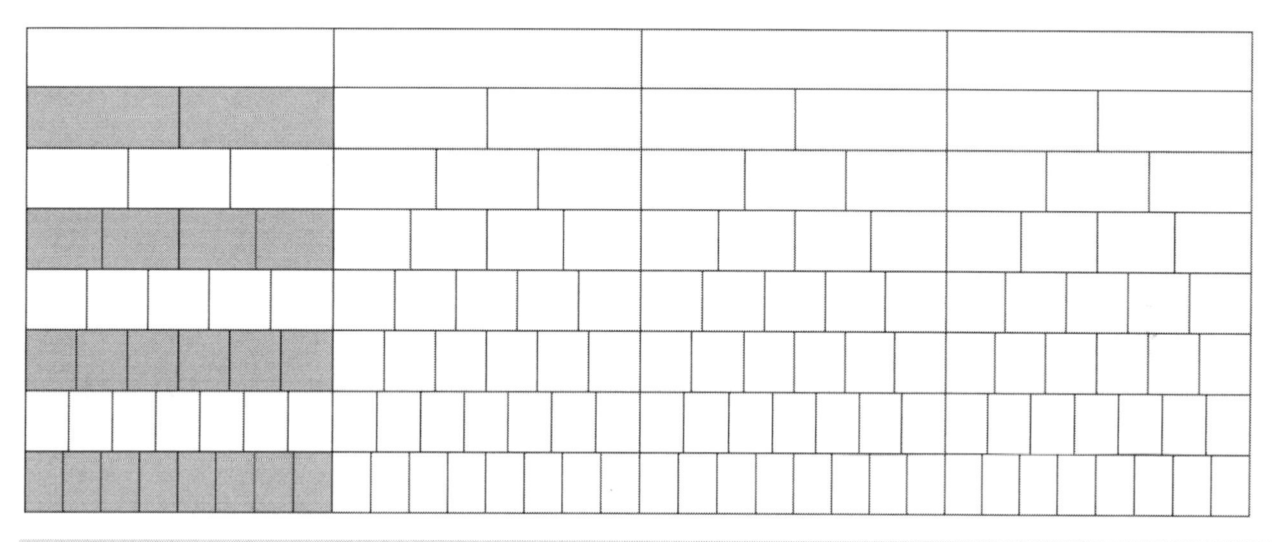

1 strip out of 4 strips is the same as 2 strips out of 8 strips, or 3 strips out of 12 strips, or 4 strips out of 16 strips, or 5 strips out of 20 strips, or 6 strips out of 24 strips, or 7 strips out of 28 strips, or 8 strips out of 32 strips - their value remains the same

$$\frac{3}{8} = \frac{6}{16} = \frac{9}{24} = \frac{12}{32}$$

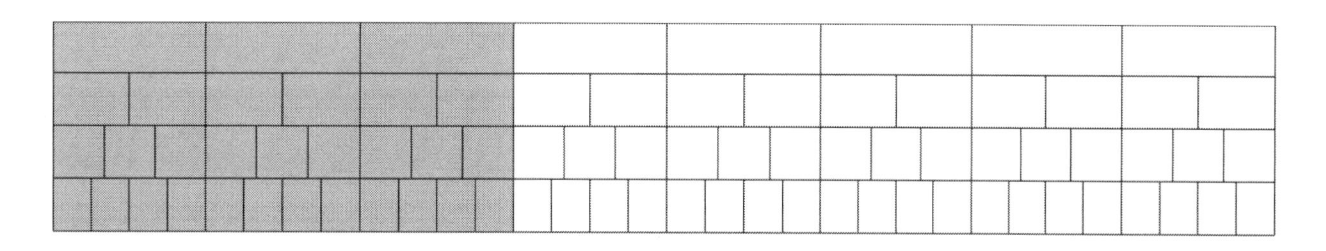

3 strips out of 8 strips are the same as 6 strips out of 16 strips, or 9 strips out of 24 strips, or 12 strips out of 32 strips - their value remains the same

Aha! The denominators are the same in these fractions!
These are common denominators.

Hint: *The denominator (8) is the numeral that tells the number of parts a whole is divided into*

$$\frac{2}{8} = \frac{4}{16} = \frac{6}{24} = \frac{8}{32}$$

$$\frac{3}{8} = \frac{6}{16} = \frac{9}{24} = \frac{12}{32}$$

Adding fractions is possible ONLY with first writing the fractions with common denominators

Step 1 is done! We changed the given fractions to equivalent fractions to find common denominators. Now we must find the least common denominator.

8 is the LEAST common denominator for

$\frac{1}{4}$ $(=\frac{2}{8})$ and $\frac{3}{8}$.

These are like fractions since they both have 8 in the denominator.

So, you can write 8 as a common denominator for both fractions.

And then, add the numerators 2 and 3 to find the sum.

Step 1: Find the Common Denominator

Step 2: Add the Numerators to find the sum

$$\frac{1}{4} + \frac{3}{8} = \frac{2}{8} + \frac{3}{8} = \frac{2+3}{8} = \frac{5}{8}$$

My problem is:

$$\frac{1}{4} + \frac{5}{6} = \ ?$$

Don't worry, we'll start slow, I may have some ideas. Find the multiples!

Hint: To find the multiples of a number, multiply this number by the counting numbers (1, 2, 3, 4, etc.) ☺

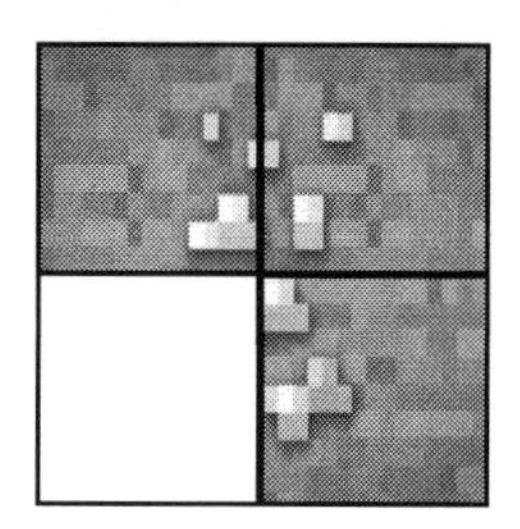

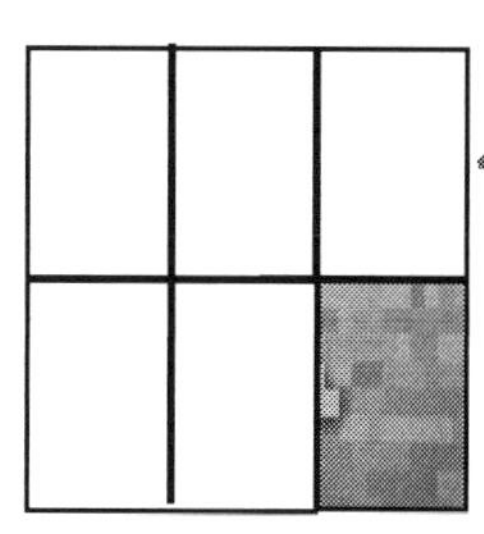

Step 1: Find the LEAST common multiple of 4 and 6.

Multiples of 4

For the denominator 4:	1 × 4 = 4	2 × 4 = 8	3 × 4 = 12
For the denominator 6:	1 × 6 = 4	2 × 6 = 12	3 × 6 = 18

Multiples of 6

The LEAST common multiple of 4 and 6 is 12:

1 × 4 = 4   2 × 4 = 8   3 × 4 = 12

1 × 6 = 4   2 × 6 = 12   3 × 6 = 18

How to find the least common multiple

Step 1: List all the multiples of the first denominator

Step 2: List all the multiples of the second denominator

Step 3: Circle the least multiple which is common to both denominators

Step 2: Draw a fractions bar, write 12 for the common denominator, and draw a slash above the numerator of each fraction:

$$\frac{1}{4} + \frac{5}{6} = \frac{\_ + \_}{12} =$$

Step 3: Divide the least common multiple by each denominator and write the quotient above each slash:

Hint: Use slashes to make calculations easy and fun 😊

$$\frac{1}{4}^{\diagdown 3} + \frac{5}{6}^{\diagdown 2} = \frac{\_ + \_}{12} =$$

$12 \div 4 = 3$

$12 \div 6 = 2$

Step 4: Multiply each numerator by the number above a slash:

$3 \times 1 = 3$

$2 \times 5 = 10$

$$\frac{1}{4}^{\diagdown 3} + \frac{5}{6}^{\diagdown 2} = \frac{(3 \times 1) + (2 \times 5)}{12} =$$

Hint: To add fractions, make sure that you add like fractions. That means they must have a common denominator 😊

You got the "new" numerators.

These are the numerators for the denominator 12:

$$\frac{1}{4} = \frac{3}{12}; \frac{5}{6} = \frac{10}{12}.$$

Step 5: Add the "new" numerators:

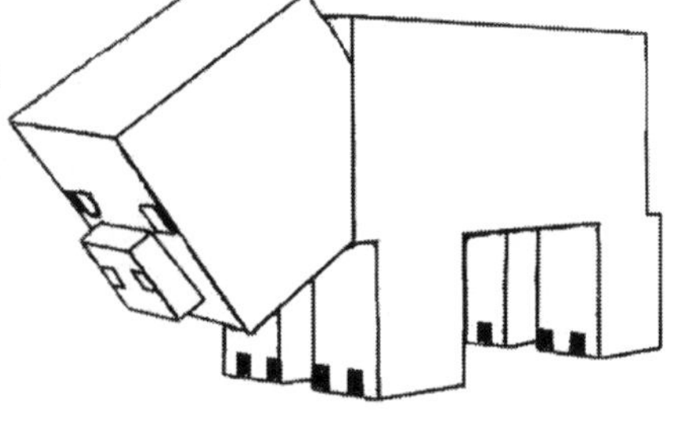

$$\frac{1}{4}^{\diagdown 3} + \frac{5}{6}^{\diagdown 2} = \frac{3 + 10}{12} = \frac{13}{12}$$

1. Add fractions. Write the missing numbers. Color the fraction of each shape. Simplify if possible.

$\frac{1}{4} + \frac{1}{3} = ?$

$\frac{3}{4} + \frac{2}{5} = ?$

Step 1: Find the least common denominator

$\frac{1}{4} = \frac{2}{\quad} = \frac{3}{\quad}$

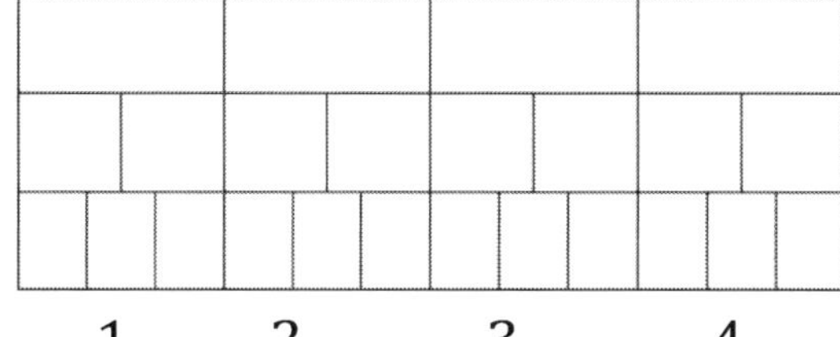

$\frac{1}{3} = \frac{2}{\quad} = \frac{3}{\quad} = \frac{4}{\quad}$

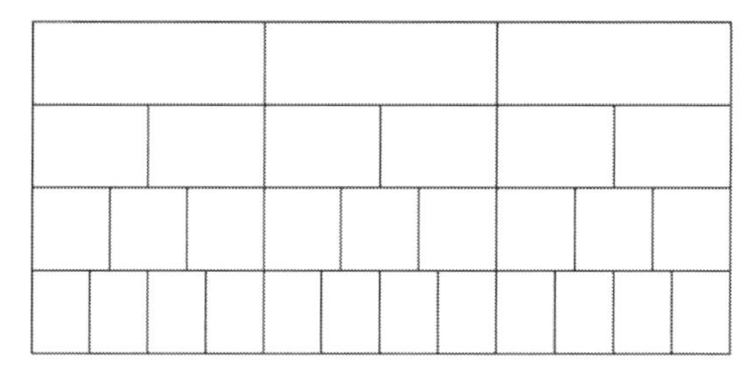

___ is the least common denominator

$\frac{3}{4} = \frac{6}{\quad} = \frac{9}{\quad} = \frac{12}{\quad} = \frac{15}{\quad}$

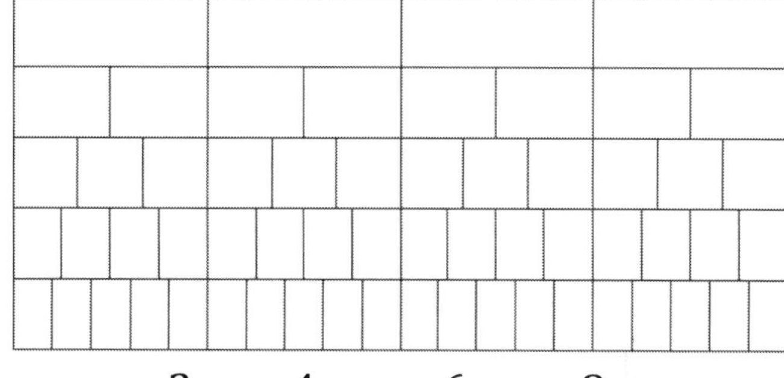

$\frac{2}{5} = \frac{4}{\quad} = \frac{6}{\quad} = \frac{8}{\quad}$

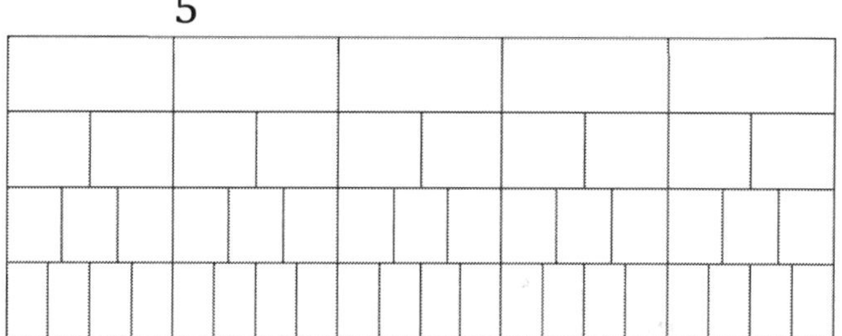

___ is the least common denominator

Step 2: Add the Numerators to find the sum

$\frac{1}{4} + \frac{1}{3} = \frac{\quad + \quad}{\quad} = \frac{\quad}{\quad} = \_\frac{\quad}{\quad}$

$\frac{3}{4} + \frac{2}{5} = \frac{\quad + \quad}{\quad} = \frac{\quad}{\quad} = \_\frac{\quad}{\quad}$

2. True or false? Explain. Circle the right answer.

The greater the numerator in like fractions the smaller the fraction.

____

a) TRUE

b) FALSE

1. Add fractions. Write the missing numbers. Color the fraction of each shape. Simplify if possible.

$\frac{2}{3} + \frac{5}{12} = ?$

$\frac{5}{6} + \frac{3}{4} = ?$

Step 1: Find the least common denominator

$\frac{2}{3} = \frac{4}{\_\_} = \frac{6}{\_\_} = \frac{8}{\_\_}$

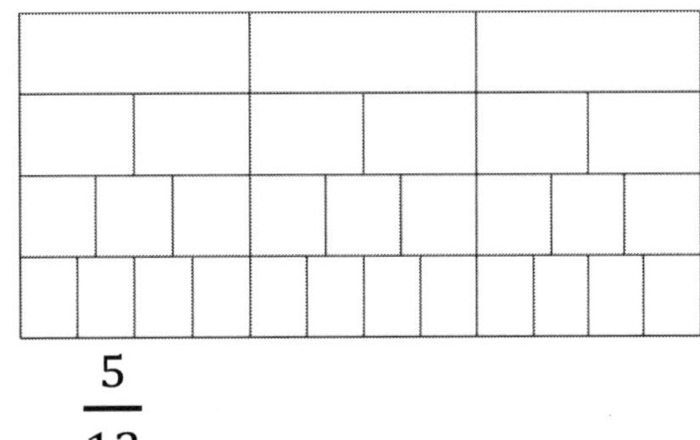

$\frac{5}{12}$

$\frac{5}{6} = \frac{10}{\_\_}$

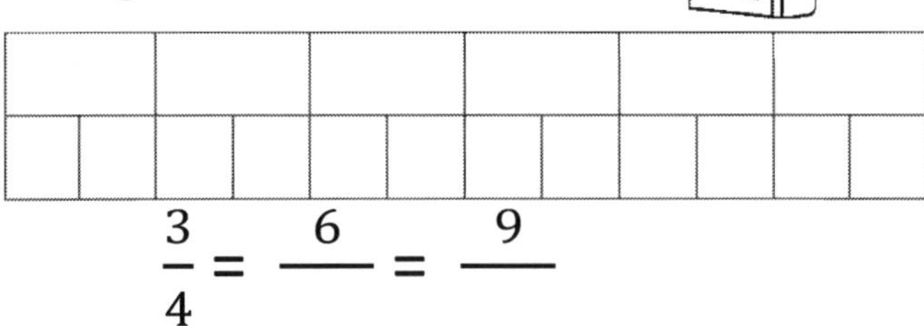

$\frac{3}{4} = \frac{6}{\_\_} = \frac{9}{\_\_}$

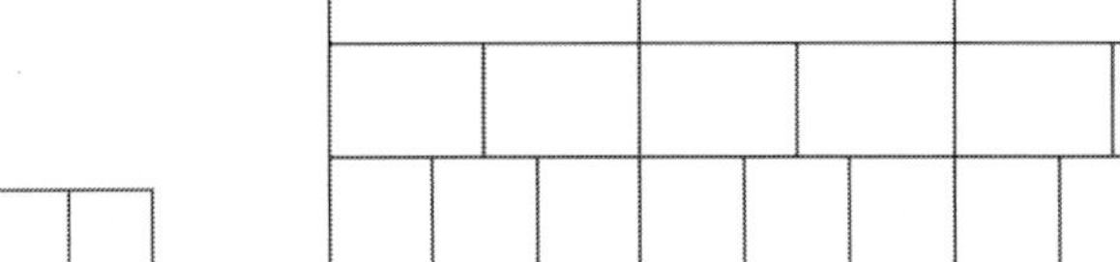

___ *is the least common denominator*

___ *is the least common denominator*

Step 2: Add the Numerators to find the sum

$\frac{2}{3} + \frac{5}{12} = \frac{\_ + \_}{\_\_} = \frac{\_}{\_} = \_\frac{\_}{\_}$

$\frac{5}{6} + \frac{3}{4} = \frac{\_ + \_}{\_\_} = \frac{\_}{\_} = \_\frac{\_}{\_}$

2. True or false? Explain. Circle the right answer.

The greater the denominator in unlike fractions the greater the fraction.

____

a) TRUE

b) FALSE

1. Add fractions. Write the missing numbers. Simplify if possible.

$\frac{2}{6} + \frac{1}{2} = \frac{\quad + \quad}{\qquad} = \frac{}{} = \_\frac{}{}$

$\frac{2}{4} + \frac{4}{5} = \frac{\quad + \quad}{\qquad} = \frac{}{} = \_\frac{}{}$

$\frac{1}{3} + \frac{1}{4} = \frac{\quad + \quad}{\qquad} = \frac{}{} = \_\frac{}{}$

$\frac{2}{3} + \frac{1}{6} = \frac{\quad + \quad}{\qquad} = \frac{}{} = \_\frac{}{}$

$\frac{4}{6} + \frac{2}{4} = \frac{\quad + \quad}{\qquad} = \frac{}{} = \_\frac{}{}$

$\frac{4}{9} + \frac{6}{18} = \frac{\quad + \quad}{\qquad} = \frac{}{} = \_\frac{}{}$

$\frac{5}{14} + \frac{3}{7} = \frac{\quad + \quad}{\qquad} = \frac{}{} = \_\frac{}{}$

$\frac{1}{6} + \frac{3}{8} = \frac{\quad + \quad}{\qquad} = \frac{}{} = \_\frac{}{}$

$\frac{3}{4} + \frac{7}{16} = \frac{\quad + \quad}{\qquad} = \frac{}{} = \_\frac{}{}$

$\frac{4}{5} + \frac{5}{6} = \frac{\quad + \quad}{\qquad} = \frac{}{} = \_\frac{}{}$

$\frac{7}{12} + \frac{7}{8} = \frac{\quad + \quad}{\qquad} = \frac{}{} = \_\frac{}{}$

$\frac{3}{10} + \frac{4}{6} = \frac{\quad + \quad}{\qquad} = \frac{}{} = \_\frac{}{}$

Hint: Find the least common denominator by multiplying the greatest given denominator by 2, then, by 3, and so on until you get a number that is exactly divided by the given denominators.

2. True or false? Explain. Circle the right answer.

The greater the denominator in unlike fractions the greater the fraction.

____

a) TRUE

b) FALSE

1. Add fractions. Write the missing numbers. Simplify if possible.

$\frac{9}{16} + \frac{11}{12} = \frac{\quad + \quad}{} = \frac{}{} = \_\frac{}{}$ $\frac{9}{18} + \frac{1}{4} = \frac{\quad + \quad}{} = \frac{}{} = \_\frac{}{}$

$\frac{7}{20} + \frac{5}{8} = \frac{\quad + \quad}{} = \frac{}{} = \_\frac{}{}$ $\frac{13}{18} + \frac{7}{10} = \frac{\quad + \quad}{} = \frac{}{} = \_\frac{}{}$

$\frac{8}{25} + \frac{4}{15} = \frac{\quad + \quad}{} = \frac{}{} = \_\frac{}{}$ $\frac{10}{28} + \frac{3}{4} = \frac{\quad + \quad}{} = \frac{}{} = \_\frac{}{}$

$\frac{9}{20} + \frac{11}{14} + \frac{2}{7} = \frac{\quad + \quad + \quad}{} = \frac{}{} = \_\frac{}{}$

$\frac{3}{7} + \frac{9}{14} + \frac{13}{21} = \frac{\quad + \quad + \quad}{} = \frac{}{} = \_\frac{}{}$

$\frac{1}{2} + \frac{7}{13} + \frac{5}{26} = \frac{\quad + \quad + \quad}{} = \frac{}{} = \_\frac{}{}$

2. True or false? Explain. Circle the right answer.

The least common denominator is the least factor for the numerator.

____

a) TRUE b) FALSE

1. Add fractions. Write the missing numbers. Simplify if possible.

$\frac{2}{13} + \frac{5}{39} = \frac{\quad + \quad}{\quad} = \frac{\quad}{\quad} = \_\frac{\quad}{\quad}$

$\frac{2}{5} + \frac{3}{10} =$

$\frac{5}{8} + \frac{5}{6} =$

$\frac{5}{6} + \frac{7}{15} =$

$\frac{7}{18} + \frac{8}{15} =$

$\frac{5}{18} + \frac{3}{12} =$

A common denominator may be found by multiplying together given denominators!

$\frac{11}{15} + \frac{7}{9} =$

$\frac{9}{13} + \frac{1}{4} =$

$\frac{13}{16} + \frac{7}{8} =$

$\frac{3}{21} + \frac{1}{6} =$

$\frac{5}{12} + \frac{3}{8} =$

$\frac{7}{16} + \frac{5}{7} =$

2. True or false? Explain. Circle the right answer.

The mixed number is the sum of a proper fraction and an improper fraction.

____

a) TRUE

b) FALSE

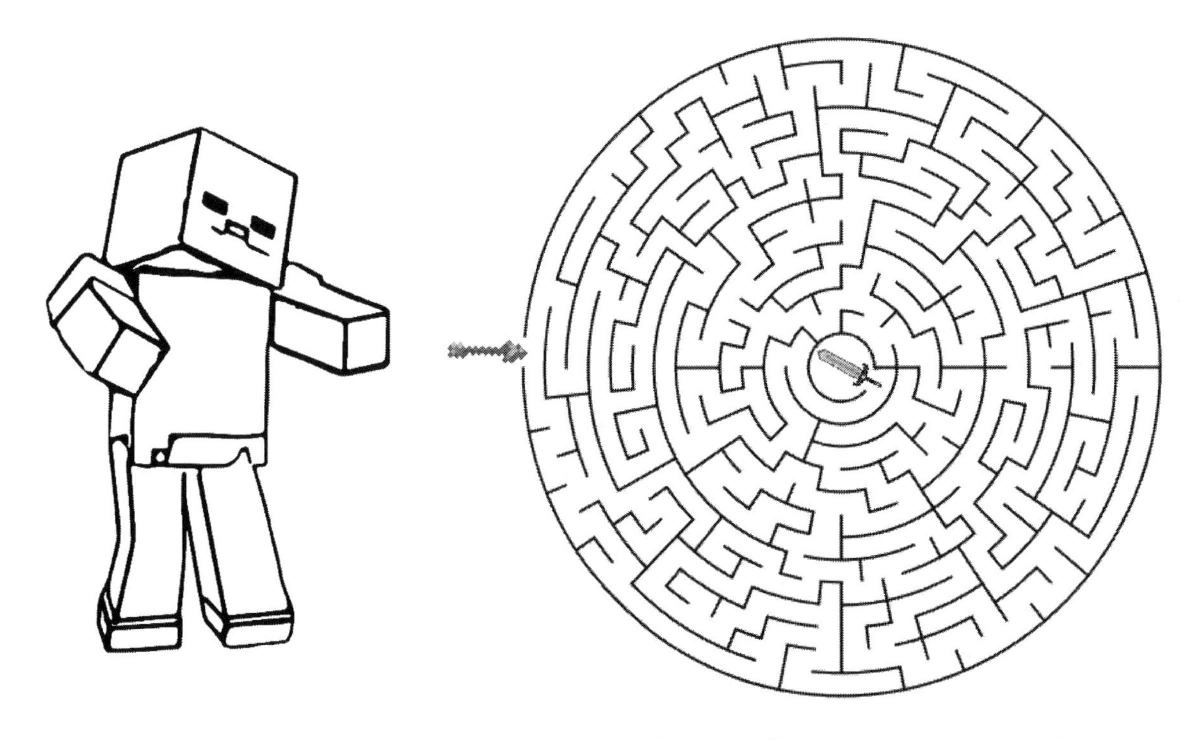

1. Find and circle or cross out the words to find out more about Minecraft.

M	E	W	L	P	O	J	X	P	U	X	G	Y
A	N	B	G	A	E	P	T	U	M	M	L	E
G	C	Z	A	A	T	I	U	H	S	E	A	G
N	L	F	K	X	D	N	G	L	T	P	V	A
I	O	P	R	N	L	C	O	A	E	D	F	G
F	S	K	I	W	Z	F	M	Z	Z	N	V	H
I	E	A	R	N	L	I	D	T	I	J	C	Q
C	M	G	H	V	X	J	O	Z	L	R	V	E
E	T	A	R	O	P	R	O	C	N	U	O	L
N	T	Z	R	R	N	R	A	R	U	Q	H	H
T	S	P	E	C	T	A	C	U	L	A	R	S
I	P	E	V	E	N	T	U	A	L	L	Y	Z
A	G	T	Q	I	V	K	R	L	G	I	P	N

OPULENCE

ENCLOSE

SPECTACULAR

MAGNIFICENT

HORIZONTAL

UNCORPORATE

EVENTUALLY

APPROXIMATELY

I need to subtract fractions:

$$\frac{3}{4} - \frac{1}{6} = ?$$

Let's try...

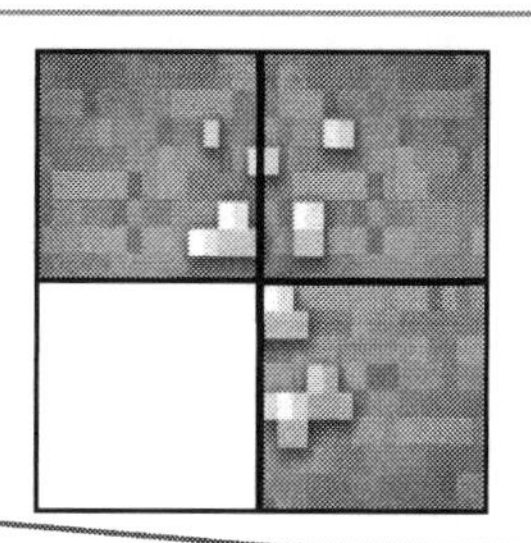

-

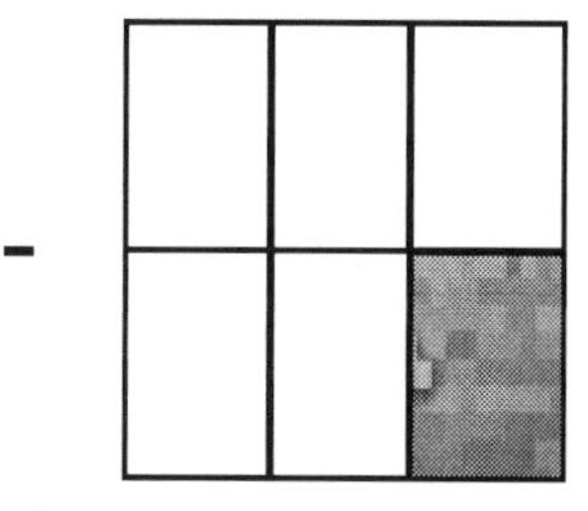

I ate $\frac{1}{6}$ out of $\frac{3}{4}$ of a Hazelnut Chocolate bar. How much is left? Can we apply the same steps to subtraction as we did in addition?

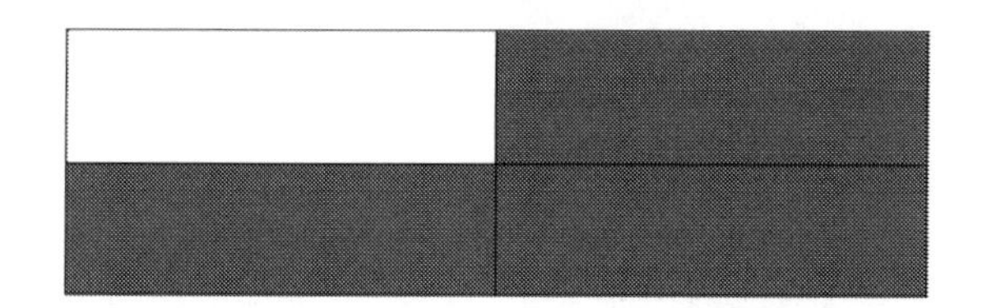

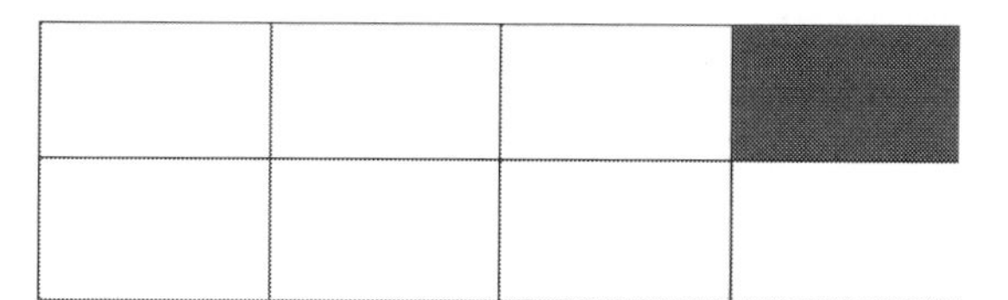

Step 1: Find out equivalent fractions to get the same common denominator (the common denominator is LCD)

Step 2: Subtract the numerators only

Step 3: Simplify where possible

Find out equivalent fractions!

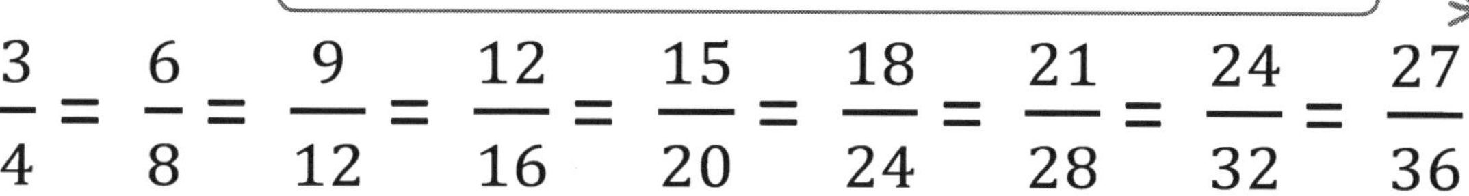

$$\frac{3}{4} = \frac{6}{8} = \frac{9}{12} = \frac{12}{16} = \frac{15}{20} = \frac{18}{24} = \frac{21}{28} = \frac{24}{32} = \frac{27}{36}$$

$$\frac{1}{6} = \frac{2}{12} = \frac{3}{18} = \frac{4}{24} = \frac{5}{30} = \frac{6}{36}$$

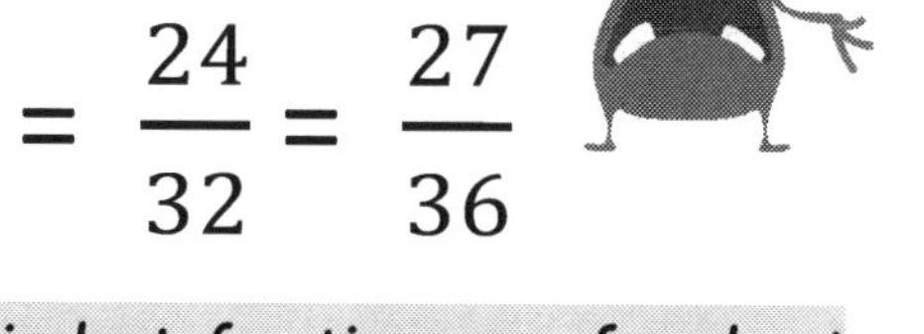

Equivalent fractions are found out by multiplying both the numerator and denominator by the same number

I have several like fractions with 12, 24, 36 in the denominator.
I colored the fractions strips grey.

$$\frac{3}{4}=\frac{6}{8}=\frac{9}{12}=\frac{12}{16}=\frac{15}{20}=\frac{18}{24}=\frac{21}{28}=\frac{24}{32}=\frac{27}{36}$$

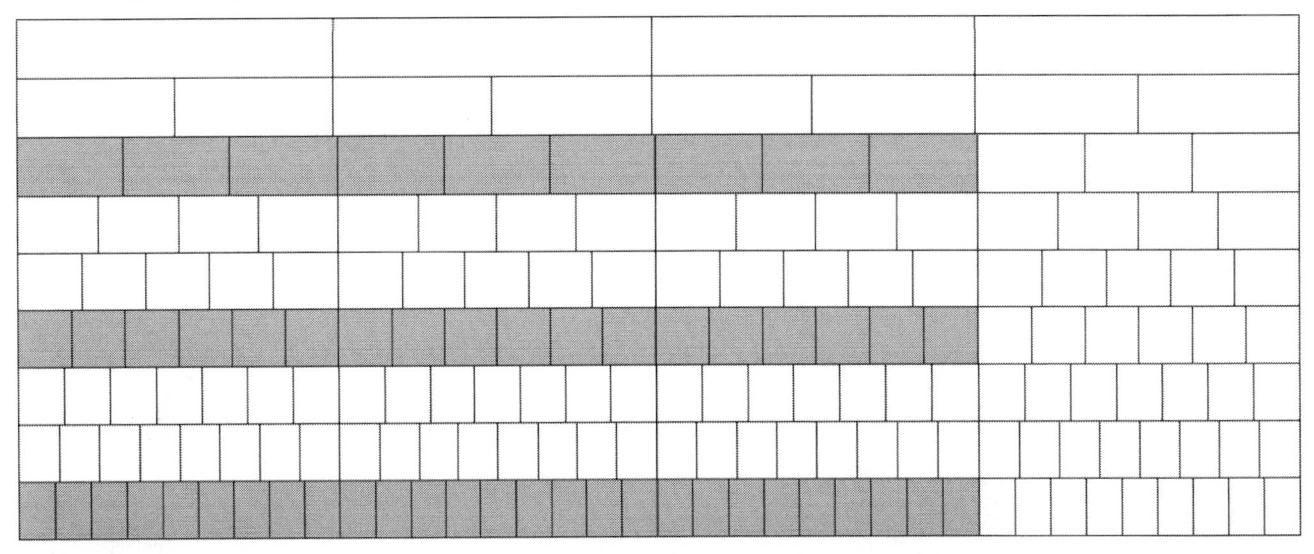

3 strips out of 4 strips are the same as 6 strips out of 8 strips, or 9 strips out of 12 strips, or 12 strips out of 16 strips, or 15 strips out of 20 strips, or 18 strips out of 24 strips, or 21 strips out of 28 strips, or 24 strips out of 32 strips, or 27 strips out of 36 strips – their value remains the same

$$\frac{1}{6}=\frac{2}{12}=\frac{3}{18}=\frac{4}{24}=\frac{5}{30}=\frac{6}{36}$$

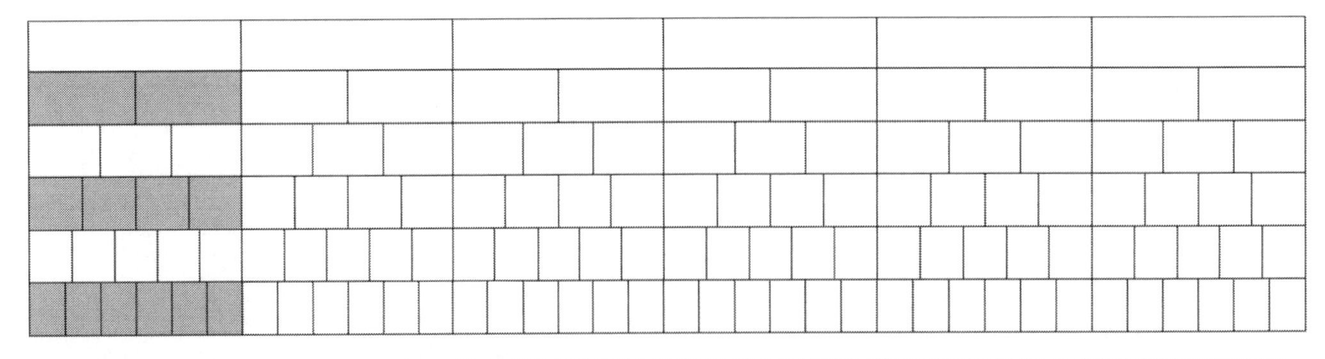

1 strip out of 6 strips is the same as 2 strips out of 12 strips, or 3 strips out of 18 strips, or 4 strips out of 24 strips, or 5 strips out of 30 strips, or 6 strips out of 36 strips – their value remains the same

Aha! The denominators are the same in these fractions! These are common denominators

$\frac{9}{12} = \frac{18}{24} = \frac{27}{36}$

$\frac{2}{12} = \frac{4}{24} = \frac{6}{36}$

Hint: You have a list of equivalent fractions 😊

Step 1 is done! We changed the given fractions to equivalent fractions to find common denominators. Now we must find the least common denominator.

12 is the LEAST common denominator for

$\frac{3}{4}$ $(=\frac{9}{12})$ and $\frac{1}{6}$ $(=\frac{2}{12})$.

Hint: The least common denominator is the denominator which is common to both denominators of the given fractions.

These are like fractions since they both have 12 in the denominator. So, you can write 12 as a common denominator for both fractions.

And then, subtract the numerators 9 and 2.

$$\frac{3}{4} - \frac{1}{6} = \frac{9}{12} - \frac{2}{12} = \frac{9-2}{12} = \frac{7}{12}$$

Subtract fractions:

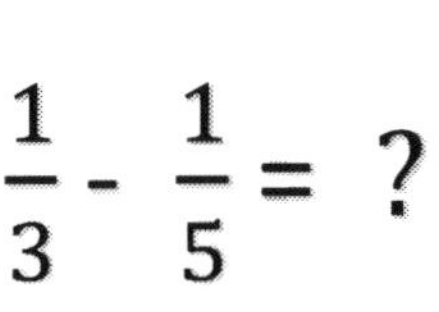

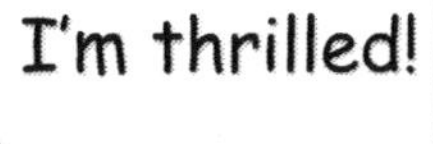

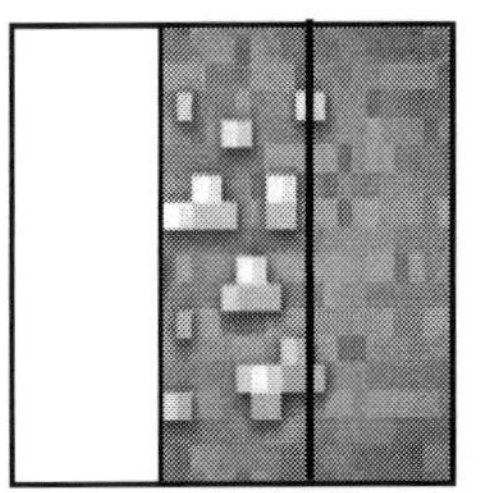 - 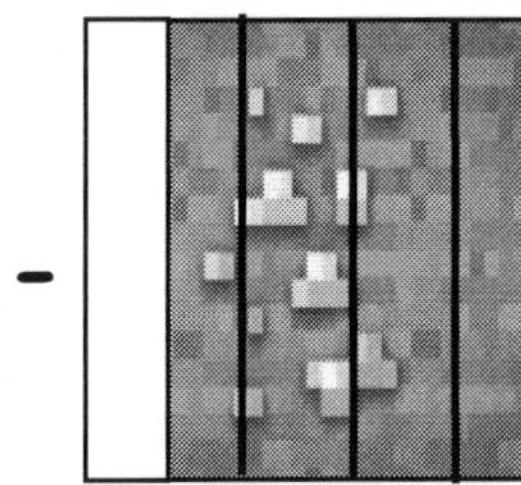

Hint: To subtract fractions, subtract LIKE fractions by simply subtracting the numerators. 😊

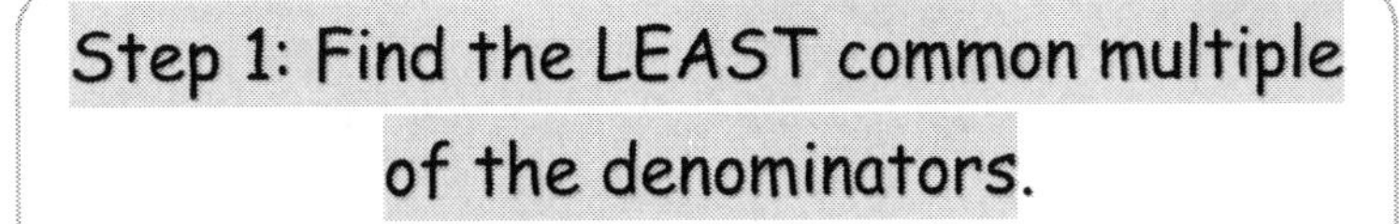

Step 1: Find the LEAST common multiple of the denominators.

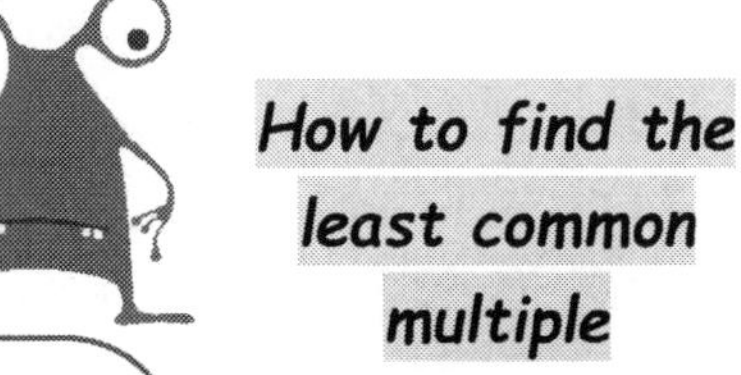

How to find the least common multiple

For the denominator 3: $1 \times 3 = 3$ $2 \times 3 = 6$

$3 \times 3 = 9$ $4 \times 3 = 12$ $5 \times 3 = 15$ $6 \times 3 = 18$

For the denominator 5: $1 \times 5 = 5$ $2 \times 5 = 10$

$3 \times 5 = 15$

Step 1: List all the multiples of the first denominator

Step 2: List all the multiples of the second denominator

Step 3: Circle the least multiple which is common to both denominators

The LEAST common multiple of 3 and 5 is 15: $5 \times 3 = 15$

$3 \times 5 = 15$

Step 2: Draw a fractions line, write 15 for the common denominator, and draw a slash above the numerator of each fraction:

$$\frac{1}{3} - \frac{1}{5} = \frac{\_ - \_}{15} =$$

Hint: To find a common denominator is to raise a fraction in higher terms to find an equivalent fraction (like $\frac{1}{3} - \frac{5}{15} = \frac{?}{15} - \frac{5}{15}$). 😊

Step 3: Divide the least common multiple by each denominator and write the quotient above each slash:

Hint: Think of how many thirds are in fifteenths? How many fifths are in fifteenths?

$$\overset{\diagdown 5}{\frac{1}{3}} - \overset{\diagdown 3}{\frac{1}{5}} = \frac{\_ - \_}{15} =$$

15 ÷ 3 = 5

15 ÷ 5 = 3

Step 4: Multiply each numerator by the number above a slash:

5 × 1 = 5

3 × 1 = 3

Hint: You are finding the "new" numerators using a common denominator of 15:

$$\overset{\diagdown 5}{\frac{1}{3}} - \overset{\diagdown 3}{\frac{1}{5}} = \frac{(5 \times 1) - (3 \times 1)}{15} =$$

You got the "new" numerators".

These are the numerators for the denominator 15:

$$\frac{1}{3} = \frac{5}{15}; \frac{1}{5} = \frac{3}{15}.$$

Step 5: Subtract the "new" numerators":

$$\overset{\diagdown 5}{\frac{1}{3}} - \overset{\diagdown 3}{\frac{1}{5}} = \frac{5 - 3}{15} = \frac{2}{15}$$

1. Subtract fractions. Write the missing numbers. Color the fraction of each shape. Simplify if possible.

$\frac{3}{4} - \frac{1}{3} = ?$

$\frac{3}{4} - \frac{1}{5} = ?$

Step 1: Find the least common denominator

$\frac{3}{4} = \frac{6}{\_\_} = \frac{9}{\_\_}$

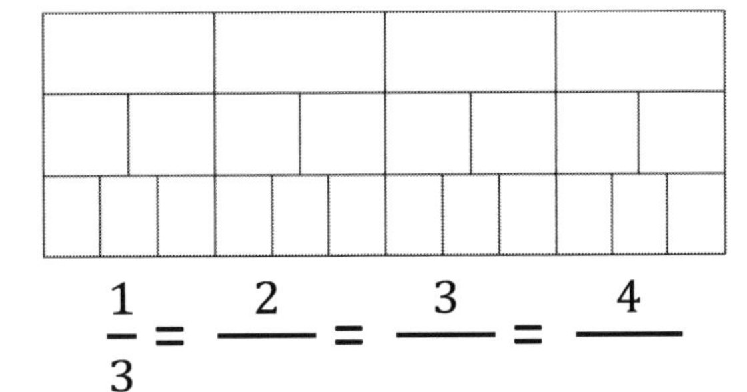

$\frac{1}{3} = \frac{2}{\_\_} = \frac{3}{\_\_} = \frac{4}{\_\_}$

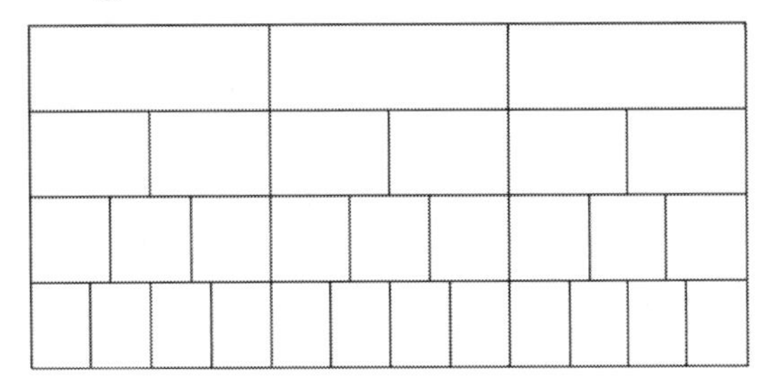

____ is the least common denominator

$\frac{3}{4} = \frac{6}{\_\_} = \frac{9}{\_\_} = \frac{12}{\_\_} = \frac{15}{\_\_}$

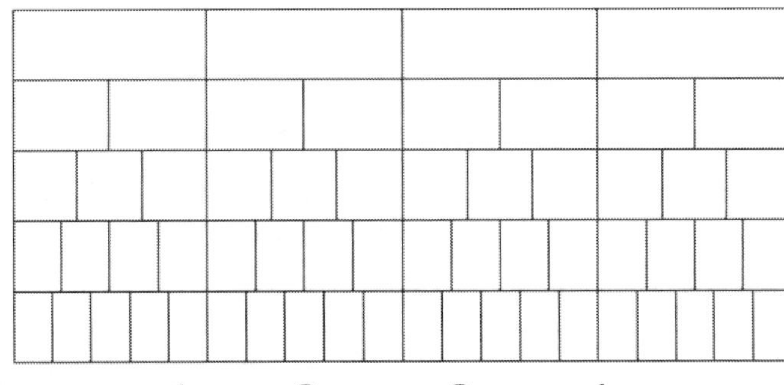

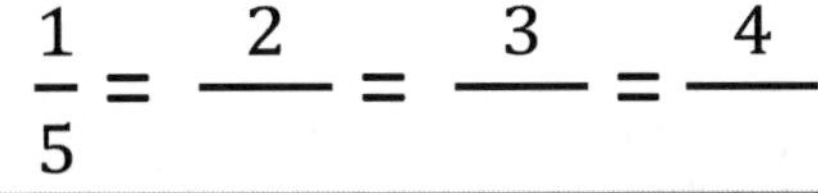

$\frac{1}{5} = \frac{2}{\_\_} = \frac{3}{\_\_} = \frac{4}{\_\_}$

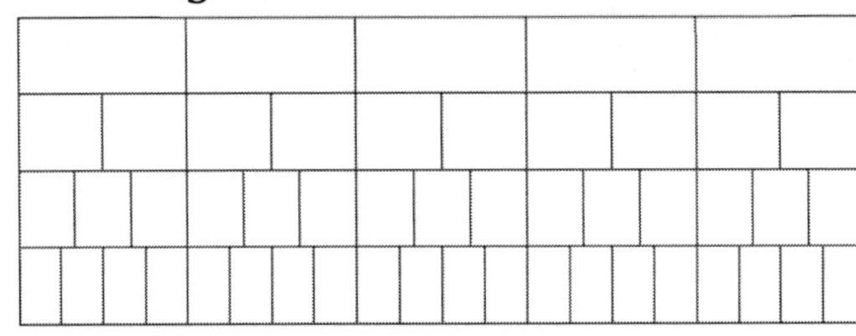

____ is the least common denominator

Step 2: Add the Numerators to find the sum

$\frac{3}{4} - \frac{1}{3} = \frac{\_ - \_}{\_\_} = \frac{\_}{\_} = \_\frac{\_}{\_}$

$\frac{3}{4} - \frac{2}{5} = \frac{\_ - \_}{\_\_} = \frac{\_}{\_} = \_\frac{\_}{\_}$

2. I had 54 bones in the beginning of the game. I used $\frac{3}{6}$ of the bones to tame one wolf and $\frac{2}{9}$ of the leftover bones to tame another wolf. How many bones did I use?

Answer: ______________________________

1. Subtract fractions. Write the missing numbers. Color the fraction of each shape. Simplify if possible.

$\frac{2}{3} - \frac{7}{12} = ?$

$\frac{1}{4} - \frac{1}{6} = ?$

Step 1: Find the least common denominator

$\frac{2}{3} = \frac{4}{\_\_} = \frac{6}{\_\_} = \frac{8}{\_\_}$

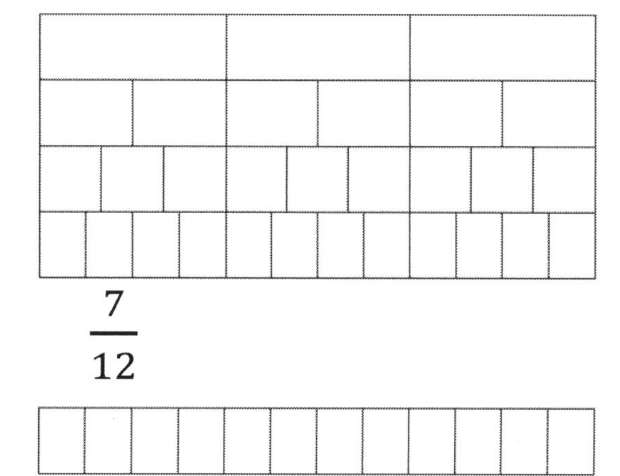

$\frac{1}{4} = \frac{2}{\_\_} = \frac{3}{\_\_}$

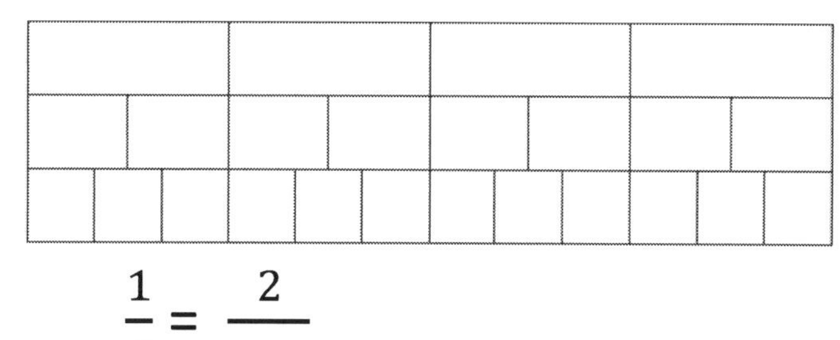

$\frac{1}{6} = \frac{2}{\_\_}$

$\frac{7}{12}$

___ *is the least common denominator*

___ *is the least common denominator*

Step 2: Add the Numerators to find the sum

$\frac{2}{3} - \frac{7}{12} = \frac{\_ - \_}{\_\_} = \frac{\_}{\_} = \_\frac{\_}{\_}$

$\frac{1}{4} - \frac{1}{6} = \frac{\_ - \_}{\_\_} = \frac{\_}{\_} = \_\frac{\_}{\_}$

2. I encountered 48 creepers in the game. I destroyed $\frac{1}{3}$ of the creepers with my iron sword and $\frac{5}{8}$ of the leftover creepers with my bow and arrow. How many creepers did I destroy?

Answer: ____

1. Subtract fractions. Write the missing numbers. Simplify if possible.

$\frac{3}{8} - \frac{1}{6} = \frac{\;-\;}{} = \frac{}{} = \_\frac{}{}$

$\frac{7}{9} - \frac{1}{3} = \frac{\;-\;}{} = \frac{}{} = \_\frac{}{}$

$\frac{5}{8} - \frac{1}{4} = \frac{\;-\;}{} = \frac{}{} = \_\frac{}{}$

$\frac{11}{15} - \frac{2}{5} = \frac{\;-\;}{} = \frac{}{} = \_\frac{}{}$

$\frac{3}{10} - \frac{1}{4} = \frac{\;-\;}{} = \frac{}{} = \_\frac{}{}$

$\frac{11}{16} - \frac{3}{8} = \frac{\;-\;}{} = \frac{}{} = \_\frac{}{}$

$\frac{8}{12} - \frac{2}{6} = \frac{\;-\;}{} = \frac{}{} = \_\frac{}{}$

$\frac{7}{8} - \frac{3}{6} = \frac{\;-\;}{} = \frac{}{} = \_\frac{}{}$

$\frac{13}{18} - \frac{2}{3} = \frac{\;-\;}{} = \frac{}{} = \_\frac{}{}$

$\frac{7}{15} - \frac{1}{3} = \frac{\;-\;}{} = \frac{}{} = \_\frac{}{}$

$\frac{5}{7} - \frac{9}{14} = \frac{\;-\;}{} = \frac{}{} = \_\frac{}{}$

$\frac{9}{10} - \frac{5}{6} = \frac{\;-\;}{} = \frac{}{} = \_\frac{}{}$

2. The villager used $\frac{1}{3}$ of his water bucket to water the carrots and $\frac{7}{24}$ of his water bucket to water the bushes, and $\frac{5}{16}$ of his water bucket to water the flowers. How much water (—) was left?

Answer: ____

1. Write in the missing numbers (add or subtract fractions).

1. Subtract fractions. Write the missing numbers. Simplify if possible.

$\frac{1}{2} - \frac{1}{6} = \frac{\quad - \quad}{\quad} = \frac{}{} = \_\frac{}{}$

$\frac{13}{30} - \frac{2}{5} = \frac{\quad - \quad}{\quad} = \frac{}{} = \_\frac{}{}$

$\frac{19}{24} - \frac{5}{8} = \frac{\quad - \quad}{\quad} = \frac{}{} = \_\frac{}{}$

$\frac{15}{28} - \frac{2}{4} = \frac{\quad - \quad}{\quad} = \frac{}{} = \_\frac{}{}$

$\frac{3}{4} - \frac{3}{12} = \frac{\quad - \quad}{\quad} = \frac{}{} = \_\frac{}{}$

$\frac{5}{7} - \frac{4}{9} = \frac{\quad - \quad}{\quad} = \frac{}{} = \_\frac{}{}$

$\frac{4}{11} - \frac{2}{22} = \frac{\quad - \quad}{\quad} = \frac{}{} = \_\frac{}{}$

$\frac{7}{10} - \frac{3}{40} = \frac{\quad - \quad}{\quad} = \frac{}{} = \_\frac{}{}$

$\frac{15}{20} - \frac{2}{5} = \frac{\quad - \quad}{\quad} = \frac{}{} = \_\frac{}{}$

$\frac{5}{6} - \frac{3}{8} = \frac{\quad - \quad}{\quad} = \frac{}{} = \_\frac{}{}$

$\frac{10}{15} - \frac{6}{12} = \frac{\quad - \quad}{\quad} = \frac{}{} = \_\frac{}{}$

$\frac{12}{36} - \frac{8}{48} = \frac{\quad - \quad}{\quad} = \frac{}{} = \_\frac{}{}$

2. The villager had some trees. He chopped down $\frac{2}{9}$ of the trees and he still had 21 trees left. How many trees did the villager have in the beginning?

Answer: ______________________________________

1. Subtract fractions. Write the missing numbers. Simplify if possible.

$\frac{5}{6} - \frac{3}{24} =$

$\frac{7}{9} - \frac{1}{3} =$

$\frac{5}{9} - \frac{1}{27} =$

$\frac{7}{16} - \frac{5}{48} =$

$\frac{23}{24} - \frac{1}{3} - \frac{1}{4} = \frac{\quad - \quad - \quad}{\qquad} = \frac{\quad}{\quad}$

$\frac{5}{6} - \frac{1}{5} - \frac{1}{15} = \frac{\quad - \quad - \quad}{\qquad} = \frac{\quad}{\quad}$

$\frac{6}{7} - \frac{2}{5} - \frac{5}{70} = \frac{\quad - \quad - \quad}{\qquad} = \frac{\quad}{\quad}$

$\frac{11}{12} - \frac{5}{48} - \frac{7}{24} = \frac{\quad - \quad - \quad}{\qquad} = \frac{\quad}{\quad}$

2. I had some diamonds. I traded $\frac{3}{8}$ of the diamonds and I still had 35 diamonds left. How many diamonds did I have in the beginning?

Answer: ______________________________

1. Find and circle or cross out the words to find out more about Minecraft.

N	U	D	R	G	Q	L	I	S	G	L	M	U	S	E
R	H	N	A	A	O	T	D	G	N	F	V	I	X	W
S	F	X	S	O	D	H	N	F	I	T	J	P	A	C
H	L	L	K	T	M	U	D	F	N	Z	E	D	Y	O
O	Y	O	N	S	E	G	Q	X	G	R	E	W	J	S
O	U	H	S	M	T	A	Y	K	I	X	J	W	R	X
T	S	J	A	H	K	N	D	M	S	B	A	E	S	H
I	D	V	D	I	H	X	E	Y	E	Q	L	E	A	Y
N	W	N	Y	P	C	N	O	M	D	L	B	Y	C	Q
G	E	L	B	A	T	C	I	D	E	R	P	N	U	S
L	P	N	I	A	B	T	L	P	R	L	S	Q	B	S
N	N	K	T	M	Y	Q	O	M	R	P	T	G	C	J
U	U	I	J	I	A	R	E	A	L	I	S	T	I	C
D	O	M	P	N	P	U	G	S	L	S	N	T	A	V
N	B	D	D	F	O	Y	D	H	M	C	A	U	W	B

LOOKOUT

PROPELLERS

EXPERIMENTATION

REALISTIC

SHOOTING

UNPREDICTABLE

BATTLEMENTS

REDESIGNING

UNSTEADY

1. Write in the missing numbers (add or subtract).

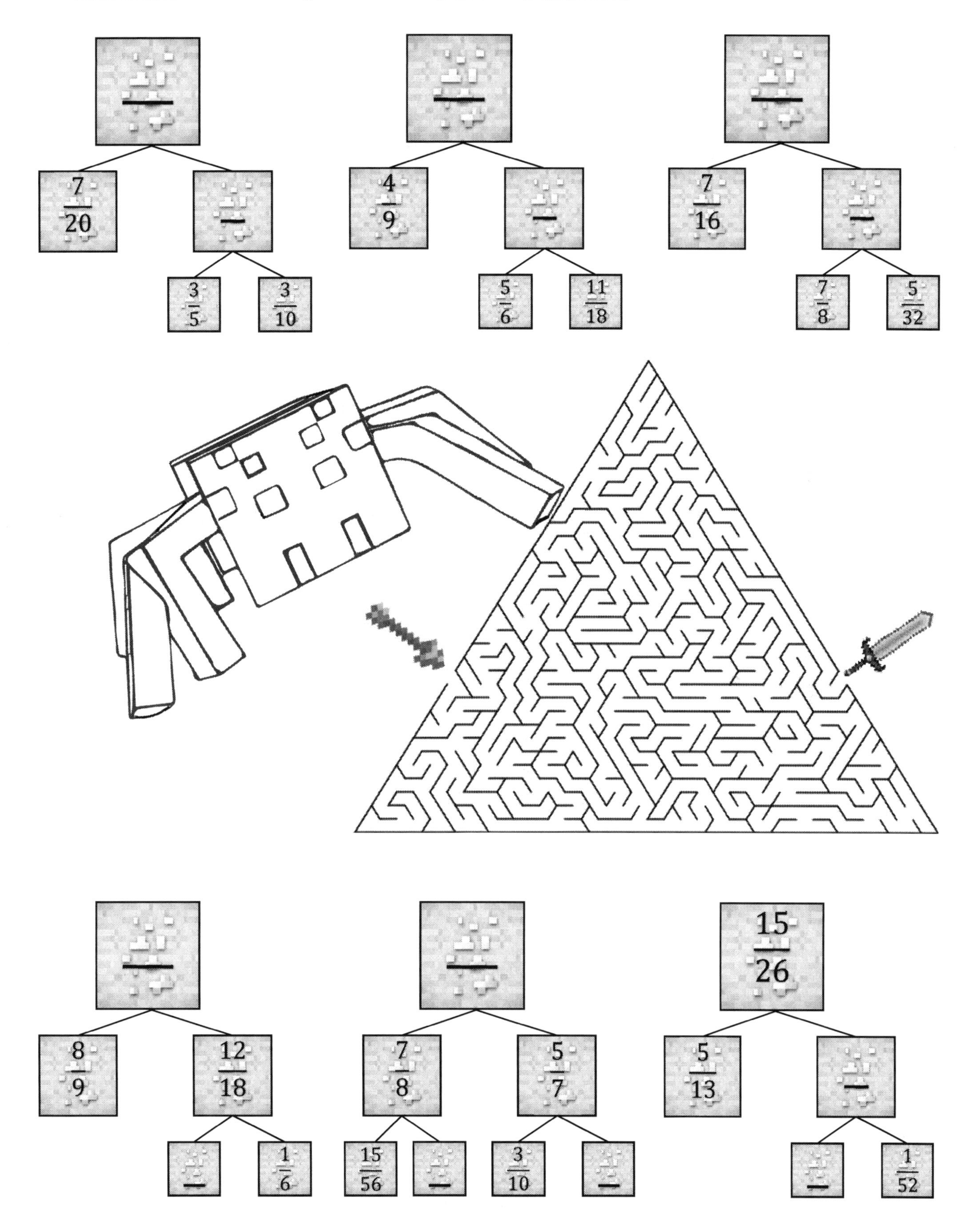

1. <u>Subtract</u> fractions. <u>Write</u> the missing numbers. <u>Simplify</u> if possible.

$1 - \frac{1}{6} = \text{— — —} = \frac{-}{\text{———}} = \text{—}$ $\quad 1 - \frac{1}{3} = \text{— — —} = \frac{-}{\text{———}} = \text{—}$

Hint: To subtract a fraction from a whole number: change a whole number to a fraction with the same denominator as that in the fraction to be subtracted

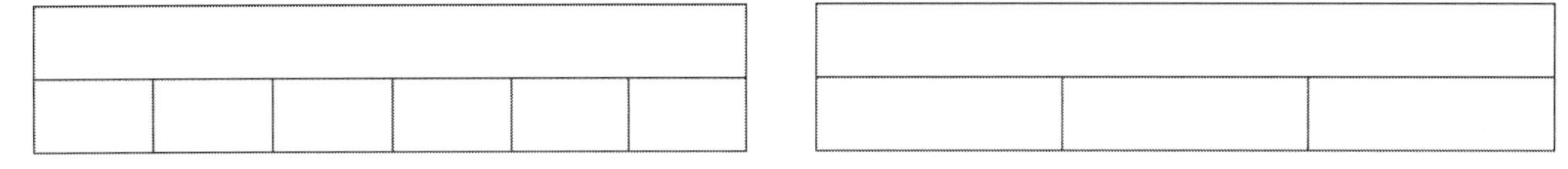

$1 - \frac{3}{4} = \text{— — —} = \frac{-}{\text{———}} = \text{—}$ $\quad 1 - \frac{2}{5} = \text{— — —} = \frac{-}{\text{———}} = \text{—}$

Hint: Rewrite a whole number to a fraction so that both fractions have a common denominator:

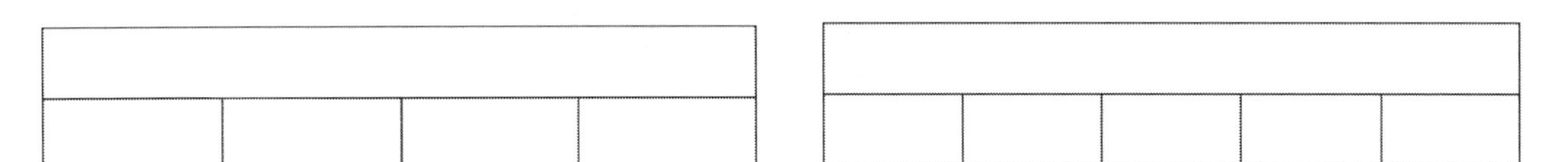

$1 - \frac{5}{7} = \text{— — —} = \frac{-}{\text{———}} = \text{—}$ $\quad 1 - \frac{3}{8} = \text{— — —} = \frac{-}{\text{———}} = \text{—}$

Hint: First, simplify if possible! Then, subtract!

$2 - \frac{2}{6} = \_\text{— —} = \frac{-}{\text{———}} = \text{—} = \_\text{—}$ $\quad 3 - \frac{3}{6} = \_\text{— —} = \frac{-}{\text{———}} = \text{—} = \_$

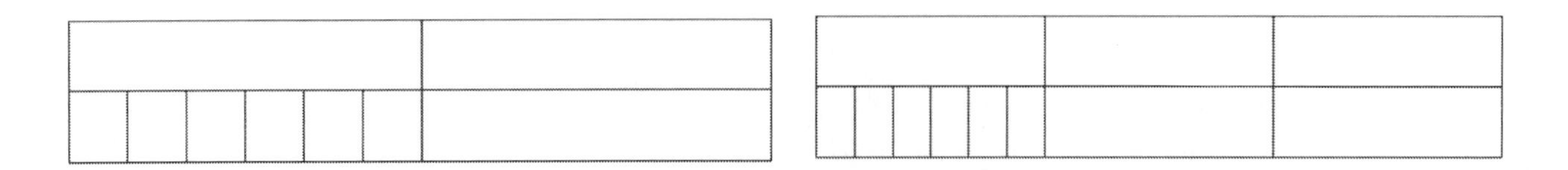

$4 - \frac{8}{12} = \_\text{— —} = \frac{-}{\text{———}} = \text{—} = \_\text{—}$ $\quad 5 - \frac{9}{15} = \_\text{— —} = \frac{-}{\text{———}} = \text{—} = \_$

1. Subtract fractions. Write the missing numbers. Simplify if possible.

Hint: To subtract a fraction from a whole number: rewrite a whole number to a mixed number: a whole number and a fraction with the same denominator (like $2\frac{9}{9}$).

$3 - \frac{4}{9} = 2\frac{9}{9} - \frac{\_}{\_} = \frac{\_ - \_}{\_} = \_\frac{\_}{\_}$ $\quad$ $6 - \frac{8}{12} = \_\frac{\_}{\_} - \frac{\_}{\_} = \frac{\_ - \_}{\_} = \_\frac{\_}{\_}$

$4 - \frac{2}{3} = \_\frac{\_}{\_} - \frac{\_}{\_} = \frac{\_ - \_}{\_} = \_\frac{\_}{\_}$ $\quad$ $5 - \frac{2}{5} = \_\frac{\_}{\_} - \frac{\_}{\_} = \frac{\_ - \_}{\_} = \_\frac{\_}{\_}$

$6 - \frac{9}{14} = \_\frac{\_}{\_} - \frac{\_}{\_} = \frac{\_ - \_}{\_} = \_\frac{\_}{\_}$ $\quad$ $7 - \frac{5}{6} = \_\frac{\_}{\_} - \frac{\_}{\_} = \frac{\_ - \_}{\_} = \_\frac{\_}{\_}$

2. Multiply and simplify if possible. Circle the right answer (letter).

Find $\frac{10}{18}$ of 54 arrows.

$\frac{10}{18} \times 54 = \frac{\_ \times \_}{\_ \times \_} = \_\_\_$

a) 50
b) 44
c) 30

Find $\frac{2}{3}$ of 72 donkeys.

Step 1: ___ ÷ ___ = ___

Step 2: ___ × ___ = ___

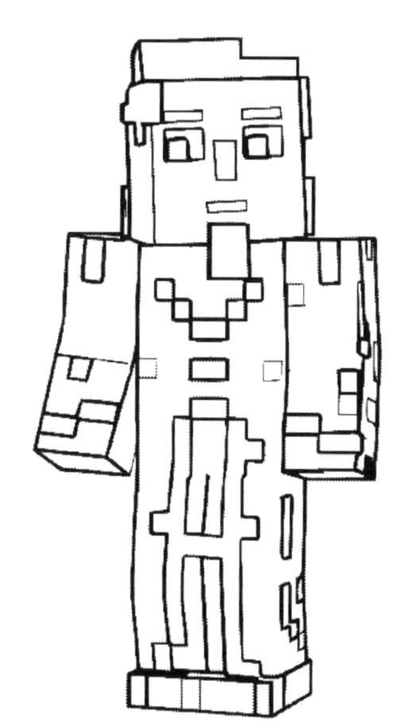

a) 48
b) 52
c) 60

Find $\frac{3}{4}$ of 56 wood planks.

Step 1: ___ ÷ ___ = ___

Step 2: ___ × ___ = ___

d) 32
e) 40
f) 42

Add mixed numbers: $2\frac{1}{3} + 1\frac{1}{4} = ?$

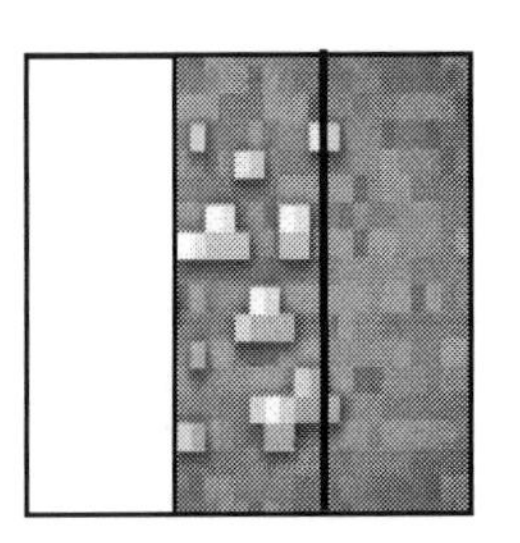

+

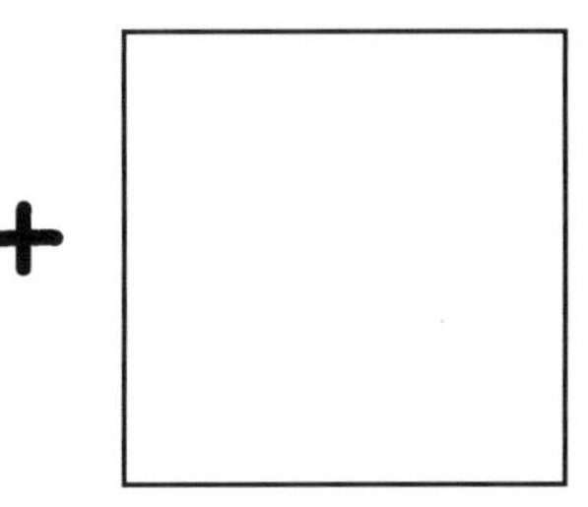
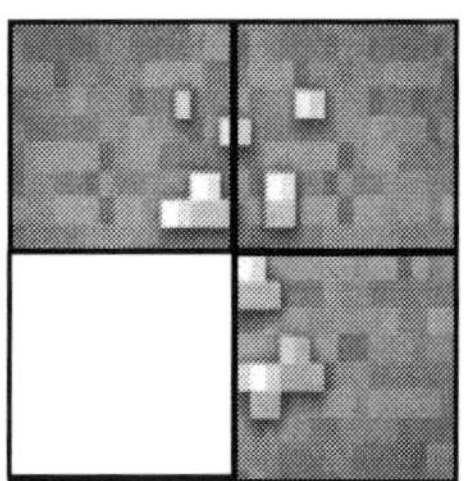

Step 1: Add the whole numbers:

2 + 1 = 3.

$$2 + 1 = 3$$

Hint: To add mixed numbers, make sure that the fractional parts have the common denominator!

Step 2: Change to equivalent fractions finding the LEAST common multiple which is equal to common denominator

For the denominator 3:	$1 \times 3 = 3$	$2 \times 3 = 6$	$3 \times 3 = 9$
	$4 \times 3 = 12$	$5 \times 3 = 15$	$6 \times 3 = 18$
For the denominator 4:	$1 \times 4 = 4$	$2 \times 4 = 8$	$3 \times 4 = 12$

The LEAST common multiple of 3 and 4 is 12:

4 × 3 = 12

3 × 4 = 12

You can say 12 is the least common denominator!

Step 3: Draw a fractions bar, write 12 for the denominator, and draw a slash above the numerator for each fraction:

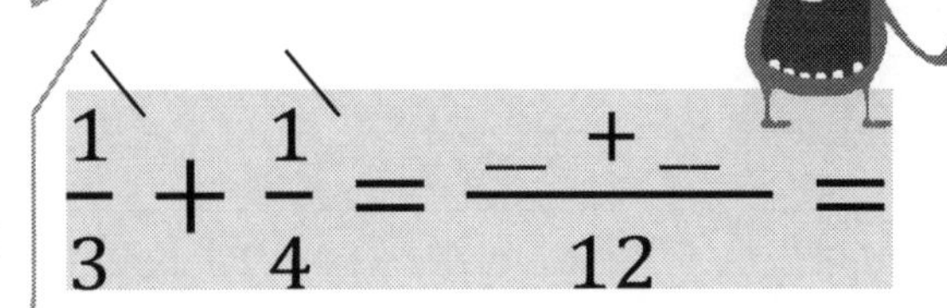

$$\frac{1}{3} + \frac{1}{4} = \frac{\_ + \_}{12} =$$

Step 4: Divide the least common multiple by each denominator and write the quotient above each slash:

Hint: Use slashes in the beginning to make the process easy and fun😊

$$\frac{1}{3}\,^{\backslash 4} + \frac{1}{4}\,^{\backslash 3} = \frac{\_ + \_}{12} =$$

$12 \div 3 = 4$

$12 \div 4 = 3$

Step 5: Multiply each numerator by the number above a slash:

$4 \times 1 = 4$

$3 \times 1 = 3$

Hint: You can also say that you're finding equivalent fractions using a common denominator of 12 😊

$$\frac{1}{3}\,^{\backslash 4} + \frac{1}{4}\,^{\backslash 3} = \frac{(4 \times 1) + (3 \times 1)}{12} =$$

You got the "new" numerators".

These are the numerators for the denominator of 12:

$$\frac{1}{3} = \frac{4}{12}; \frac{1}{4} = \frac{3}{12}.$$

Step 6: Add the "new" numerators":

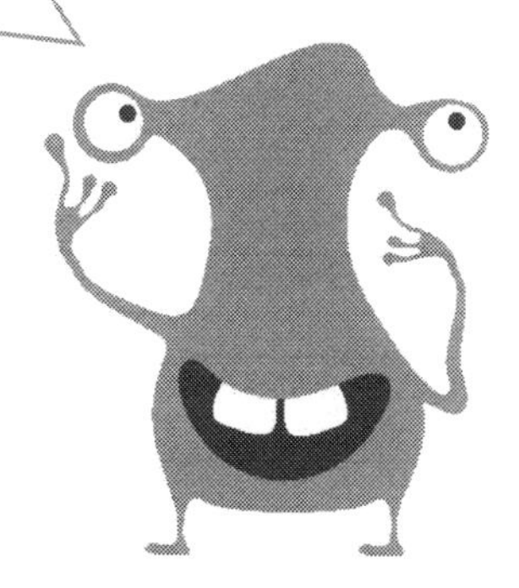

$$\frac{1}{3}\,^{\backslash 4} + \frac{1}{4}\,^{\backslash 3} = \frac{4 + 3}{12} = \frac{7}{12}$$

Step 7: Add the whole number and a fractional part:

$$3 + \frac{7}{12} = 3\frac{7}{12}$$

1. Add fractions. Write the missing numbers. Simplify if possible.

$2\frac{3}{8} + 4\frac{5}{9} = \_\,\frac{}{}$

1) Add whole numbers

2) Add fractions

1)

2)

$3\frac{3}{4} + 5\frac{3}{5} = \_\,\frac{}{}$

1)

2)

$6\frac{3}{10} + 4\frac{5}{6} = \_\,\frac{}{}$

1)

2)

2. Change the mixed number to an improper faction. Circle the correct answer.

$3\frac{7}{13} = \frac{}{}$

a) $\frac{39}{13}$ b) $\frac{23}{13}$ c) $\frac{46}{13}$

$8\frac{5}{9} = \frac{}{}$

a) $\frac{72}{9}$ b) $\frac{75}{9}$ c) $\frac{77}{9}$

Subtract mixed numbers: $2\frac{1}{3} - 1\frac{2}{8} = ?$

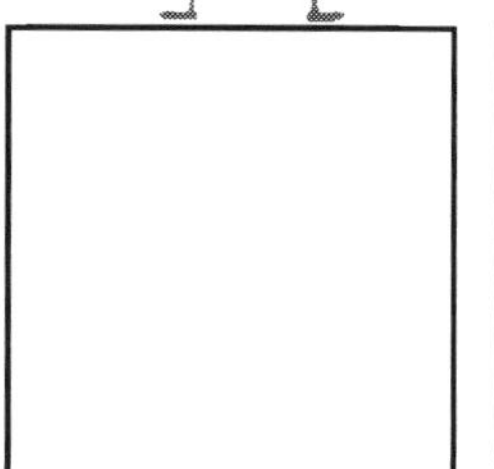  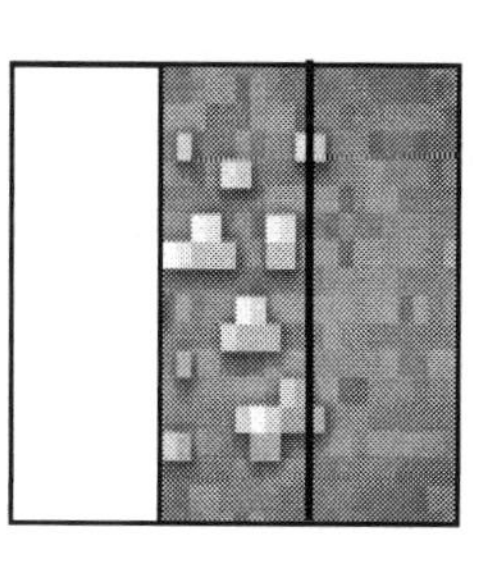 - 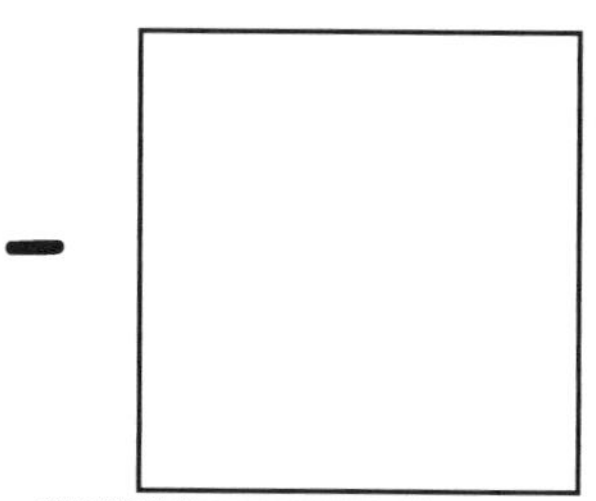 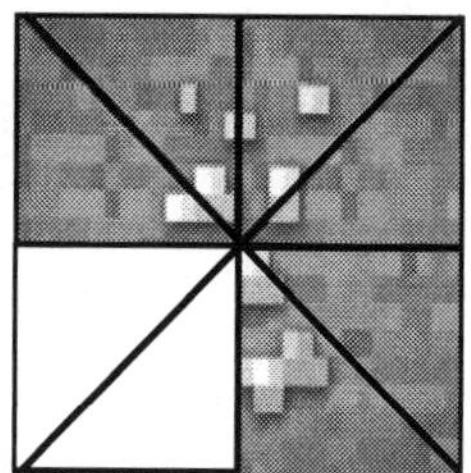

$2 - 1 = 1$

Step 1: Subtract the whole numbers: 2 - 1 = 1.

Hint: To subtract mixed numbers, make sure that the fractional parts have the common denominator!

Step 2: Change to equivalent fractions finding the LEAST common multiple which is equal to common denominator.

For the denominator 3:	$1 \times 3 = 3$	$2 \times 3 = 6$	$3 \times 3 = 9$
	$4 \times 3 = 12$	$5 \times 3 = 15$	$6 \times 3 = 18$
	$7 \times 3 = 21$	$8 \times 3 = 24$	
For the denominator 4:	$1 \times 8 = 8$	$2 \times 8 = 16$	$3 \times 8 = 24$

The LEAST common multiple of 3 and 8 is 24:

8 × 3 = 24

3 × 8 = 24

You can say 24 is the least common denominator!

Step 3: Draw a fractions bar, write 12 for the denominator, and draw a slash above the numerator for each fraction:

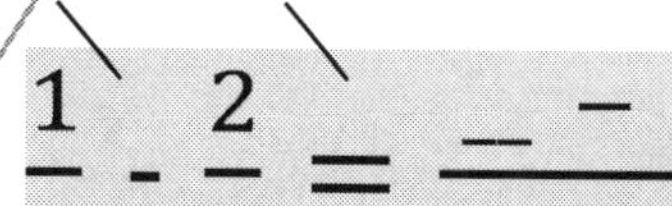

$\frac{1}{3} - \frac{2}{8} = \frac{\_ - \_}{24} =$

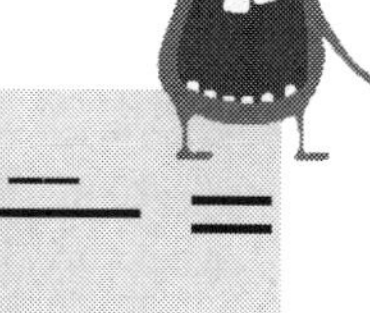

Step 4: Divide the least common multiple by each denominator and write the quotient above each slash:

Hint: Use slashes in the beginning to make the process easy and fun☺

$$\frac{1}{3}\overset{8}{\diagdown} - \frac{2}{8}\overset{3}{\diagdown} = \frac{\_\ -\ \_}{24} =$$

$$24 \div 3 = 8$$
$$24 \div 8 = 3$$

Step 5: Multiply each numerator by the number above a slash:

$$8 \times 1 = 8$$
$$3 \times 2 = 6$$

Hint: You can also say that you're finding the equivalent fractions using a common denominator of 24 ☺

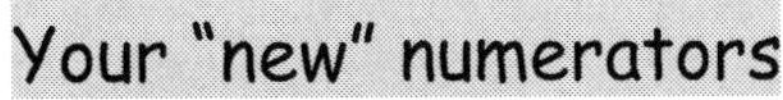
Your "new" numerators

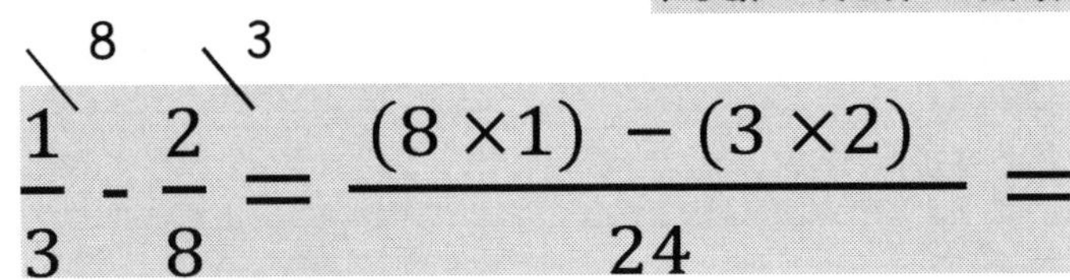

$$\frac{1}{3}\overset{8}{\diagdown} - \frac{2}{8}\overset{3}{\diagdown} = \frac{(8 \times 1) - (3 \times 2)}{24} =$$

You got the "new" numerators".

These are the numerators for the denominator 12:

$$\frac{1}{3} = \frac{8}{24}; \frac{2}{8} = \frac{6}{24}.$$

Step 6: Add the "new" numerators" and simplify:

$$\frac{1}{3}\overset{8}{\diagdown} - \frac{2}{8}\overset{3}{\diagdown} = \frac{8 - 6}{24} = \frac{2}{24} = \frac{1}{12}$$

Step 7: Add the whole number and a fractional part:

$$1 + \frac{1}{12} = 1\frac{1}{12}$$

1. Subtract fractions. Write the missing numbers. Simplify if possible.

$3\frac{5}{8} - 1\frac{5}{6} = \_\frac{}{}$

1) Subtract whole numbers

2) Subtract fractions

To subtract a greater fraction from a smaller fraction: Step 1: borrow one from a whole number; Step 2: rewrite the borrowed whole number to a fraction: $3\frac{1}{3} - 1\frac{3}{8} = 1\frac{23}{24}$

Step 1: 3-1=~~2~~ 1

Step 2: $\frac{1}{3} - \frac{3}{8} = \frac{8-9}{24} = \frac{8+24-9}{24} = \frac{23}{24}$

1)

2)

$8\frac{7}{10} - 2\frac{3}{4} = \_\frac{}{}$

1)

2)

$9\frac{1}{15} - 3\frac{7}{9} = \_\frac{}{}$

1)

2)

2. Change the improper faction to a mixed number. Circle the correct answer.

$\frac{32}{5} = \frac{}{}$

a) $3\frac{2}{5}$ b) $6\frac{1}{5}$ c) $6\frac{2}{5}$

$\frac{49}{11} = \frac{}{}$

a) $11\frac{5}{11}$ b) $1\frac{38}{11}$ c) $4\frac{5}{11}$

1. Multiply fractions. Write the missing numbers. Simplify if possible.

Step 1: Change the mixed number to an improper fraction

Step 2: Multiply the numerators and denominators

$$3 \times 2\frac{3}{15} = \frac{3}{1} \times \frac{33}{15} = \frac{3}{1} \times \frac{11}{5} = \frac{11+11+11}{5} = \_\frac{\_}{\_}$$

Hint: Simplify before multiplying where you can! (cancel the 3s from 33 and 15)

Hint: Change the mixed number to improper fractions!

$$6 \times 3\frac{5}{12} = \frac{\_}{\_} \times \frac{\_}{\_} = \frac{\_ \times \_}{\_ \times \_} = \frac{\_}{\_} = \_\frac{\_}{\_}$$

Hint: Cancel or cross out the common factors of the numerator and the denominator to simplify where it's possible 😊

$$5 \times 2\frac{7}{20} = \frac{\_}{\_} \times \frac{\_}{\_} = \frac{\_ \times \_}{\_ \times \_} = \frac{\_}{\_} = \_\frac{\_}{\_}$$

$$4 \times 5\frac{5}{16} = \frac{\_}{\_} \times \frac{\_}{\_} = \frac{\_ \times \_}{\_ \times \_} = \frac{\_}{\_} = \_\frac{\_}{\_}$$

2. Multiply. Express your answer in simplest form.

$$\frac{9}{16} \times \frac{20}{36}$$

a) $\frac{15}{36}$ b) $\frac{5}{16}$ c) $\frac{180}{432}$ d) $2\frac{2}{5}$

1. Find and circle or cross out the words to find out more about Minecraft.

O	H	V	N	H	F	Z	V	X	V	C	W	F	V	E	E	R
S	C	A	I	O	E	V	Q	G	O	V	M	S	X	W	K	E
Z	S	Y	R	H	I	W	Y	O	H	V	T	A	S	X	I	D
R	B	E	G	D	H	T	R	C	Z	B	K	O	A	S	R	W
G	X	Y	N	X	C	D	N	W	W	C	X	R	V	U	T	O
U	M	I	D	T	E	O	N	E	I	C	W	J	A	M	S	P
V	Q	Q	M	N	H	E	R	P	T	V	R	O	N	S	G	N
J	V	B	A	O	M	G	D	E	V	X	M	G	N	W	N	U
T	P	T	O	R	F	N	I	B	M	V	E	H	A	W	I	G
S	E	C	E	C	O	Z	H	R	E	O	X	Q	B	B	N	R
S	E	D	T	M	I	I	C	Q	B	X	D	H	I	L	T	E
U	N	Z	A	F	G	K	A	C	E	N	N	E	O	I	H	M
E	O	I	C	Q	K	K	Y	G	H	L	E	Q	M	Z	G	F
T	D	V	G	C	M	X	P	M	Z	O	N	E	E	M	I	D
V	E	R	T	I	C	A	L	T	O	R	C	H	R	X	L	E
T	Z	M	S	I	N	A	H	C	E	M	S	T	Z	C	M	Z
R	A	V	I	N	E	S	Y	N	O	Q	O	J	O	O	S	L

MECHANISM	SCREEN BRIGHTNESS
VERTICAL TORCH	DIAMOND PICKAXE
EXTENTION	GUNPOWDER
HARDCORE MODE	RAVINES
ENDERMEN	SAVANNA BIOME
COORDINATES	LIGHTNING STRIKE

1. Multiply fractions. Write the missing numbers. Simplify if possible.

Step 1: Change the mixed number to an improper fraction

Step 2: Multiply the numerators and denominators

$4\frac{1}{5} \times 4\frac{2}{7} = \frac{}{} \times \frac{}{} = \frac{\times}{\times} = \frac{}{} = \_\frac{}{}$

Hint: Factor the numerator and the denominator by the same number to reduce the fraction to lowest terms! 😊

$2\frac{1}{4} \times 3\frac{4}{6} = \frac{}{} \times \frac{}{} = \frac{\times}{\times} = \frac{}{} = \_\frac{}{}$

$3\frac{1}{2} \times 2\frac{2}{7} = \frac{}{} \times \frac{}{} = \frac{\times}{\times} = \frac{}{} = \_\frac{}{}$

$9\frac{5}{8} \times 3\frac{7}{11} = \frac{}{} \times \frac{}{} = \frac{\times}{\times} = \frac{}{} = \_\frac{}{}$

2. Multiply. Express your answer in simplest form.

$\frac{5}{12} \times \frac{36}{55}$

a) $\frac{3}{10}$ b) $\frac{36}{60}$ c) $\frac{3}{11}$ d) $1\frac{2}{5}$

1. Write in the missing numbers (practice adding and subtracting mixed numbers).

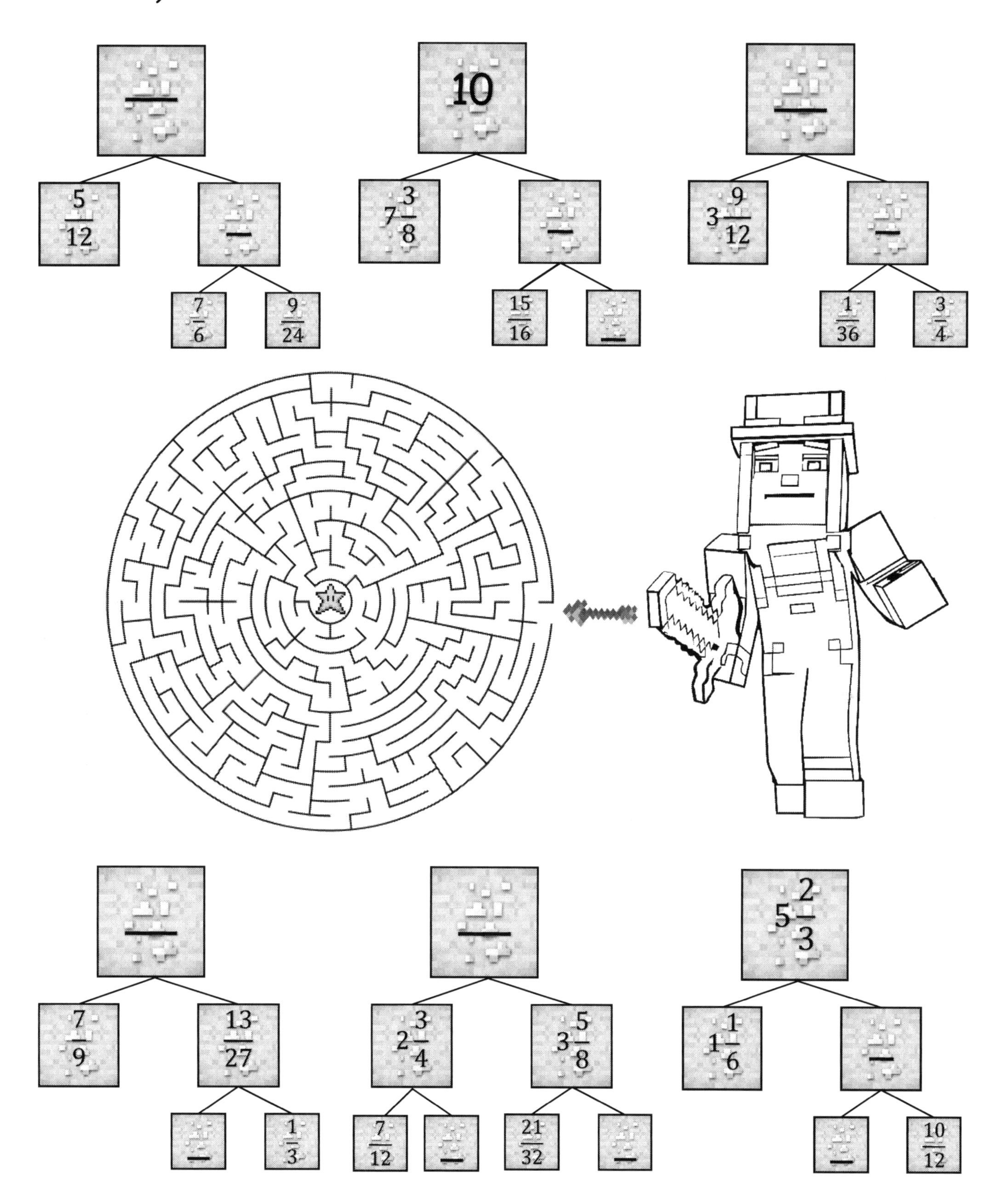

1. Divide fractions. Write the missing numbers. Simplify if possible.

Divide fractions! That's impossible! I cannot divide 3 blocks by, say, $\frac{5}{9}$!

$3 \div \frac{5}{9}$ = **no** answer, no solution, let's skip it.

Your problem means: "How many $\frac{5}{9}$'s are there in 3?" or "How many $\frac{5}{9}$'s are contained in 3?", or "How many $\frac{5}{9}$'s in 3?".

Step 1: Change 3 to $\frac{1}{9}$'s. Use a common denominator of 9.

$$3 = \frac{27}{9}$$

Step 2: Divide the numerators, then, divide the denominators, then, divide the numerator by the denominator.

$$\text{So, } 3 \div \frac{5}{9} = \frac{27}{9} \div \frac{5}{9} = \frac{(27 \div 5)}{(9 \div 9)} = \frac{(27 \div 5)}{1} = \frac{27}{5} = 5\frac{2}{5}$$

```
       5   =the whole number
   5 | 27  =Numerator
Denominator
     25
     ---
      2    = Remainder
```

Dividend = Numerator

Divisor = Denominator

Quotient = The Whole Number

Remainder = Numerator of a Fraction Part

$$5 \div \frac{2}{7} = \frac{35}{7} \div \frac{2}{7} = \frac{(\quad \div \quad)}{(\quad \div \quad)} = \frac{\quad}{\quad} = \_\_ \frac{\quad}{\quad}$$

$$4 \div \frac{2}{3} = \frac{\quad}{\quad} \div \frac{\quad}{\quad} = \frac{(\quad \div \quad)}{(\quad \div \quad)} = \frac{\quad}{\quad} = \_\_ \frac{\quad}{\quad}$$

One whole →

One out of 2 equal parts →

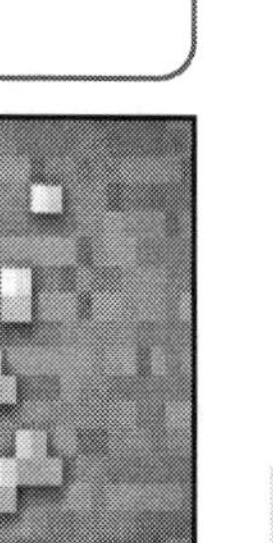

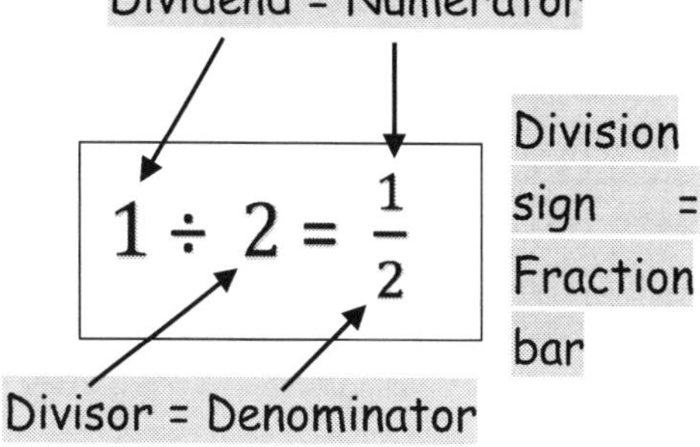

Hint: **Ask yourself, how many 2s are contained in a half? By using blocks, it is easy to see that when you divide a block by 2, you get one-fourth.**

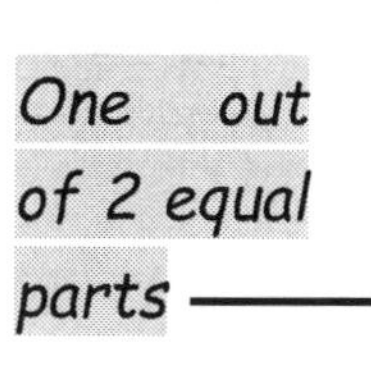

One out of 2 equal parts →

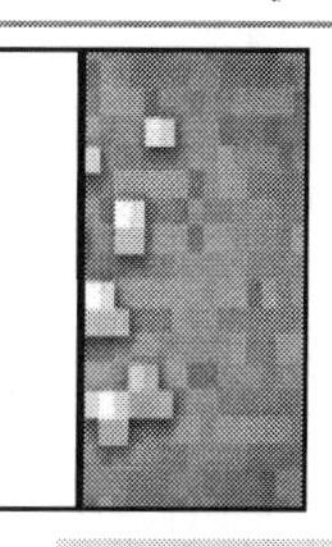

One out of 4 equal parts →

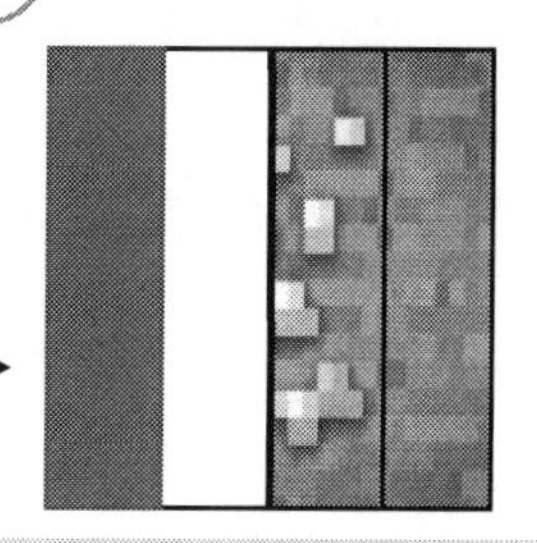

$$\frac{1}{2} \div 2 = \frac{1}{4}$$

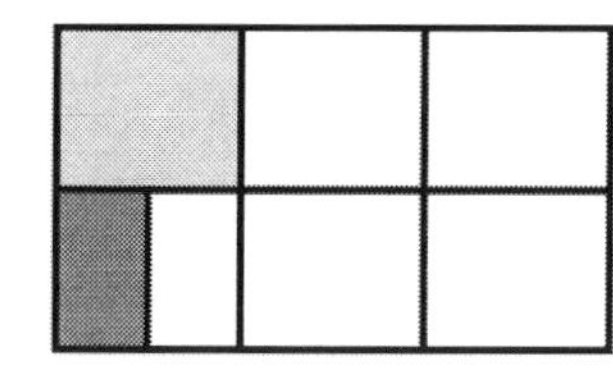

Hint: **Ask yourself, how many 2s are contained in one-third? By using blocks, it is easy to see that when you divide one-third by 2, you get one-sixth.**

$$\frac{1}{3} \div 2 = \frac{1}{6}$$

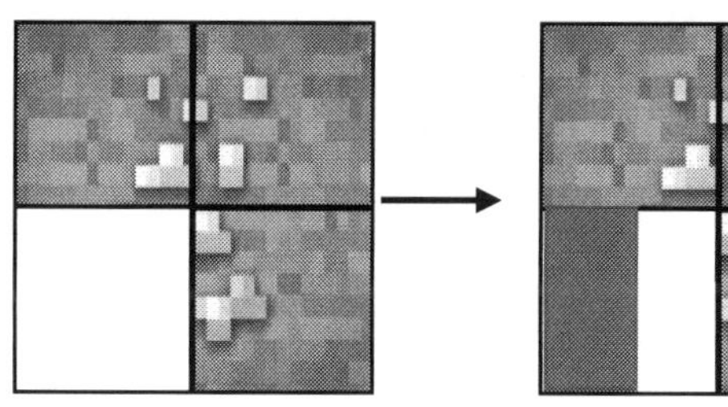

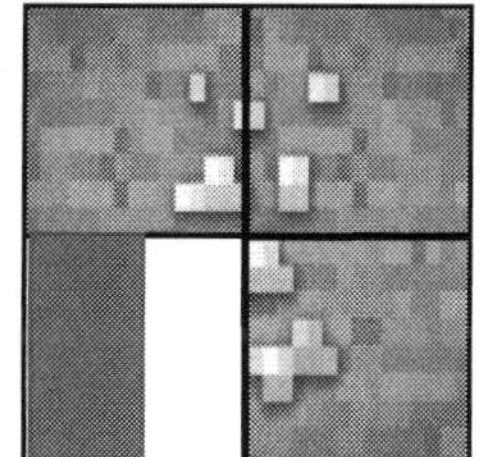

Hint: **Ask yourself, how many 2s are contained in a quarter? By using blocks, it is easy to see that when you divide a quarter by 2, you get one-eighth.**

$$\frac{1}{4} \div 2 = \frac{1}{8}$$

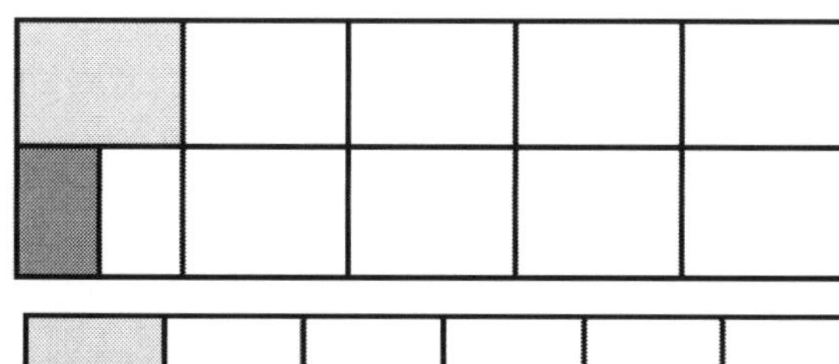

Hint: **By using blocks, it is easy to see that when you divide one-fifth by 2, you get one-tenth.**

$$\frac{1}{5} \div 2 = \frac{1}{10}$$

Hint: **By using blocks, it is easy to see that when you divide one-fifth by 2, you get one-tenth.**

$$\frac{1}{6} \div 2 = \frac{1}{12}$$

Aha... For all whole numbers (like 3 and 5) we can rewrite division as:

$$3 \div 5 = \frac{3}{5}$$

Ask yourself, "How many groups of size 5 are in 3?"

The answer is: three fifths.

I need to divide:

$$\frac{3}{4} \div 3 = ?$$

How many 3s are contained in $\frac{3}{4}$?

1			
$\frac{3}{4}$			
$\frac{1}{4}$			

By using blocks, it is easy to see that when you divide three quarters by 3, you get one quarter.

$$\frac{3}{4} \div 3 = \frac{1}{4}$$

I need to divide:

$$\frac{1}{2} \div 4 = ?$$

How many 4s are contained in $\frac{1}{2}$?

1				
$\frac{1}{2}$				
$\frac{1}{8}$				

By using blocks, it is easy to see that when you divide a half by 4, you get one-eighth.

$$\frac{1}{2} \div 4 = \frac{1}{8}$$

1. How many 8s are contained in $\frac{2}{4}$?

Write the missing numbers. Fill in the fractions in the table.

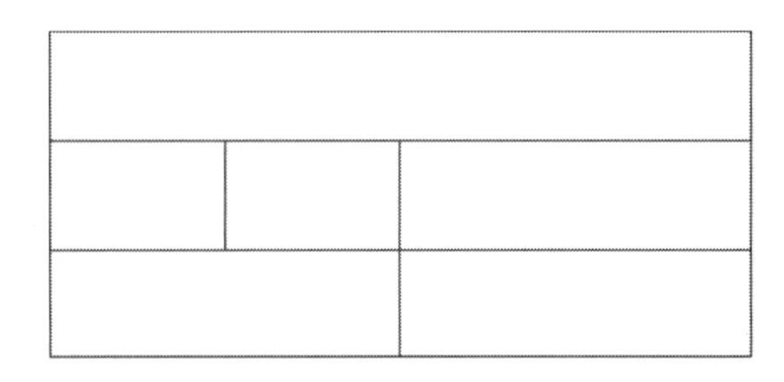

$$\frac{2}{4} \div 8 = \frac{\quad}{\quad}$$

Can you show me some rules to do division problems easier?

Turn a fraction upside down. It is often called taking the RECIPROCAL. Reciprocal is a number (fraction) where you interchange the numerator and the denominator.

Write the missing numbers to get a reciprocal.

$\frac{1}{4} \rightarrow \frac{4}{1}$ $\frac{3}{8} \rightarrow$ — $\frac{5}{9} \rightarrow$ — $\frac{6}{11} \rightarrow$ —

Hint: To find a reciprocal you need to invert the fraction so that the numerator and the denominator switched their places. ☺

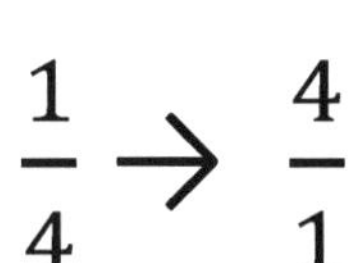

Find the reciprocal for 1: $1 = \frac{1}{1} \rightarrow$ —

Write the missing numbers to get a reciprocal.

First, find an improper fraction, then, a reciprocal:

$12 \rightarrow \frac{1}{12}$ $\frac{7}{23} \rightarrow$ — $9\frac{7}{8} \rightarrow$ —

Hint: If you multiply a number by its reciprocal, you get 1:

$$\frac{4}{9} \times \frac{9}{4} = \frac{4 \times 9}{9 \times 4} = \frac{1 \times 1}{1 \times 1} = 1$$

Hint: Cancel 9s from 9 and 9. Cancel 4s from 4 and 4.

I draw the arrows to help myself.

Multiply the given fractions and invert the divisor!

$$\frac{5}{8} \div 10 = \frac{5}{8} \div \frac{10}{1} = \frac{5}{8} \times \frac{1}{10} = \frac{5 \times 1}{8 \times 10_{2}} = \frac{1}{16}$$

1. Divide fractions. Write the missing numbers. Simplify if possible.

$\frac{10}{3} \div 2 = \frac{10}{3} \div \frac{2}{1} = \frac{10}{3} \times \frac{1}{2} = \frac{\quad \times \quad}{\quad \times \quad} = \_\frac{\quad}{\quad}$

$\frac{8}{15} \div 4 = \frac{\quad}{\quad} \div \frac{\quad}{\quad} = \frac{\quad}{\quad} \times \frac{\quad}{\quad} = \frac{\quad \times \quad}{\quad \times \quad} = \frac{\quad}{\quad}$

$\frac{2}{3} \div 4 = \frac{\quad}{\quad} \div \frac{\quad}{\quad} = \frac{\quad}{\quad} \times \frac{\quad}{\quad} = \frac{\quad \times \quad}{\quad \times \quad} = \frac{\quad}{\quad}$

$\frac{5}{6} \div 10 =$

$\frac{13}{18} \div 26 =$

2. My brother and my sister had a total of 56 swords at first. After she lost 8 of her swords and he lost $\frac{1}{2}$ of his swords, they had an equal number of swords. How many swords did she have at first?

Answer: ______________________________

1. Divide fractions. Write the missing numbers. Simplify if possible.

$\frac{49}{50} \div 7 = \frac{}{} \div \frac{}{} = \frac{}{} \times \frac{}{} = \frac{\times}{\times} = \frac{}{}$

$\frac{14}{18} \div 2 = \frac{}{} \div \frac{}{} = \frac{}{} \times \frac{}{} = \frac{\times}{\times} = \frac{}{}$

$\frac{8}{27} \div 18 =$ $\frac{6}{14} \div 21 =$

$\frac{21}{25} \div 14 =$ $\frac{12}{13} \div 24 =$

2. My brother and my sister had a total of 20 wood planks at first. After she used $\frac{3}{4}$ of her planks and he used $\frac{5}{6}$ of his planks, they had an equal number of wood planks. How many wood planks did he have at first?

Answer: ______________________________

1. Divide fractions. Write the missing numbers. Simplify if possible.

$\frac{9}{17} \div 18 = \text{—} \div \text{—} = \text{—} \times \text{—} = \frac{\ \times\ }{\ \times\ } = \text{—}$

$\frac{24}{35} \div 3 = \text{—} \div \text{—} = \text{—} \times \text{—} = \frac{\ \times\ }{\ \times\ } = \text{—}$

$\frac{49}{50} \div x = \frac{7}{50}$

$x = \frac{49}{50} \div \frac{7}{50} = \frac{49}{50} \times \frac{50}{7} = \text{—}$

$\frac{15}{16} \div x = \frac{3}{64}$

$x =$

$\frac{36}{41} \div x = \frac{4}{41}$

$x =$

$\frac{25}{50} \div x = \frac{1}{30}$

$x =$

2. My brother destroyed $\frac{2}{5}$ as many spiders as my sister at first. After he destroyed 9 more spiders and she destroyed 6 more spiders, they had an equal number of spiders destroyed in the end. How many spiders did he destroy at first?

Answer: ______________________________

1. Find and circle or cross out the words to find out more about Minecraft.

A C T I V A T O R O D B
V Z D A M A G E P P O O
L C Y V A J H O G P X O
P U M Y J M Z U B O J K
A G G R A V A T I N G S
O L A O C R A H C E R H
Q Q K H D C R I T N E E
E U Q I N H C E T T B L
H K A F E X C Z L S M V
G N D P U S T X N M A E
T F Z J M B P F I S H S
A Z Z Y H S K E G Z C O

TECHNIQUE
OPPONENT
DAMAGE
CHAMBER
GUARDIAN
AGGRAVATING
BOOKSHELVES
CHARCOAL
ACTIVATOR

I need to divide: $1 \div \frac{1}{2} = ?$

Think: <u>How many halves</u> are there in one whole?

$$1 \div \frac{1}{2} = 2$$

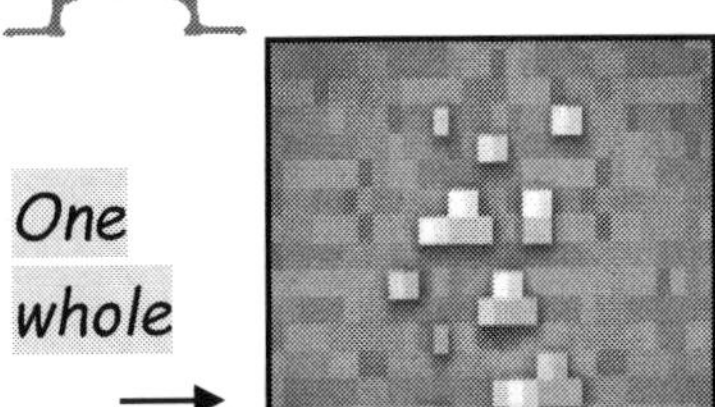

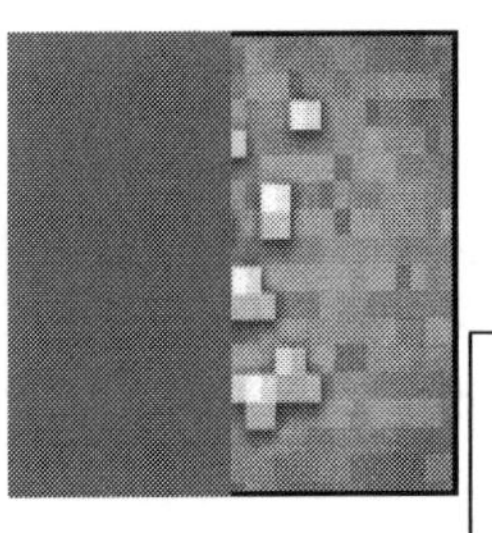

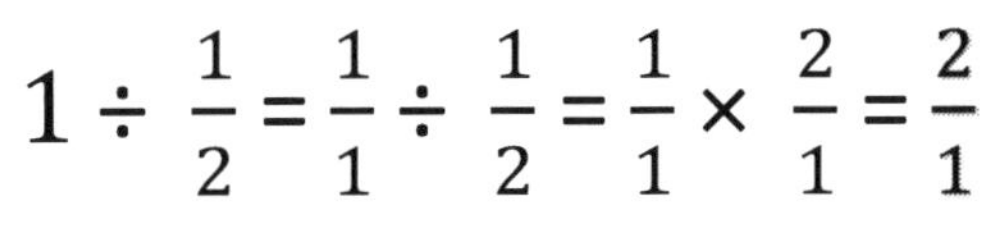

$$1 \div \frac{1}{2} = \frac{1}{1} \div \frac{1}{2} = \frac{1}{1} \times \frac{2}{1} = \frac{2}{1}$$

I divide: $2 \div \frac{1}{2} = ?$

Think: <u>How many halves</u> are there in two wholes?

There are 2 halves in 1 whole. Therefore, there must be 4 halves in 2 wholes.

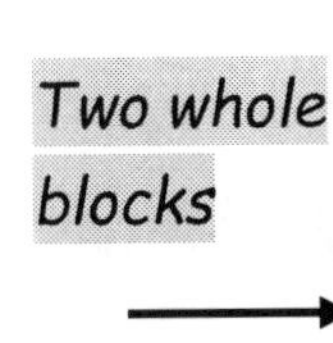

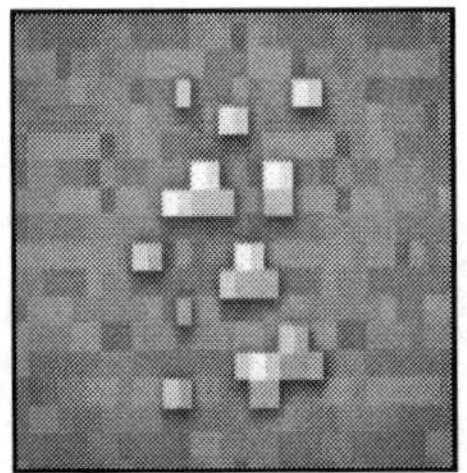

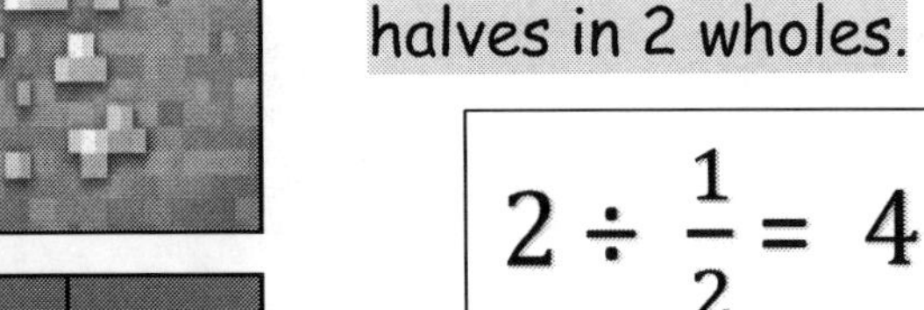

$$2 \div \frac{1}{2} = 4$$

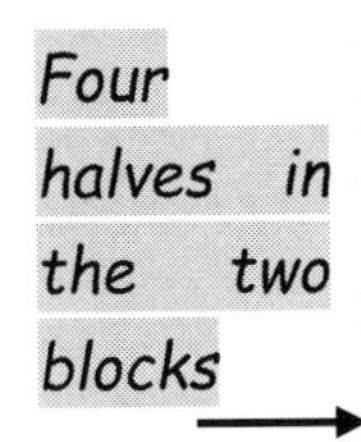

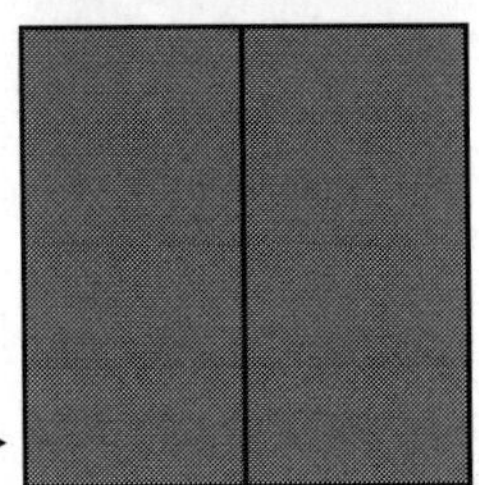

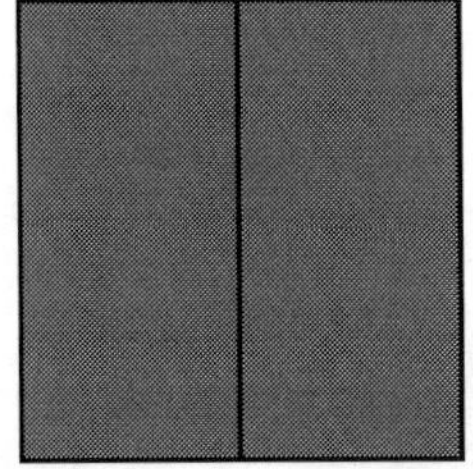

$$2 \div \frac{1}{2} = \frac{2}{1} \div \frac{1}{2} = \frac{2}{1} \times \frac{2}{1} = \frac{4}{1}$$

$$4 \div \frac{1}{3} = ?$$

Think: <u>How many thirds</u> are there in 4?

<u>To divide a whole number by a fraction</u>:

1) change a whole number so that both fractions have a common denominator
2) or turn the second fraction upside down (= use the reciprocal).

$$4 \div \frac{1}{3} = \frac{12}{3} \div \frac{1}{3} = \frac{(12 \div 1)}{(3 \div 3)} = \frac{12}{1}$$

$$4 \div \frac{1}{3} = \frac{4}{1} \div \frac{1}{3} = \frac{4}{1} \times \frac{3}{1} = \frac{12}{1}$$

1			1			1			1		
$\frac{1}{3}$	$\frac{1}{3}$	$\frac{1}{3}$	$\frac{1}{3}$	$\frac{1}{3}$	$\frac{1}{3}$	$\frac{1}{3}$	$\frac{1}{3}$	$\frac{1}{3}$	$\frac{1}{3}$	$\frac{1}{3}$	$\frac{1}{3}$

1. Divide fractions. Write the missing numbers. Simplify if possible.

$8 \div \frac{2}{7} = \frac{8}{1} \div \frac{2}{7} = \frac{8}{1} \times \frac{7}{2} = \_\frac{}{\_}$

$6 \div \frac{12}{13} = - \div - = - \times - = \_\frac{}{\_}$

$5 \div \frac{10}{17} = - \div - = - \times - = \_\frac{}{\_}$

$4 \div \frac{12}{21} = - \div - = - \times - = \_\frac{}{\_}$

$9 \div \frac{3}{8} = - \div - = - \times - = \_\frac{}{\_}$

2. I came across x spiders and destroyed $\frac{3}{4}$ of them with my iron sword. How many spiders did I come across if I destroyed 12 spiders?

$\frac{3}{4}x = 12$

$x = 12 \div \frac{3}{4} = - \div - = - \times - = \frac{\times}{\times} = \_\frac{}{\_}$

Answer: ______________________________

1. Divide fractions. Write the missing numbers. Simplify if possible.

$3 \div \frac{21}{25} = - \div - = - \times - = \_ -$

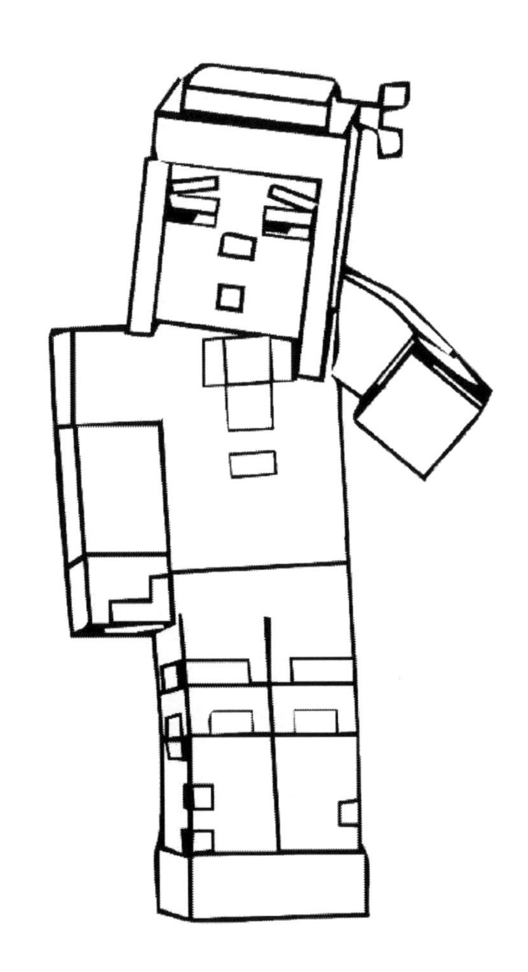

$7 \div \frac{14}{17} = - \div - = - \times - = \_ -$

$3 \div \frac{6}{11} = - \div - = - \times - = \_ -$

$6 \div \frac{24}{25} =$ $9 \div \frac{18}{22} =$

$3 \div \frac{15}{18} =$ $10 \div \frac{10}{15} =$

2. The villager planted x flowers on his farm and $\frac{3}{5}$ of them were beautiful 30 lilacs. How many flowers did he plant?

$\frac{3}{5}x = 30$

$x = \_\_\_ \div - = - \div - = - \times - = \_ -$

Answer: ____________________________________

1. Divide or multiply. Write the missing numbers. Simplify if possible.

$x \div \frac{5}{7} = 7$

$x = — \times — = \frac{\times}{\times} = —$

$x \div \frac{4}{15} = 22\frac{1}{2}$

Hint: Change the mixed number to an improper fraction, then divide or multiply!

$x =$

$x \div \frac{24}{25} = 8\frac{1}{3}$

$x =$

$x \div \frac{18}{22} = 11$

$x =$

$x \div \frac{24}{35} = 8\frac{3}{4}$

$x =$

$x \div \frac{15}{36} = 28\frac{4}{5}$

$x =$

2. I met $\boxed{x}$ animals and $\boxed{\frac{5}{12}}$ of them were $\boxed{35}$ black horses. How many animals did I meet?

$— x = \_\_\_$

$x = \_\_\_ \div — = — \div — = — \times — = \_ —$

Answer: ____________________________________

I need to divide: $\frac{2}{3} \div \frac{1}{6} =$?

Think: <u>How many sixths</u> are there in two thirds?

$$\frac{2}{3} \div \frac{1}{6} = \frac{2}{\cancel{3}} \times \frac{\cancel{6}^{2}}{1} = \frac{4}{1}$$

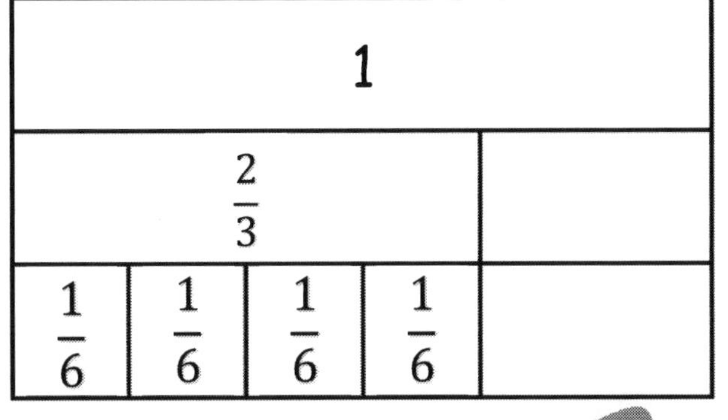

I need to divide $\frac{5}{12} \div \frac{1}{4} =$

I'll show you the algorithm.

Step 1: Rewrite the first fraction:

$$\frac{5}{12} \div \frac{1}{4} = \frac{5}{12}$$

Step 2: Change the operation (÷ → ×):

$$\frac{5}{12} \div \frac{1}{4} = \frac{5}{12} \times$$

Step 3: Write the reciprocal of the second fraction - switch the numerator and the denominator:

$$\frac{5}{12} \div \frac{1}{4} = \frac{5}{12} \times \frac{4}{1}$$

Step 4: Multiply the numerators and denominators and simplify where possible:

$$\frac{5}{12} \div \frac{1}{4} = \frac{5}{\cancel{12}_{3}} \times \frac{4}{1}$$

Step 5: Change to mixed fraction where possible:

$$\frac{5}{12} \div \frac{1}{4} = \frac{5}{\cancel{12}_{3}} \times \frac{4}{1} = 6\frac{2}{3}$$

1. Divide fractions. Write the missing numbers. Simplify if possible.

$\frac{1}{12} \div \frac{5}{6} = \frac{}{} \times \frac{}{} = \_\frac{}{}$

$\frac{8}{9} \div \frac{24}{45} = \frac{}{} \times \frac{}{} = \_\frac{}{}$

$\frac{5}{18} \div \frac{8}{9} = \frac{}{} \times \frac{}{} = \_\frac{}{}$

First, change the mixed number to an improper fraction, then, divide or multiply!

$6\frac{5}{6} \div 2\frac{22}{30} = \frac{}{} \div \frac{}{} = \frac{}{} \times \frac{}{} = \_\frac{}{}$

$1\frac{8}{9} \div 2\frac{14}{27} =$

$4\frac{3}{7} \div 4\frac{6}{14} =$

2. I spent $\frac{1}{3}$ of an hour to find a cave. Then, I spent $\frac{5}{6}$ of an hour to dig out 2 diamonds. Then, I came back to my tiny house, found the recipe of a diamond sword, and spent $\frac{1}{4}$ of an hour to make it. How much time did I spend in all?

Answer: ____________________________________

1. Divide fractions. Write the missing numbers. Simplify if possible.

$\frac{14}{18} \div \frac{2}{9} = \text{—} \times \text{—} = \_\text{—}$

$\frac{6}{7} \div \frac{24}{49} =$

$3\frac{5}{6} \div 4\frac{2}{7} =$

$\frac{21}{25} \div \frac{14}{15} =$

$2\frac{3}{8} \div 2\frac{1}{3} =$

$1\frac{8}{9} \div 2\frac{2}{33} =$

$4\frac{3}{7} \div 3\frac{6}{14} =$

$8\frac{3}{4} \div 3\frac{6}{8} =$

$10\frac{2}{3} \div 5\frac{4}{6} =$

2. I had $\frac{2}{7}$ as many emeralds as my brother. After my brother used 8 of his emeralds, and I used 6 emeralds, I have $\frac{1}{4}$ as many emeralds as my brother. How many emeralds do we have in the end?

Answer: ______________________________

1. Divide fractions. Write the missing numbers. Simplify if possible.

$\frac{16}{25} \div \frac{32}{55} = \frac{}{} \times \frac{}{} = \_\frac{}{}$

$\frac{6}{7} \div \frac{24}{49} =$

$5\frac{1}{7} \div 4\frac{1}{2} =$

$\frac{21}{25} \div \frac{14}{15} =$

$11\frac{3}{7} \div 4\frac{10}{14} =$

$15\frac{1}{8} \div 2\frac{1}{5} =$

$4\frac{1}{5} \div 2\frac{1}{3} =$

$x \div 3\frac{4}{8} = 2\frac{2}{7}$

$x \div 7\frac{1}{7} = 3\frac{9}{25}$

$x =$

$x =$

2. I crafted $\frac{3}{5}$ as many wooden sticks as my sister. After my sister used 8 of her wooden sticks to make weapons, and I crafted 12 more wooden sticks, I have $\frac{3}{4}$ as many wooden sticks as my sister. How many wooden sticks does she have in the end?

Answer: ______________________________________

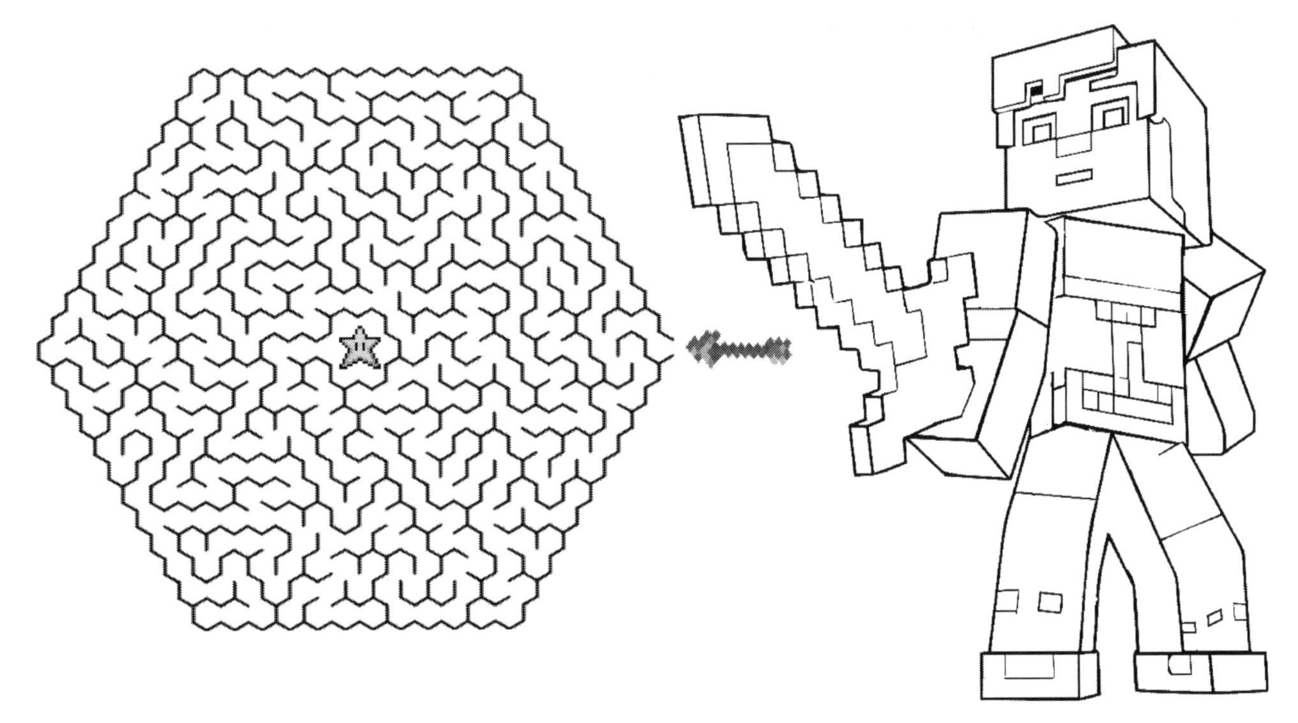

1. Find and circle or cross out the words to find out more about Minecraft.

E	V	I	D	E	T	Y	U	A	E	O	P
W	M	A	N	S	Q	X	M	C	S	E	H
M	G	O	A	V	E	J	N	A	R	I	S
X	M	H	I	X	E	E	H	I	E	P	I
X	G	G	Q	B	I	N	M	O	P	J	F
E	H	U	I	R	E	E	T	R	E	E	R
K	A	E	E	Z	T	L	B	O	E	B	E
F	J	P	S	E	D	F	G	C	R	R	V
V	X	P	R	W	X	Y	I	N	C	Y	L
E	S	R	E	T	S	N	O	M	U	X	I
E	N	D	E	R	M	I	T	E	S	J	S
K	W	L	D	D	J	Y	Q	Y	O	Z	Y

SILVERFISH
MONSTERS
INVENTORY
JUNGLE BIOME
EXPERIENCE
ENDERMITES
PERIMETER
GHAST
CREEPERS

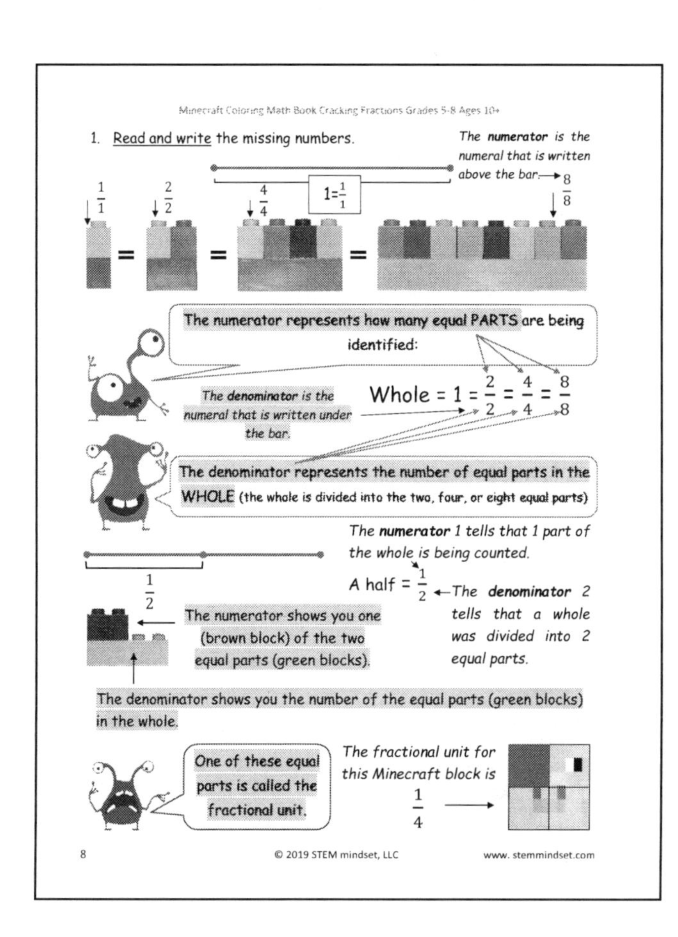

1. Read and write the missing numbers.

*The **numerator** is the numeral that is written above the bar.*

$\frac{1}{1}$ = $\frac{2}{2}$ = $\frac{4}{4}$ = $\frac{8}{8}$ $1=\frac{1}{1}$

The numerator represents how many equal PARTS are being identified:

*The **denominator** is the numeral that is written under the bar.*

Whole = $1 = \frac{2}{2} = \frac{4}{4} = \frac{8}{8}$

The denominator represents the number of equal parts in the WHOLE (the whole is divided into the two, four, or eight equal parts)

*The **numerator** 1 tells that 1 part of the whole is being counted.*

A half = $\frac{1}{2}$ ← *The **denominator** 2 tells that a whole was divided into 2 equal parts.*

$\frac{1}{2}$ The numerator shows you one (brown block) of the two equal parts (green blocks).

The denominator shows you the number of the equal parts (green blocks) in the whole.

One of these equal parts is called the fractional unit.

The fractional unit for this Minecraft block is $\frac{1}{4}$ →

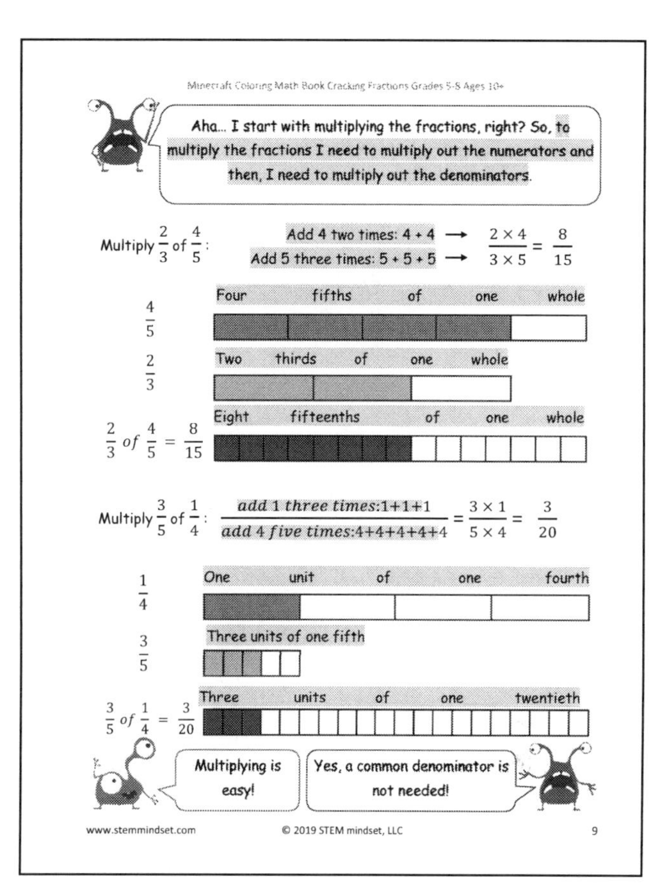

Aha... I start with multiplying the fractions, right? So, to multiply the fractions I need to multiply out the numerators and then, I need to multiply out the denominators.

Multiply $\frac{2}{3}$ of $\frac{4}{5}$: Add 4 two times: 4 + 4 → Add 5 three times: 5 + 5 + 5 → $\frac{2 \times 4}{3 \times 5} = \frac{8}{15}$

$\frac{4}{5}$ Four fifths of one whole

$\frac{2}{3}$ Two thirds of one whole

$\frac{2}{3}$ of $\frac{4}{5} = \frac{8}{15}$ Eight fifteenths of one whole

Multiply $\frac{3}{5}$ of $\frac{1}{4}$: $\frac{\textit{add 1 three times:}1+1+1}{\textit{add 4 five times:}4+4+4+4+4} = \frac{3 \times 1}{5 \times 4} = \frac{3}{20}$

$\frac{1}{4}$ One unit of one fourth

$\frac{3}{5}$ Three units of one fifth

$\frac{3}{5}$ of $\frac{1}{4} = \frac{3}{20}$ Three units of one twentieth

Multiplying is easy!

Yes, a common denominator is not needed!

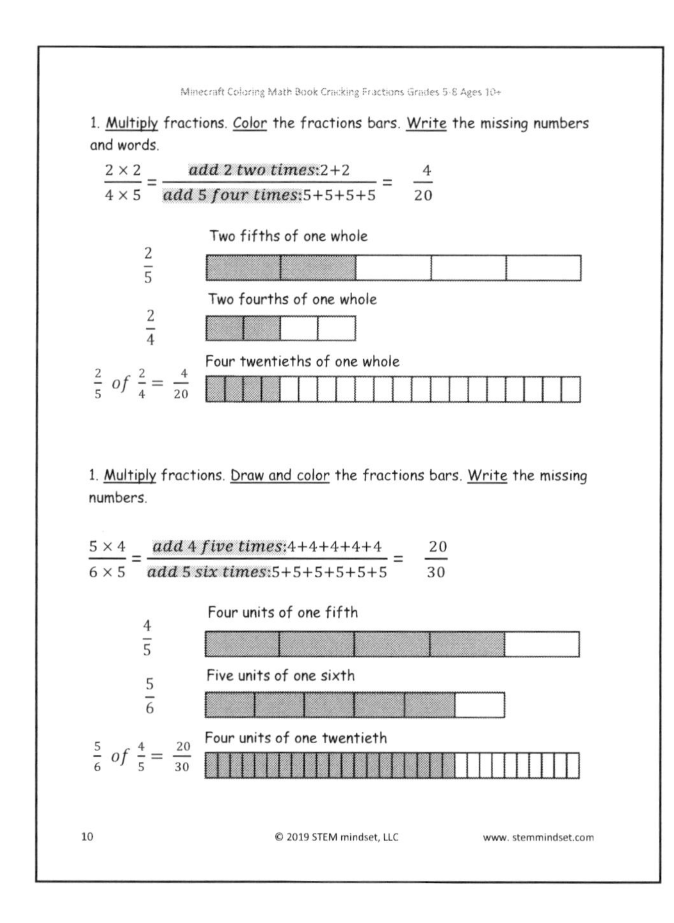

1. Multiply fractions. Color the fractions bars. Write the missing numbers and words.

$\frac{2 \times 2}{4 \times 5} = \frac{\textit{add 2 two times:}2+2}{\textit{add 5 four times:}5+5+5+5} = \frac{4}{20}$

$\frac{2}{5}$ Two fifths of one whole

$\frac{2}{4}$ Two fourths of one whole

$\frac{2}{5}$ of $\frac{2}{4} = \frac{4}{20}$ Four twentieths of one whole

1. Multiply fractions. Draw and color the fractions bars. Write the missing numbers.

$\frac{5 \times 4}{6 \times 5} = \frac{\textit{add 4 five times:}4+4+4+4+4}{\textit{add 5 six times:}5+5+5+5+5+5} = \frac{20}{30}$

$\frac{4}{5}$ Four units of one fifth

$\frac{5}{6}$ Five units of one sixth

$\frac{5}{6}$ of $\frac{4}{5} = \frac{20}{30}$ Four units of one twentieth

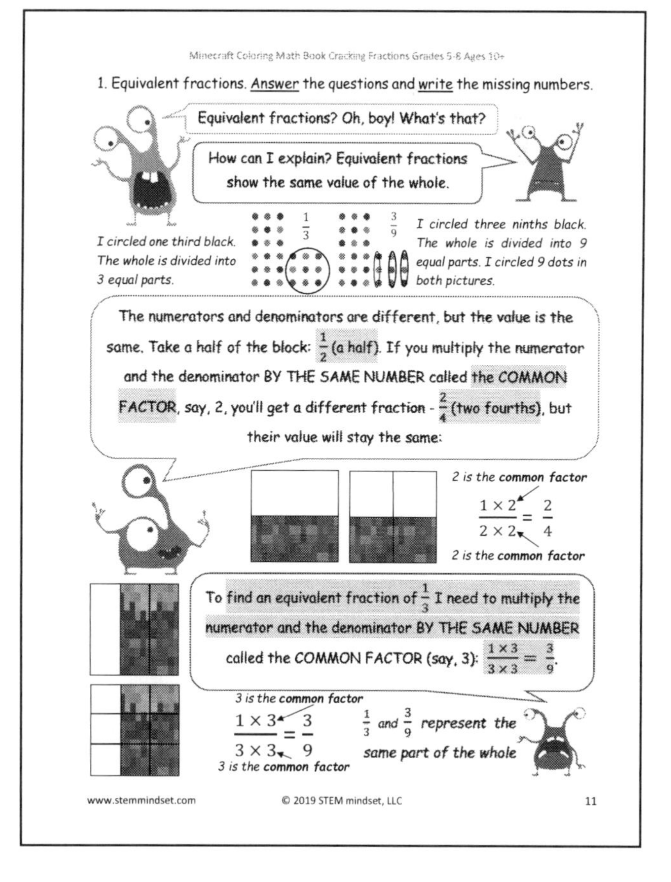

1. Equivalent fractions. Answer the questions and write the missing numbers.

Equivalent fractions? Oh, boy! What's that?

How can I explain? Equivalent fractions show the same value of the whole.

I circled one third black. The whole is divided into 3 equal parts. $\frac{1}{3}$ $\frac{3}{9}$ *I circled three ninths black. The whole is divided into 9 equal parts. I circled 9 dots in both pictures.*

The numerators and denominators are different, but the value is the same. Take a half of the block: $\frac{1}{2}$ (a half). If you multiply the numerator and the denominator BY THE SAME NUMBER called the COMMON FACTOR, say, 2, you'll get a different fraction - $\frac{2}{4}$ (two fourths), but their value will stay the same:

2 is the common factor

$\frac{1 \times 2}{2 \times 2} = \frac{2}{4}$

2 is the common factor

To find an equivalent fraction of $\frac{1}{3}$ I need to multiply the numerator and the denominator BY THE SAME NUMBER called the COMMON FACTOR (say, 3): $\frac{1 \times 3}{3 \times 3} = \frac{3}{9}$.

3 is the common factor

$\frac{1 \times 3}{3 \times 3} = \frac{3}{9}$

3 is the common factor

$\frac{1}{3}$ *and* $\frac{3}{9}$ *represent the same part of the whole*

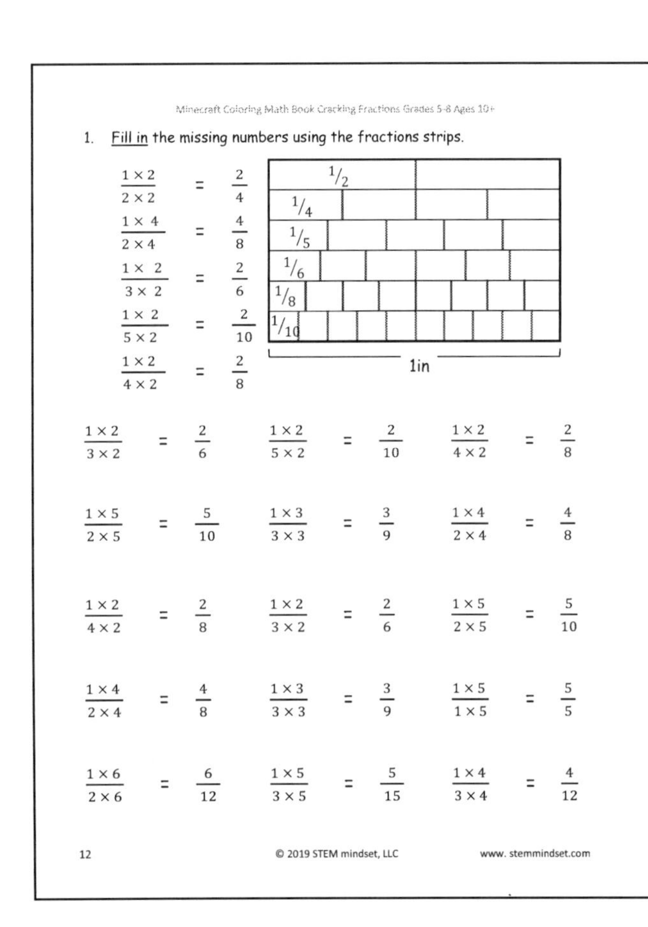

1. Fill in the missing numbers using the fractions strips.

$\frac{1\times2}{2\times2} = \frac{2}{4}$

$\frac{1\times4}{2\times4} = \frac{4}{8}$

$\frac{1\times2}{3\times2} = \frac{2}{6}$

$\frac{1\times2}{5\times2} = \frac{2}{10}$

$\frac{1\times2}{4\times2} = \frac{2}{8}$

$\frac{1\times2}{3\times2} = \frac{2}{6}$ $\frac{1\times2}{5\times2} = \frac{2}{10}$ $\frac{1\times2}{4\times2} = \frac{2}{8}$

$\frac{1\times5}{2\times5} = \frac{5}{10}$ $\frac{1\times3}{3\times3} = \frac{3}{9}$ $\frac{1\times4}{2\times4} = \frac{4}{8}$

$\frac{1\times2}{4\times2} = \frac{2}{8}$ $\frac{1\times2}{3\times2} = \frac{2}{6}$ $\frac{1\times5}{2\times5} = \frac{5}{10}$

$\frac{1\times4}{2\times4} = \frac{4}{8}$ $\frac{1\times3}{3\times3} = \frac{3}{9}$ $\frac{1\times5}{1\times5} = \frac{5}{5}$

$\frac{1\times6}{2\times6} = \frac{6}{12}$ $\frac{1\times5}{3\times5} = \frac{5}{15}$ $\frac{1\times4}{3\times4} = \frac{4}{12}$

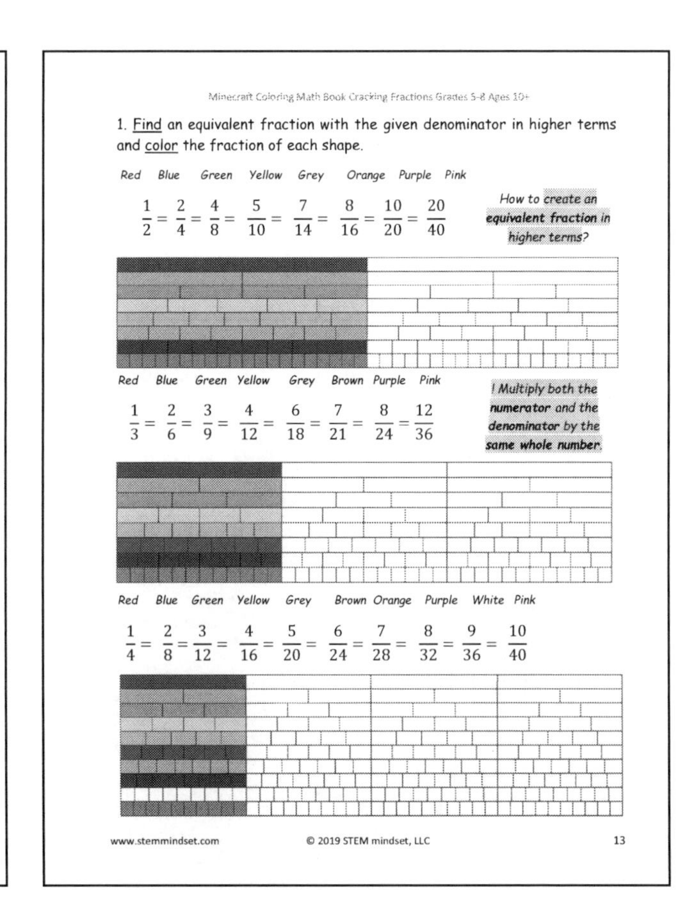

1. Find an equivalent fraction with the given denominator in higher terms and color the fraction of each shape.

Red Blue Green Yellow Grey Orange Purple Pink

$\frac{1}{2} = \frac{2}{4} = \frac{4}{8} = \frac{5}{10} = \frac{7}{14} = \frac{8}{16} = \frac{10}{20} = \frac{20}{40}$

*How to create an **equivalent fraction** in higher terms?*

Red Blue Green Yellow Grey Brown Purple Pink

$\frac{1}{3} = \frac{2}{6} = \frac{3}{9} = \frac{4}{12} = \frac{6}{18} = \frac{7}{21} = \frac{8}{24} = \frac{12}{36}$

*! Multiply both the **numerator** and the **denominator** by the **same whole number**.*

Red Blue Green Yellow Grey Brown Orange Purple White Pink

$\frac{1}{4} = \frac{2}{8} = \frac{3}{12} = \frac{4}{16} = \frac{5}{20} = \frac{6}{24} = \frac{7}{28} = \frac{8}{32} = \frac{9}{36} = \frac{10}{40}$

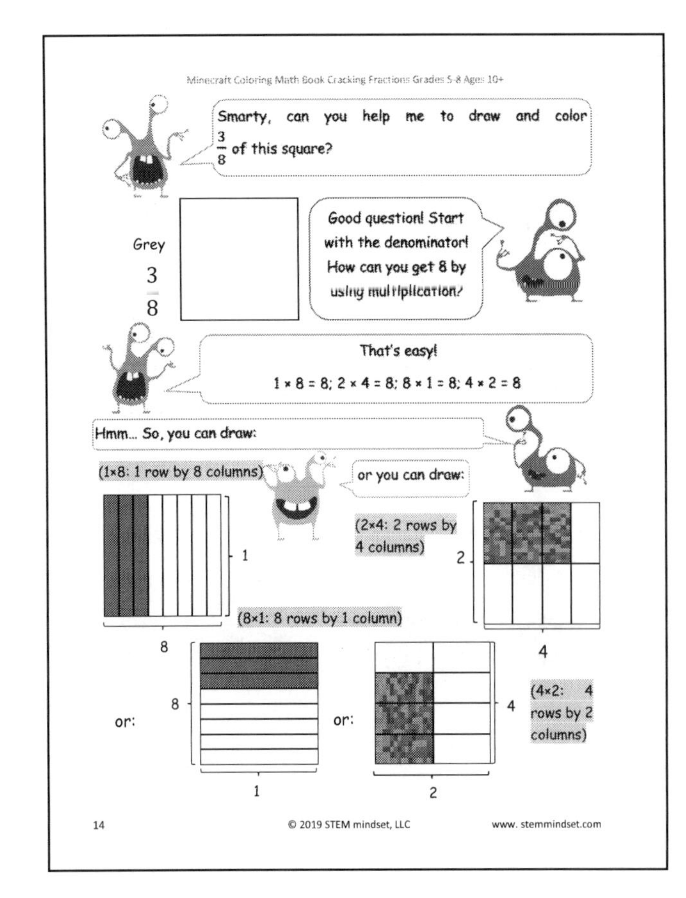

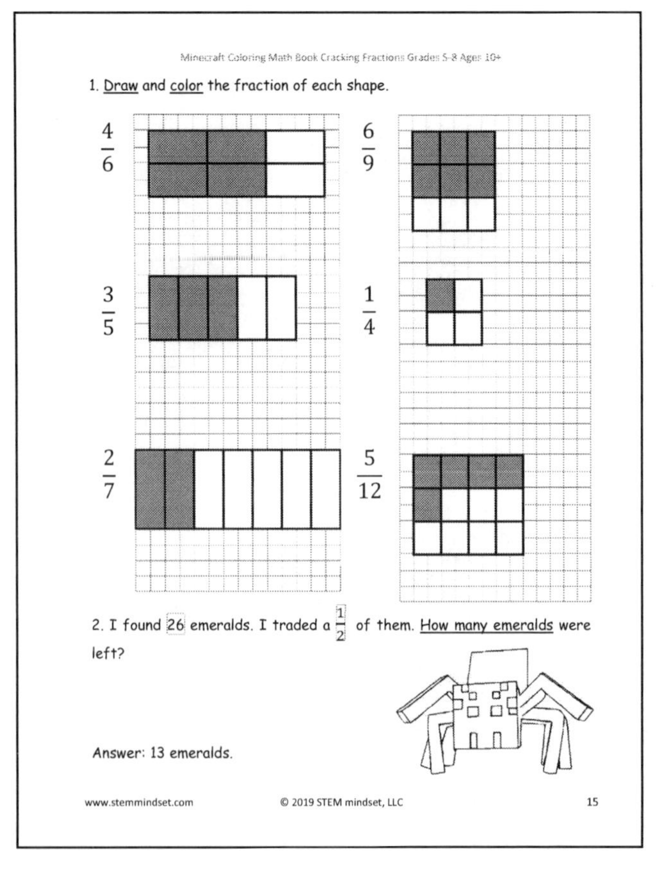

1. Draw and color the fraction of each shape.

2. I found 26 emeralds. I traded a $\frac{1}{2}$ of them. How many emeralds were left?

Answer: 13 emeralds.

1. Find an equivalent fraction with the given numerator.

Step 1: Use the given fraction $(\frac{1}{2})$.

Step 2: The numerator 1 has been multiplied by 2 to get 2.

Step 3: To get an equivalent fraction, the denominator 2 must also be multiplied by 2 to get 4.

Therefore $\frac{1}{2} = \frac{2}{4}$.

$$\frac{1}{2}=\frac{2}{4}=\frac{11}{22}=\frac{4}{8}=\frac{9}{18}=\frac{6}{12}=\frac{10}{20}=\frac{8}{16}=\frac{5}{10}=\frac{7}{14}=\frac{3}{6}=\frac{12}{24}$$

$$\frac{1}{3}=\frac{2}{6}=\frac{12}{36}=\frac{4}{12}=\frac{8}{24}=\frac{11}{33}=\frac{7}{21}=\frac{5}{15}=\frac{9}{27}=\frac{10}{30}=\frac{6}{18}=\frac{3}{9}$$

$$\frac{1}{4}=\frac{9}{36}=\frac{3}{12}=\frac{12}{48}=\frac{5}{20}=\frac{8}{32}=\frac{11}{44}=\frac{6}{24}=\frac{2}{8}=\frac{10}{40}=\frac{7}{28}=\frac{4}{16}$$

$$\frac{1}{5}=\frac{8}{40}=\frac{3}{15}=\frac{12}{60}=\frac{2}{10}=\frac{10}{50}=\frac{4}{20}=\frac{5}{25}=\frac{9}{45}=\frac{6}{30}=\frac{11}{55}=\frac{7}{35}$$

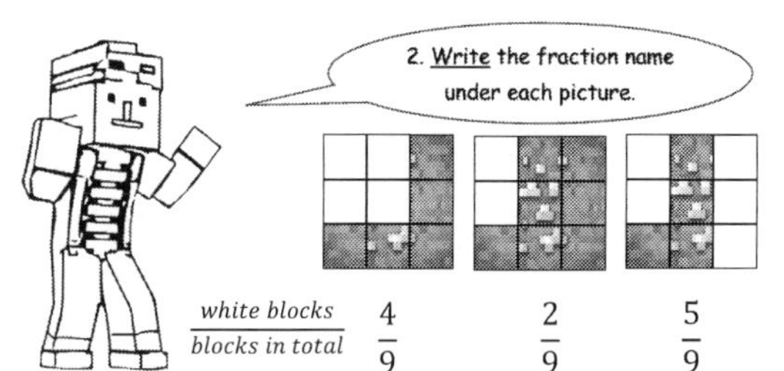
2. Write the fraction name under each picture.

$\frac{\text{white blocks}}{\text{blocks in total}}$ $\frac{4}{9}$ $\frac{2}{9}$ $\frac{5}{9}$

1. Find an equivalent fraction with the given denominator.

Step 1: Use the given fraction $(\frac{1}{2})$.

Step 2: The denominator 2 has been multiplied by 9 to get 18.

Step 3: To get an equivalent fraction, the numerator 1 must also be multiplied by 9 to get 9.

Therefore $\frac{1}{2} = \frac{9}{18}$.

$$\frac{1}{2}=\frac{9}{18}=\frac{2}{4}=\frac{6}{12}=\frac{11}{22}=\frac{8}{16}=\frac{4}{8}=\frac{14}{28}=\frac{20}{40}=\frac{3}{6}=\frac{5}{10}=\frac{17}{34}$$

$$\frac{1}{3}=\frac{7}{21}=\frac{3}{9}=\frac{20}{60}=\frac{5}{15}=\frac{9}{27}=\frac{4}{12}=\frac{2}{6}=\frac{6}{18}=\frac{12}{36}=\frac{16}{48}=\frac{8}{24}$$

$$\frac{1}{4}=\frac{6}{24}=\frac{10}{40}=\frac{4}{16}=\frac{8}{32}=\frac{3}{12}=\frac{12}{48}=\frac{2}{8}=\frac{9}{36}=\frac{5}{20}=\frac{7}{28}=\frac{14}{56}$$

$$\frac{1}{5}=\frac{7}{35}=\frac{3}{15}=\frac{9}{45}=\frac{4}{20}=\frac{6}{30}=\frac{13}{65}=\frac{5}{25}=\frac{14}{70}=\frac{10}{50}=\frac{2}{10}=\frac{8}{40}$$

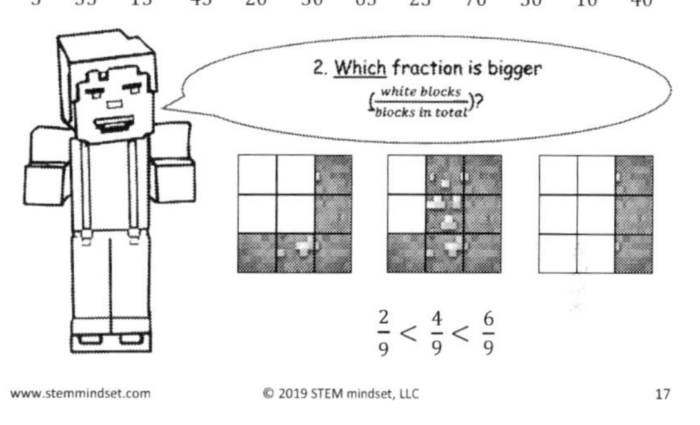
2. Which fraction is bigger
(white blocks / blocks in total)?

$$\frac{2}{9} < \frac{4}{9} < \frac{6}{9}$$

1. Find an equivalent fraction with the given numerator.

Step 1: Use the given fraction $(\frac{1}{6})$.

Step 2: The numerator 1 has been multiplied by 2 to get 2.

Step 3: To get an equivalent fraction, the denominator 6 must also be multiplied by 2 to get 12.

Therefore $\frac{1}{6} = \frac{2}{12}$.

$$\frac{1}{6}=\frac{2}{12}=\frac{9}{54}=\frac{4}{24}=\frac{12}{72}=\frac{6}{36}=\frac{10}{60}=\frac{8}{48}=\frac{3}{18}=\frac{7}{42}=\frac{11}{66}=\frac{5}{30}$$

$$\frac{1}{7}=\frac{2}{14}=\frac{7}{49}=\frac{12}{84}=\frac{5}{35}=\frac{9}{63}=\frac{3}{21}=\frac{8}{56}=\frac{6}{42}=\frac{10}{70}=\frac{11}{77}=\frac{4}{28}$$

$$\frac{1}{8}=\frac{2}{16}=\frac{6}{48}=\frac{4}{32}=\frac{9}{72}=\frac{3}{24}=\frac{11}{88}=\frac{8}{64}=\frac{5}{40}=\frac{10}{80}=\frac{7}{56}=\frac{12}{96}$$

$$\frac{1}{9}=\frac{5}{45}=\frac{3}{27}=\frac{9}{81}=\frac{2}{18}=\frac{11}{99}=\frac{7}{63}=\frac{12}{108}=\frac{4}{36}=\frac{10}{90}=\frac{6}{54}=\frac{8}{72}$$

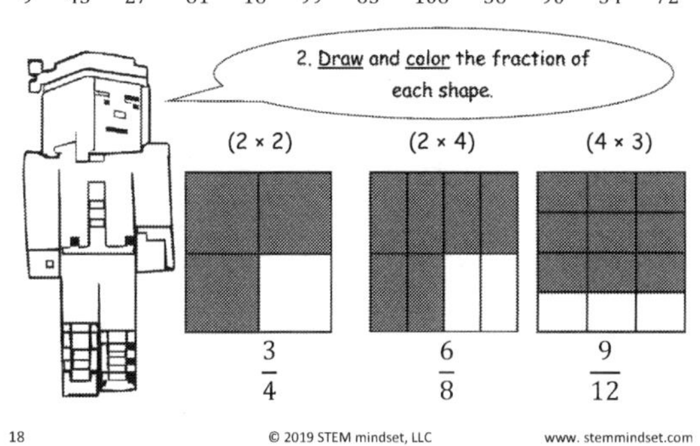
2. Draw and color the fraction of each shape.
(2 × 2)
(2 × 4)
(4 × 3)

$\frac{3}{4}$ $\frac{6}{8}$ $\frac{9}{12}$

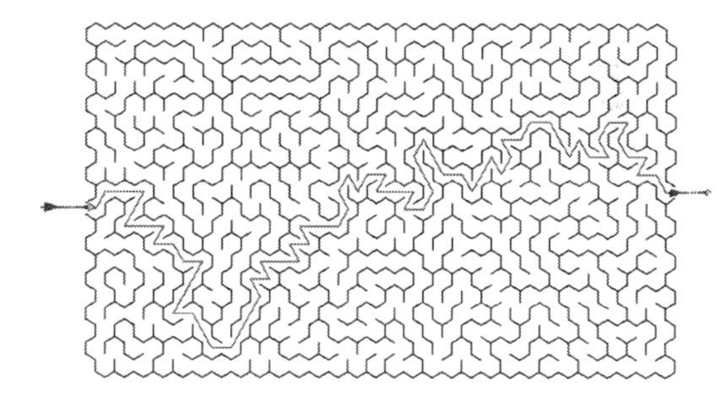

1. Find and circle or cross out the words to find out more about Minecraft.

C	A	E	A	W	S	F	D	Q	D	D	P	R
L	O	W	V	F	J	J	G	E	K	O	B	E
A	T	N	A	I	M	Z	T	S	S	E	G	Q
V	H	C	S	C	T	A	I	S	R	G	X	U
I	F	G	B	T	L	A	E	M	W	J	Z	I
V	H	M	P	O	R	S	E	O	H	W	R	R
R	Y	L	S	P	S	U	D	R	P	K	H	E
U	Q	I	U	I	A	X	C	J	C	X	D	M
S	S	D	O	P	L	X	K	T	L	Z	C	E
T	E	N	T	R	A	N	C	E	I	W	L	N
P	S	M	A	E	G	N	G	W	G	O	M	T
E	N	O	T	S	E	L	B	B	O	C	N	S
R	C	L	E	V	V	T	G	C	Y	V	X	S

CREATIVE
POSSESSIONS
SURVIVAL
REQUIREMENTS
ISOLATED
CONSTRUCTIONS
ENTRANCE
COBBLESTONE

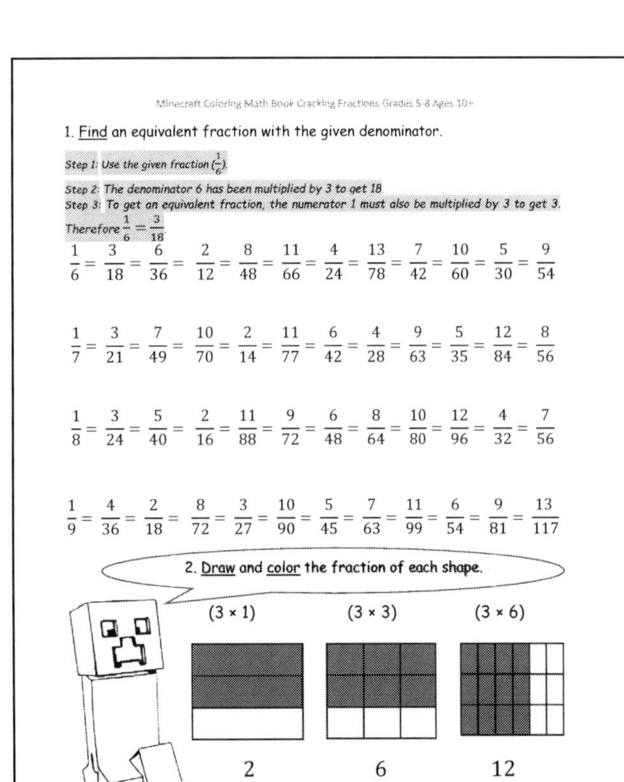
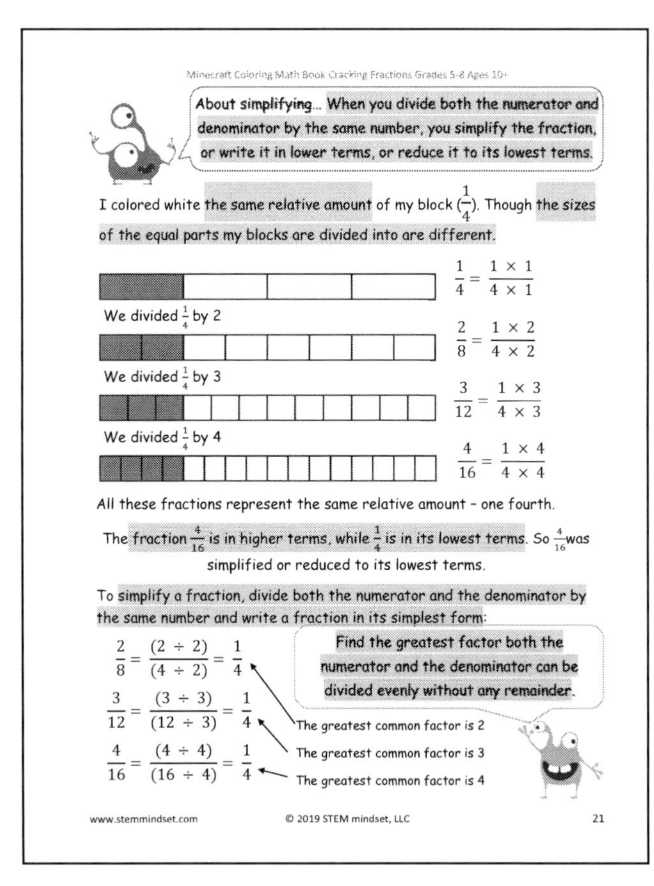
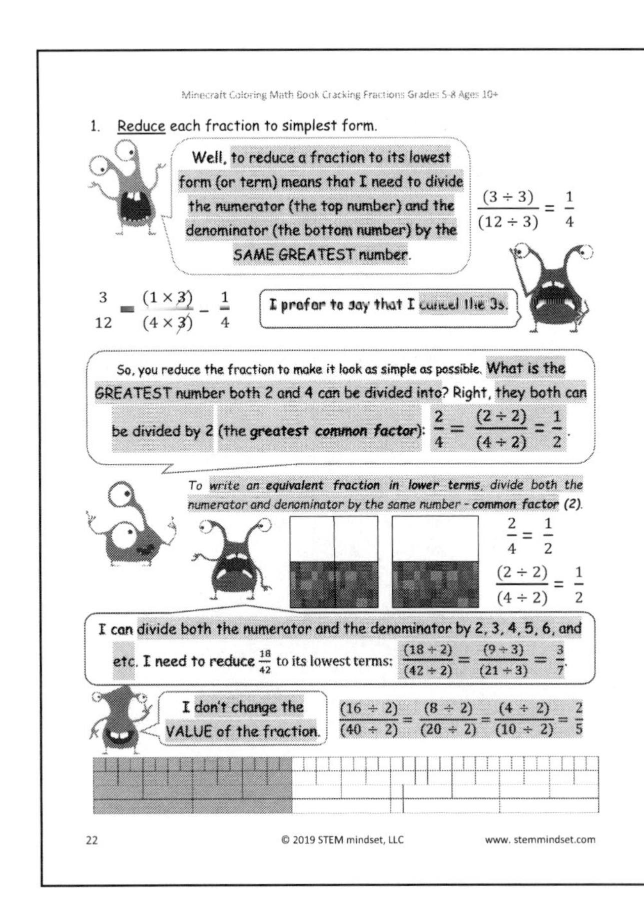
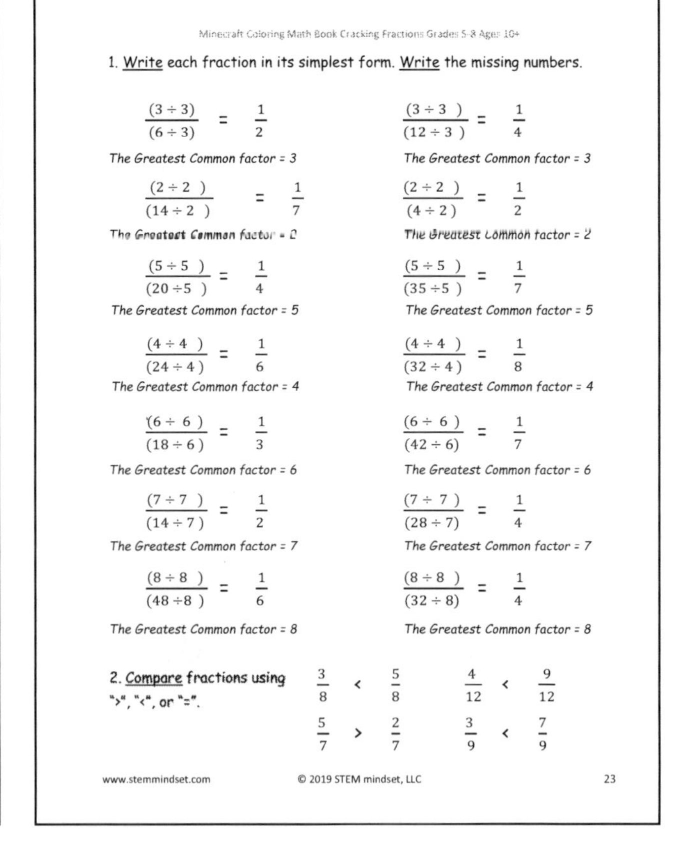

1. Reduce each fraction to its lowest terms and color the fraction of each shape. Write the missing numbers.

Blue Green Red

$$\frac{16}{28} = \frac{(16 \div 2)}{(28 \div 2)} = \frac{(8 \div 2)}{(14 \div 2)} = \frac{4}{7}$$

The Greatest common factor is the greatest factor that divides both the numerator and the denominator evenly into a number, and there is no remainder.

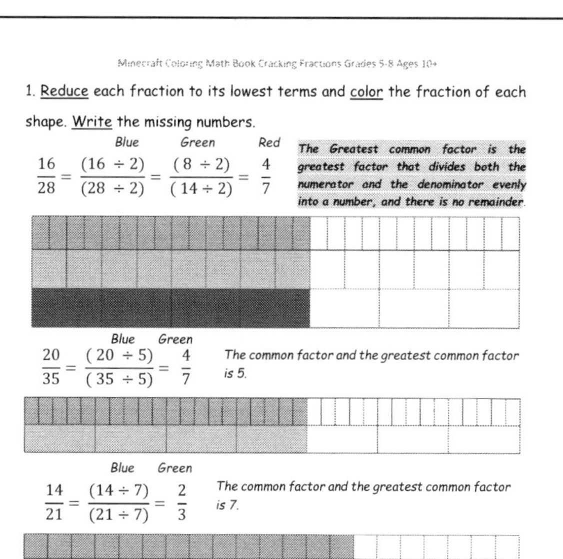

Blue Green

$$\frac{20}{35} = \frac{(20 \div 5)}{(35 \div 5)} = \frac{4}{7}$$

The common factor and the greatest common factor is 5.

Blue Green

$$\frac{14}{21} = \frac{(14 \div 7)}{(21 \div 7)} = \frac{2}{3}$$

The common factor and the greatest common factor is 7.

Blue Green Red

$$\frac{36}{40} = \frac{(36 \div 2)}{(40 \div 2)} = \frac{(18 \div 2)}{(20 \div 2)} = \frac{9}{10}$$

***Factors of 36**: 1, 2, 3, 4, 6, 9, 12, 18, 36.*

***Factors of 40**: 1, 2, 4, 5, 8, 10, 20, 40.*

***The greatest common factor (GCF)** is 4.*

1. Reduce each fraction to its simplest form and color the fraction of each shape. Write the missing numbers.

Blue Green Red Yellow

$$\frac{18}{36} = \frac{(18 \div 2)}{(36 \div 2)} = \frac{(9 \div 3)}{(18 \div 3)} = \frac{(3 \div 3)}{(6 \div 3)} = \frac{1}{2}$$

2 × 3 × 3 = 18

*18 is the **greatest common factor***

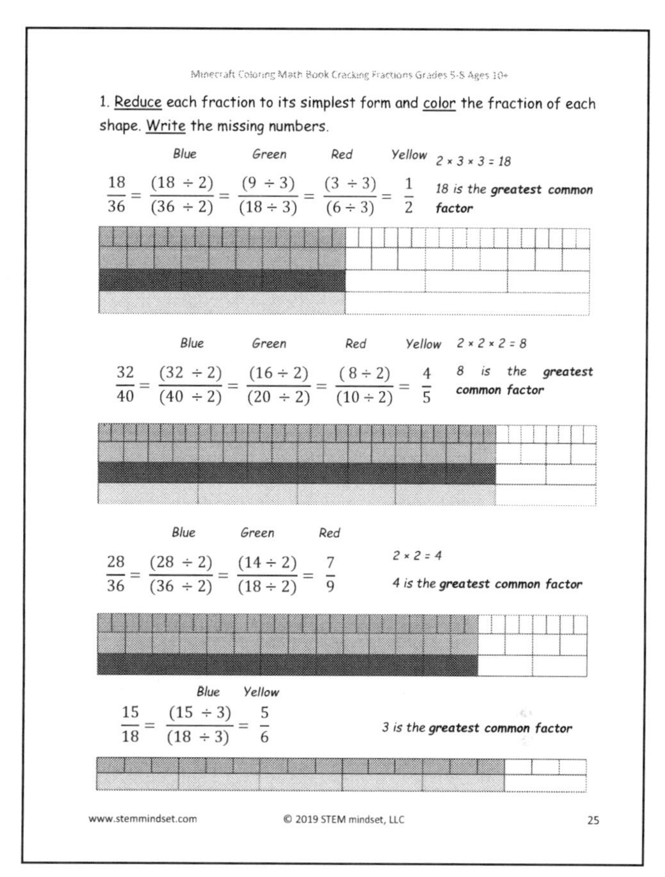

Blue Green Red Yellow

$$\frac{32}{40} = \frac{(32 \div 2)}{(40 \div 2)} = \frac{(16 \div 2)}{(20 \div 2)} = \frac{(8 \div 2)}{(10 \div 2)} = \frac{4}{5}$$

2 × 2 × 2 = 8

*8 is the **greatest common factor***

Blue Green Red

$$\frac{28}{36} = \frac{(28 \div 2)}{(36 \div 2)} = \frac{(14 \div 2)}{(18 \div 2)} = \frac{7}{9}$$

2 × 2 = 4

*4 is the **greatest common factor***

Blue Yellow

$$\frac{15}{18} = \frac{(15 \div 3)}{(18 \div 3)} = \frac{5}{6}$$

*3 is the **greatest common factor***

1. Reduce each fraction to its simplest form. Write the missing numbers.

$$\frac{32}{48} = \frac{(32 \div 2)}{(48 \div 2)} = \frac{(16 \div 2)}{(24 \div 2)} = \frac{(8 \div 2)}{(12 \div 2)} = \frac{(4 \div 2)}{(6 \div 2)} = \frac{2}{3}$$

$$\frac{(32 \div 16)}{(48 \div 16)} = \frac{2}{3}$$

2×2×2×2=16 *is the **greatest common factor***

$$\frac{12}{28} = \frac{(12 \div 2)}{(28 \div 2)} = \frac{(6 \div 2)}{(14 \div 2)} = \frac{3}{7} \qquad \frac{(12 \div 4)}{(28 \div 4)} = \frac{3}{7}$$

2×2=4 *is the **greatest common factor***

$$\frac{36}{48} = \frac{(36 \div 2)}{(48 \div 2)} = \frac{(18 \div 2)}{(24 \div 2)} = \frac{(9 \div 3)}{(12 \div 3)} = \frac{3}{4} \qquad \frac{(36 \div 12)}{(48 \div 12)} = \frac{3}{4}$$

2×2×3=12 *is the **greatest common factor***

$$\frac{50}{75} = \frac{(50 \div 5)}{(75 \div 5)} = \frac{(10 \div 5)}{(15 \div 5)} = \frac{2}{3} \qquad \frac{(50 \div 25)}{(75 \div 25)} = \frac{2}{3}$$

5×5=25 *is the **greatest common factor***

2. Write the fraction name under each picture. Compare the fractions.

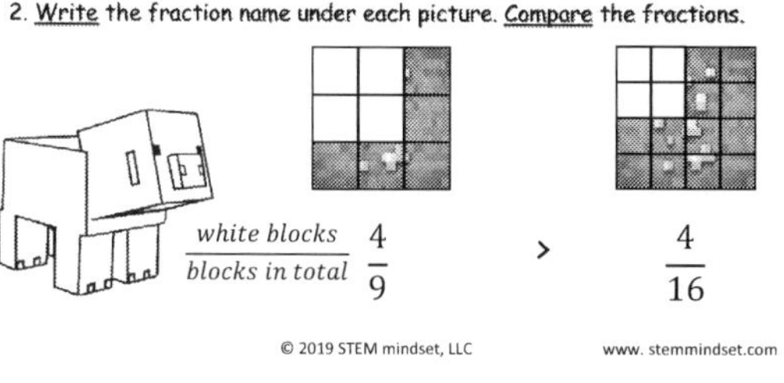

$\frac{\text{white blocks}}{\text{blocks in total}}$ $\frac{4}{9}$ > $\frac{4}{16}$

1. Reduce each fraction to its simplest form. Write the missing numbers.

$$\frac{56}{72} = \frac{(56 \div 2)}{(72 \div 2)} = \frac{(28 \div 2)}{(36 \div 2)} = \frac{(14 \div 2)}{(18 \div 2)} = \frac{7}{9} \qquad \frac{(56 \div 8)}{(72 \div 8)} = \frac{7}{9}$$

2×2×2=8 *is the **greatest common factor***

$$\frac{36}{81} = \frac{(36 \div 3)}{(81 \div 3)} = \frac{(12 \div 3)}{(27 \div 3)} = \frac{4}{9} \qquad \frac{(36 \div 9)}{(81 \div 9)} = \frac{4}{9}$$

3×3=9 *is the **greatest common factor***

$$\frac{60}{96} = \frac{(60 \div 2)}{(96 \div 2)} = \frac{(30 \div 2)}{(48 \div 2)} = \frac{(15 \div 3)}{(24 \div 2)} = \frac{5}{8} \qquad \frac{(60 \div 12)}{(96 \div 12)} = \frac{5}{8}$$

2×2×3=12 *is the **greatest common factor***

$$\frac{48}{96} = \frac{(48 \div 2)}{(96 \div 2)} = \frac{(24 \div 2)}{(48 \div 2)} = \frac{(12 \div 2)}{(24 \div 2)} = \frac{(6 \div 2)}{(12 \div 2)} = \frac{(3 \div 3)}{(6 \div 3)} = \frac{1}{2}$$

$$\frac{(48 \div 48)}{(96 \div 48)} = \frac{1}{2}$$

2×2×2×2×3=48 *is the **greatest common factor***

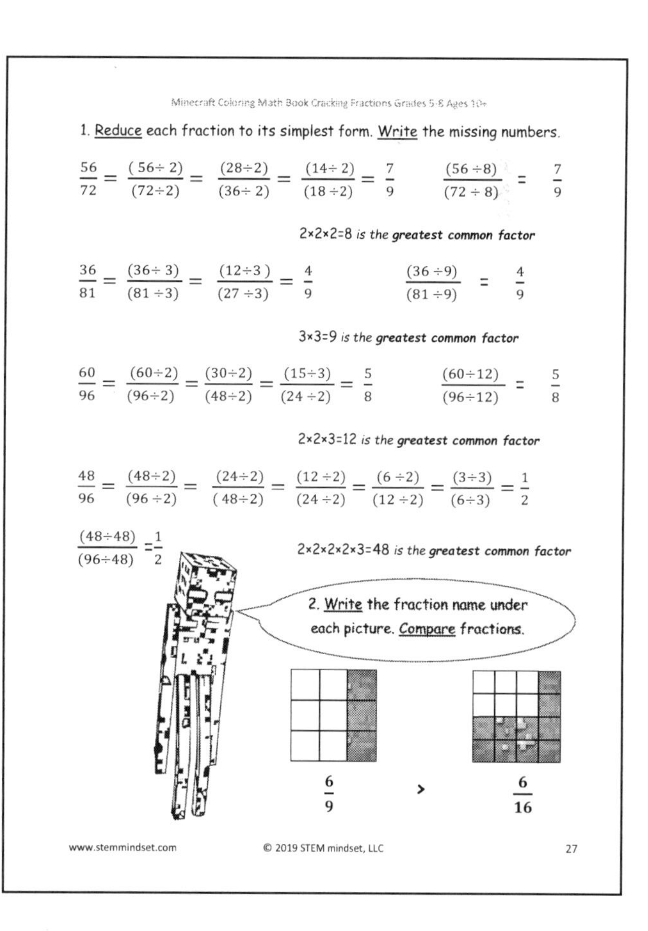
2. Write the fraction name under each picture. Compare fractions.

$\frac{6}{9}$ > $\frac{6}{16}$

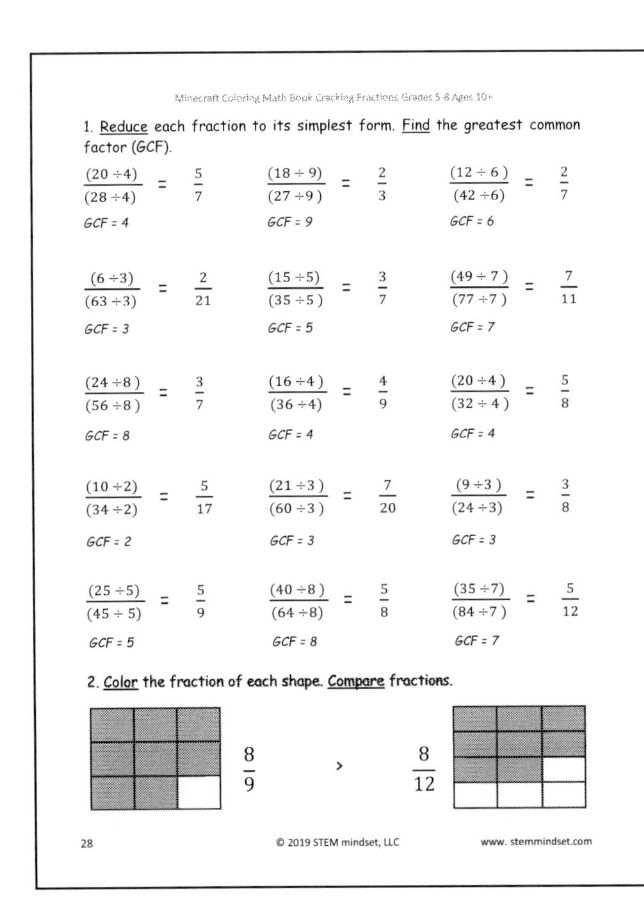

1. Reduce each fraction to its simplest form. Find the greatest common factor (GCF).

$\frac{(20 \div 4)}{(28 \div 4)} = \frac{5}{7}$ GCF = 4 | $\frac{(18 \div 9)}{(27 \div 9)} = \frac{2}{3}$ GCF = 9 | $\frac{(12 \div 6)}{(42 \div 6)} = \frac{2}{7}$ GCF = 6

$\frac{(6 \div 3)}{(63 \div 3)} = \frac{2}{21}$ GCF = 3 | $\frac{(15 \div 5)}{(35 \div 5)} = \frac{3}{7}$ GCF = 5 | $\frac{(49 \div 7)}{(77 \div 7)} = \frac{7}{11}$ GCF = 7

$\frac{(24 \div 8)}{(56 \div 8)} = \frac{3}{7}$ GCF = 8 | $\frac{(16 \div 4)}{(36 \div 4)} = \frac{4}{9}$ GCF = 4 | $\frac{(20 \div 4)}{(32 \div 4)} = \frac{5}{8}$ GCF = 4

$\frac{(10 \div 2)}{(34 \div 2)} = \frac{5}{17}$ GCF = 2 | $\frac{(21 \div 3)}{(60 \div 3)} = \frac{7}{20}$ GCF = 3 | $\frac{(9 \div 3)}{(24 \div 3)} = \frac{3}{8}$ GCF = 3

$\frac{(25 \div 5)}{(45 \div 5)} = \frac{5}{9}$ GCF = 5 | $\frac{(40 \div 8)}{(64 \div 8)} = \frac{5}{8}$ GCF = 8 | $\frac{(35 \div 7)}{(84 \div 7)} = \frac{5}{12}$ GCF = 7

2. Color the fraction of each shape. Compare fractions.

$\frac{8}{9} > \frac{8}{12}$

  www. stemmindset.com

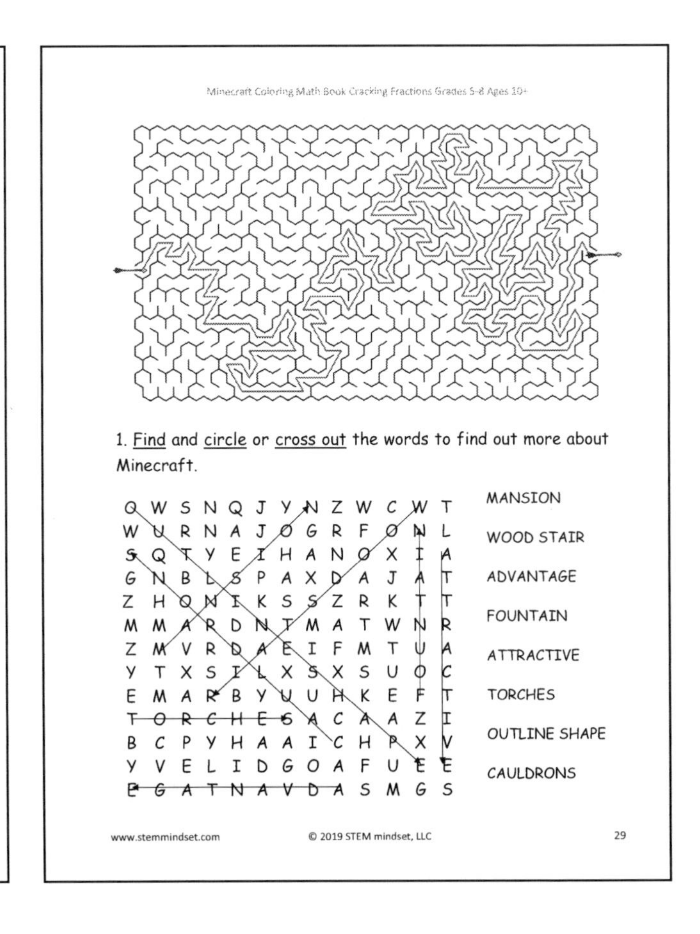

1. Find and circle or cross out the words to find out more about Minecraft.

Q	W	S	N	Q	J	Y	N	Z	W	C	W	T
W	U	R	N	A	J	O	G	R	F	O	N	L
S	Q	T	Y	E	I	H	A	N	O	X	I	A
G	N	B	L	S	P	A	X	D	A	J	A	T
Z	H	O	N	I	K	S	S	Z	R	K	T	T
M	M	A	R	D	N	T	M	A	T	W	N	R
Z	M	V	R	D	A	E	I	F	M	T	U	A
Y	T	X	S	I	L	X	S	X	S	U	O	C
E	M	A	R	B	Y	U	U	H	K	E	F	T
T	O	R	C	H	E	S	A	C	A	A	Z	I
B	C	P	Y	H	A	A	I	C	H	P	X	V
Y	V	E	L	I	D	G	O	A	F	U	E	E
E	G	A	T	N	A	V	D	A	S	M	G	S

MANSION
WOOD STAIR
ADVANTAGE
FOUNTAIN
ATTRACTIVE
TORCHES
OUTLINE SHAPE
CAULDRONS

www.stemmindset.com

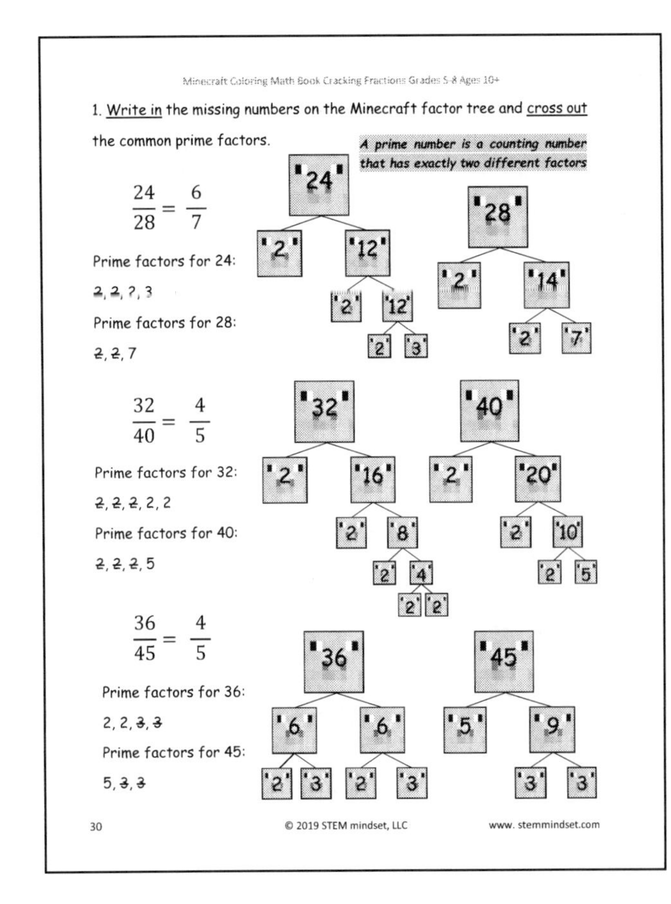

1. Write in the missing numbers on the Minecraft factor tree and cross out the common prime factors.

A prime number is a counting number that has exactly two different factors

$\frac{24}{28} = \frac{6}{7}$

Prime factors for 24: ~~2~~, ~~2~~, 2, 3

Prime factors for 28: ~~2~~, ~~2~~, 7

$\frac{32}{40} = \frac{4}{5}$

Prime factors for 32: ~~2~~, ~~2~~, ~~2~~, 2, 2

Prime factors for 40: ~~2~~, ~~2~~, ~~2~~, 5

$\frac{36}{45} = \frac{4}{5}$

Prime factors for 36: 2, 2, ~~3~~, ~~3~~

Prime factors for 45: 5, ~~3~~, ~~3~~

  www. stemmindset.com

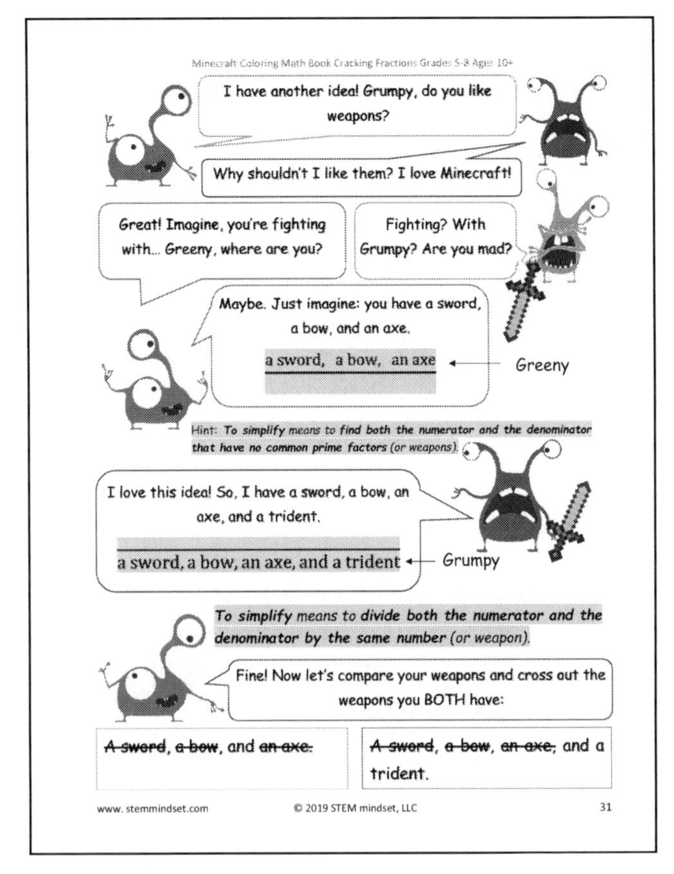

www.stemmindset.com

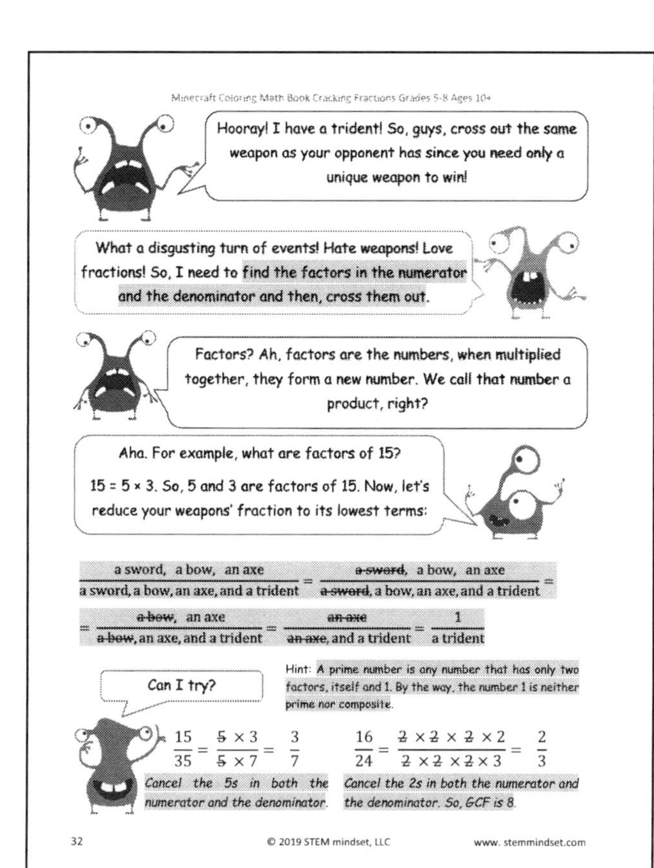

$$\frac{\text{a sword, a bow, an axe}}{\text{a sword, a bow, an axe, and a trident}} = \frac{\cancel{\text{a sword}}\text{, a bow, an axe}}{\cancel{\text{a sword}}\text{, a bow, an axe, and a trident}} = \frac{\cancel{\text{a bow}}\text{, an axe}}{\cancel{\text{a bow}}\text{, an axe, and a trident}} = \frac{\cancel{\text{an axe}}}{\cancel{\text{an axe}}\text{, and a trident}} = \frac{1}{\text{a trident}}$$

Hint: A prime number is any number that has only two factors, itself and 1. By the way, the number 1 is neither prime nor composite.

$$\frac{15}{35} = \frac{\cancel{5} \times 3}{\cancel{5} \times 7} = \frac{3}{7}$$

Cancel the 5s in both the numerator and the denominator.

$$\frac{16}{24} = \frac{\cancel{2} \times \cancel{2} \times \cancel{2} \times 2}{\cancel{2} \times \cancel{2} \times \cancel{2} \times 3} = \frac{2}{3}$$

Cancel the 2s in both the numerator and the denominator. So, GCF is 8.

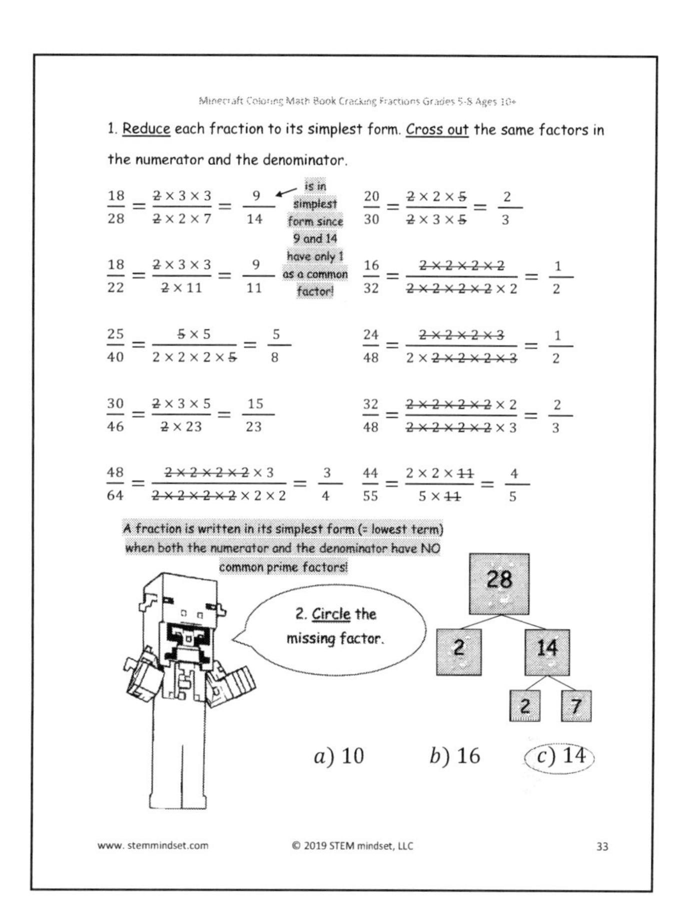

1. Reduce each fraction to its simplest form. Cross out the same factors in the numerator and the denominator.

$$\frac{18}{28} = \frac{\cancel{2} \times 3 \times 3}{\cancel{2} \times 2 \times 7} = \frac{9}{14}$$

← is in simplest form since 9 and 14 have only 1 as a common factor!

$$\frac{20}{30} = \frac{\cancel{2} \times 2 \times \cancel{5}}{\cancel{2} \times 3 \times \cancel{5}} = \frac{2}{3}$$

$$\frac{18}{22} = \frac{\cancel{2} \times 3 \times 3}{\cancel{2} \times 11} = \frac{9}{11}$$

$$\frac{16}{32} = \frac{\cancel{2 \times 2 \times 2 \times 2}}{\cancel{2 \times 2 \times 2 \times 2} \times 2} = \frac{1}{2}$$

$$\frac{25}{40} = \frac{\cancel{5} \times 5}{2 \times 2 \times 2 \times \cancel{5}} = \frac{5}{8}$$

$$\frac{24}{48} = \frac{\cancel{2 \times 2 \times 2 \times 3}}{2 \times \cancel{2 \times 2 \times 2 \times 3}} = \frac{1}{2}$$

$$\frac{30}{46} = \frac{\cancel{2} \times 3 \times 5}{\cancel{2} \times 23} = \frac{15}{23}$$

$$\frac{32}{48} = \frac{\cancel{2 \times 2 \times 2 \times 2} \times 2}{\cancel{2 \times 2 \times 2 \times 2} \times 3} = \frac{2}{3}$$

$$\frac{48}{64} = \frac{\cancel{2 \times 2 \times 2 \times 2} \times 3}{\cancel{2 \times 2 \times 2 \times 2} \times 2 \times 2} = \frac{3}{4}$$

$$\frac{44}{55} = \frac{2 \times 2 \times \cancel{11}}{5 \times \cancel{11}} = \frac{4}{5}$$

A fraction is written in its simplest form (= lowest term) when both the numerator and the denominator have NO common prime factors!

$a)$ 10 $b)$ 16 $c)$ 14

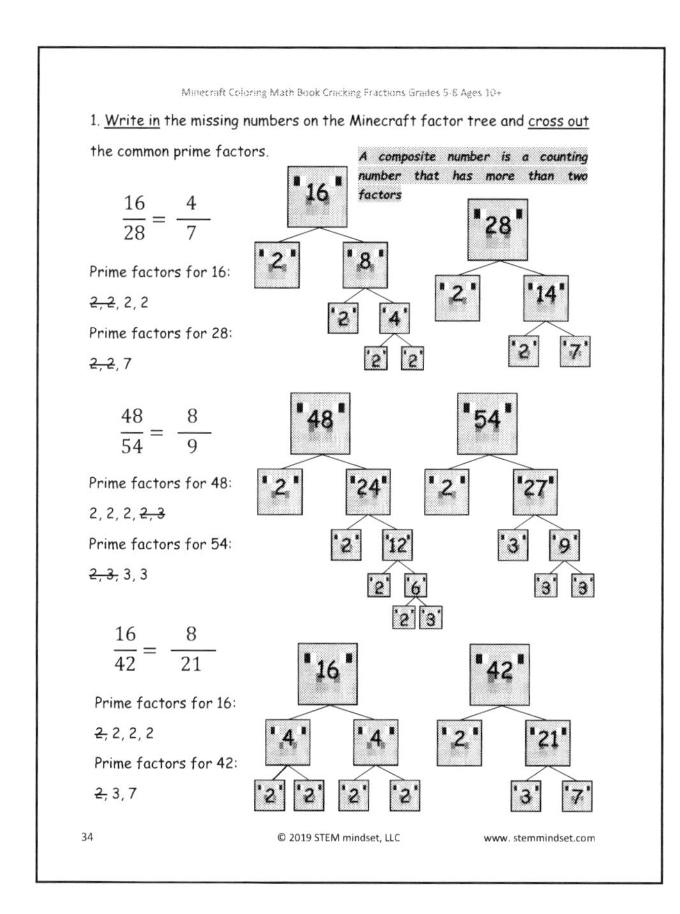

1. Write in the missing numbers on the Minecraft factor tree and cross out the common prime factors.

A composite number is a counting number that has more than two factors

$$\frac{16}{28} = \frac{4}{7}$$

Prime factors for 16:

~~2, 2~~, 2, 2

Prime factors for 28:

~~2, 2~~, 7

$$\frac{48}{54} = \frac{8}{9}$$

Prime factors for 48:

2, 2, 2, ~~2, 3~~

Prime factors for 54:

~~2, 3~~, 3, 3

$$\frac{16}{42} = \frac{8}{21}$$

Prime factors for 16:

~~2~~, 2, 2, 2

Prime factors for 42:

~~2~~, 3, 7

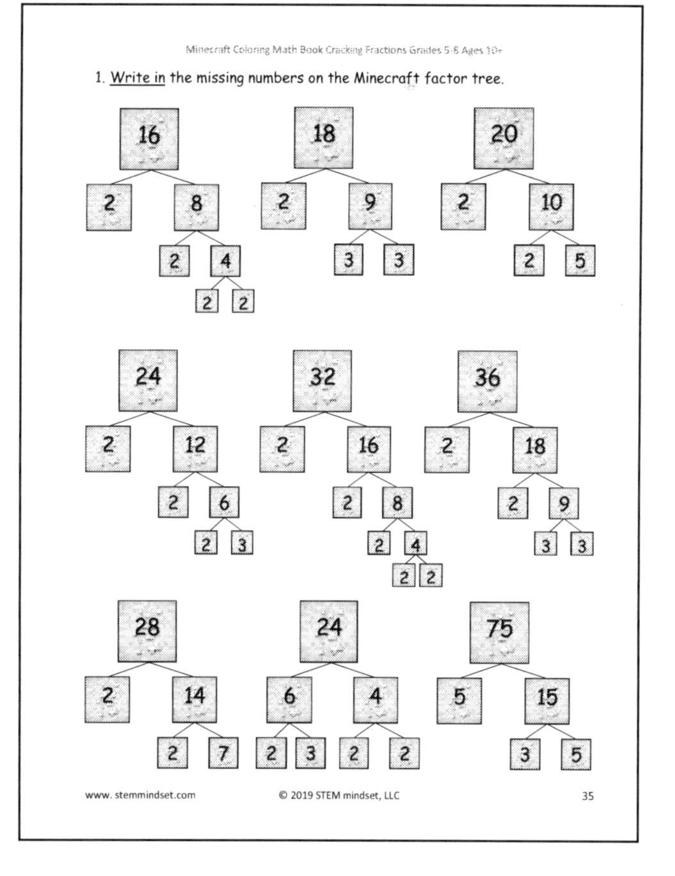

1. Write in the missing numbers on the Minecraft factor tree.

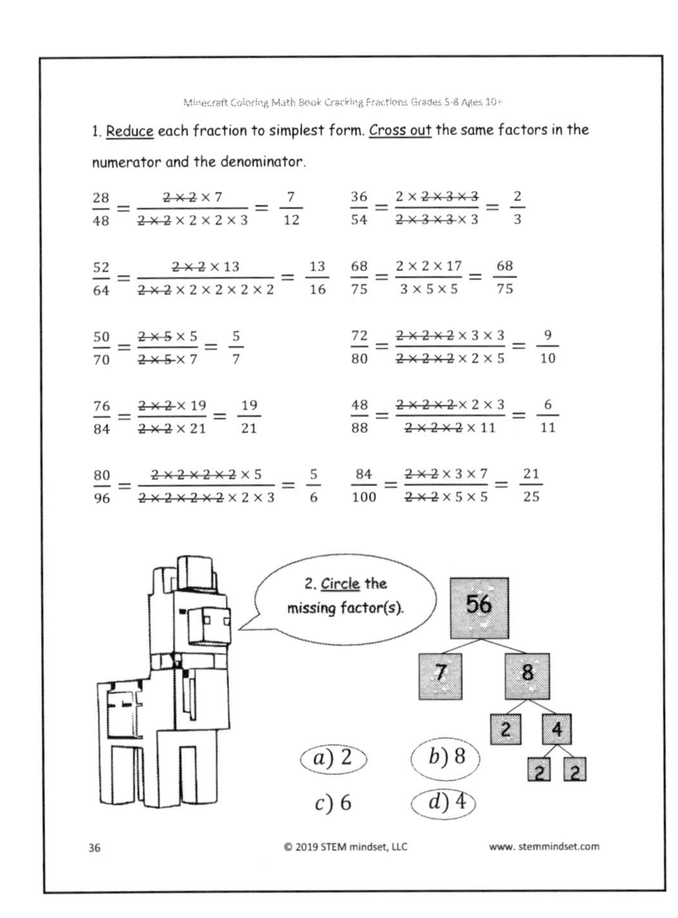

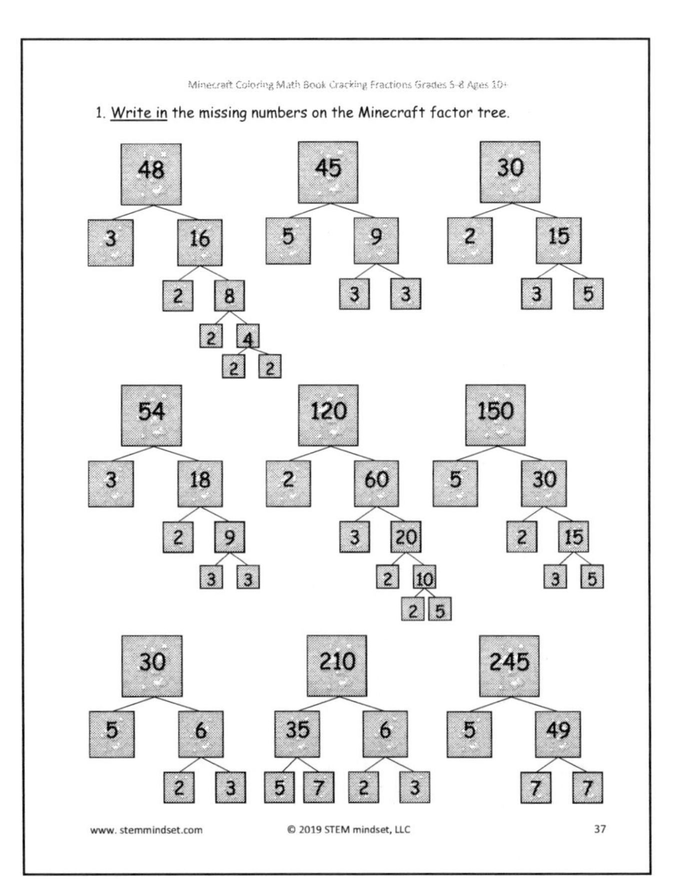

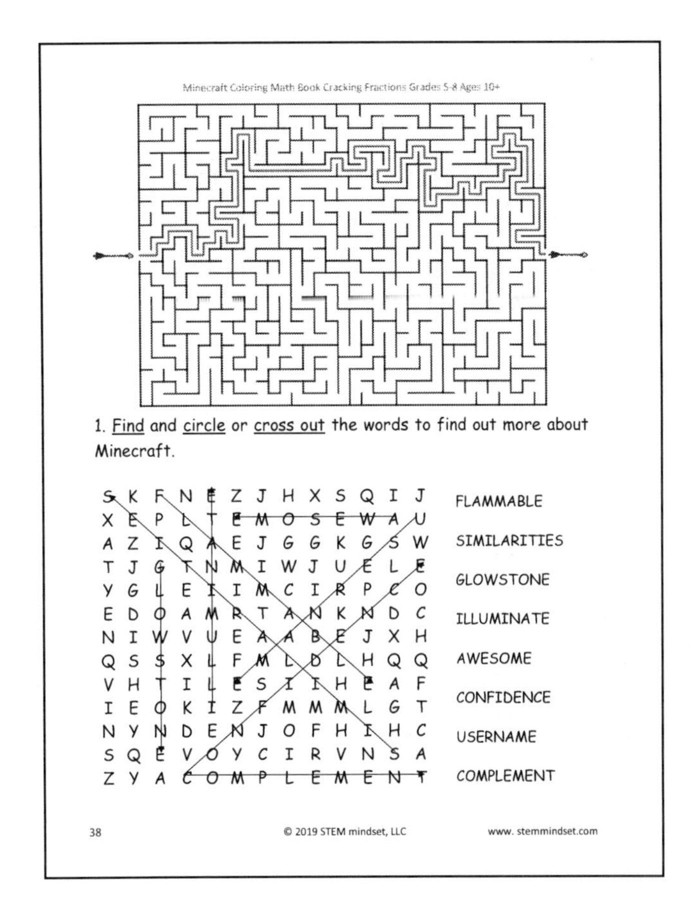

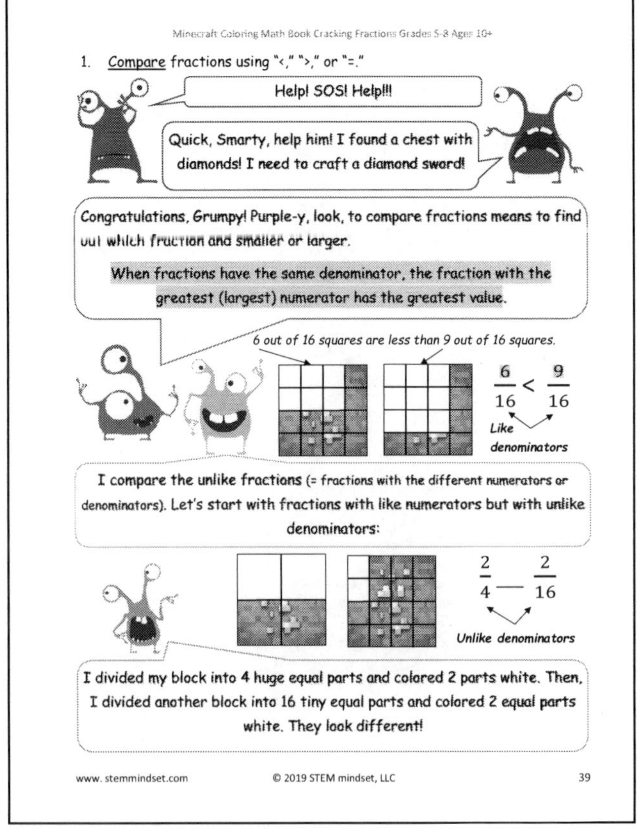

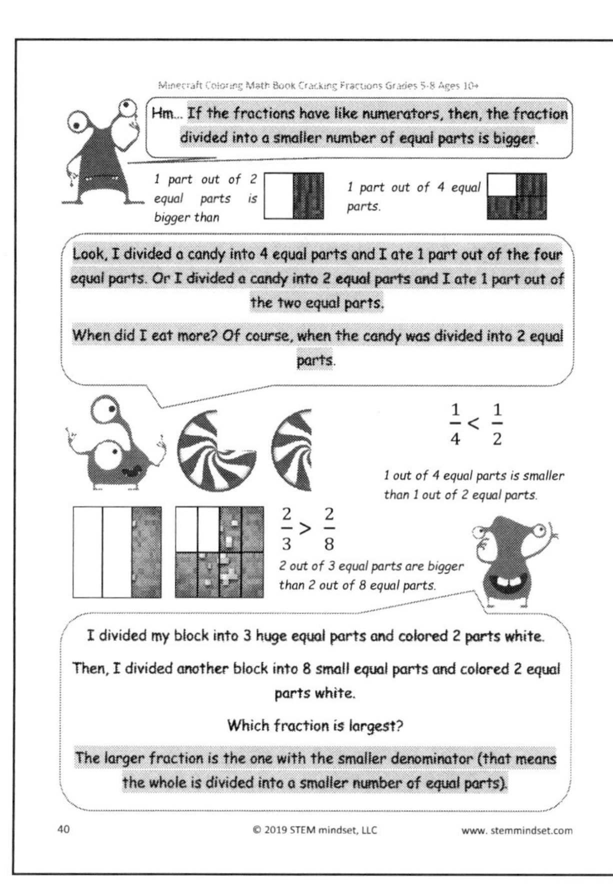

Hm... If the fractions have like numerators, then, the fraction divided into a smaller number of equal parts is bigger.

1 part out of 2 equal parts is bigger than *1 part out of 4 equal parts.*

Look, I divided a candy into 4 equal parts and I ate 1 part out of the four equal parts. Or I divided a candy into 2 equal parts and I ate 1 part out of the two equal parts.

When did I eat more? Of course, when the candy was divided into 2 equal parts.

$\frac{1}{4} < \frac{1}{2}$

1 out of 4 equal parts is smaller than 1 out of 2 equal parts.

$\frac{2}{3} > \frac{2}{8}$

2 out of 3 equal parts are bigger than 2 out of 8 equal parts.

I divided my block into 3 huge equal parts and colored 2 parts white.

Then, I divided another block into 8 small equal parts and colored 2 equal parts white.

Which fraction is largest?

The larger fraction is the one with the smaller denominator (that means the whole is divided into a smaller number of equal parts).

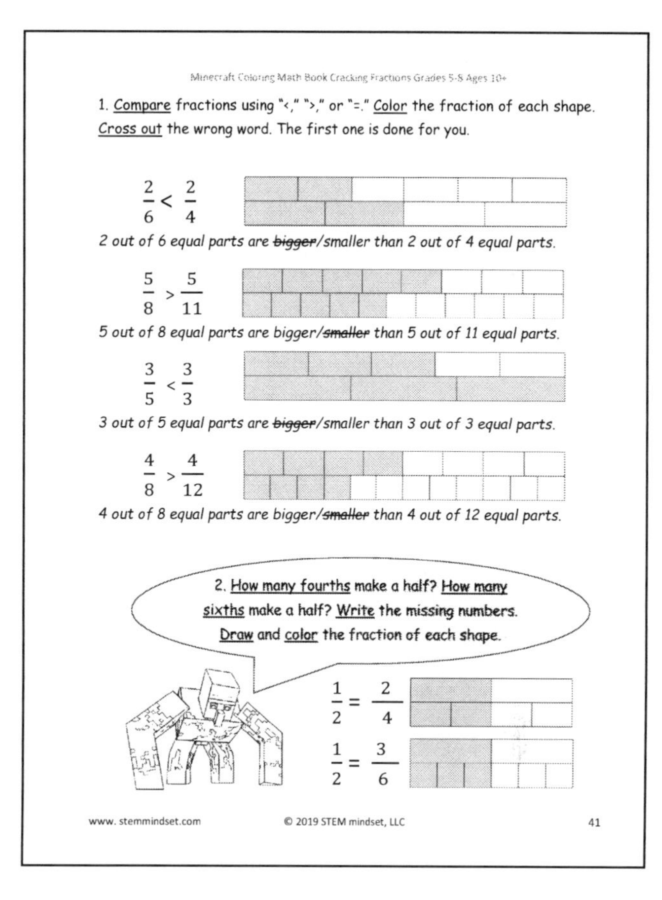

1. Compare fractions using "<," ">," or "=." Color the fraction of each shape. Cross out the wrong word. The first one is done for you.

$\frac{2}{6} < \frac{2}{4}$

2 out of 6 equal parts are ~~bigger~~/smaller than 2 out of 4 equal parts.

$\frac{5}{8} > \frac{5}{11}$

5 out of 8 equal parts are bigger/~~smaller~~ than 5 out of 11 equal parts.

$\frac{3}{5} < \frac{3}{3}$

3 out of 5 equal parts are ~~bigger~~/smaller than 3 out of 3 equal parts.

$\frac{4}{8} > \frac{4}{12}$

4 out of 8 equal parts are bigger/~~smaller~~ than 4 out of 12 equal parts.

2. How many fourths make a half? How many sixths make a half? Write the missing numbers. Draw and color the fraction of each shape.

$\frac{1}{2} = \frac{2}{4}$

$\frac{1}{2} = \frac{3}{6}$

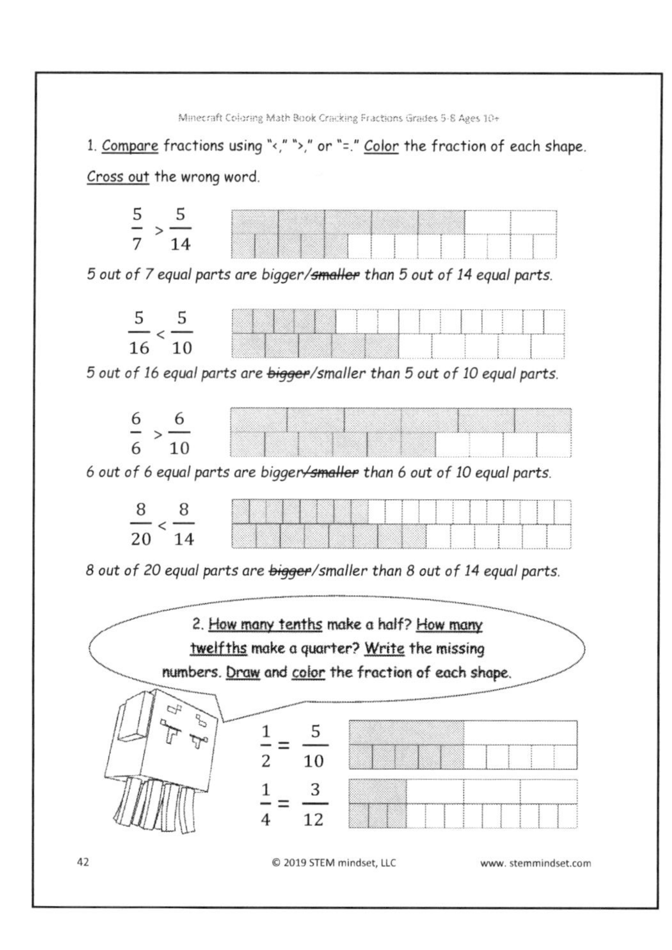

1. Compare fractions using "<," ">," or "=." Color the fraction of each shape. Cross out the wrong word.

$\frac{5}{7} > \frac{5}{14}$

5 out of 7 equal parts are bigger/~~smaller~~ than 5 out of 14 equal parts.

$\frac{5}{16} < \frac{5}{10}$

5 out of 16 equal parts are ~~bigger~~/smaller than 5 out of 10 equal parts.

$\frac{6}{6} > \frac{6}{10}$

6 out of 6 equal parts are bigger/~~smaller~~ than 6 out of 10 equal parts.

$\frac{8}{20} < \frac{8}{14}$

8 out of 20 equal parts are ~~bigger~~/smaller than 8 out of 14 equal parts.

2. How many tenths make a half? How many twelfths make a quarter? Write the missing numbers. Draw and color the fraction of each shape.

$\frac{1}{2} = \frac{5}{10}$

$\frac{1}{4} = \frac{3}{12}$

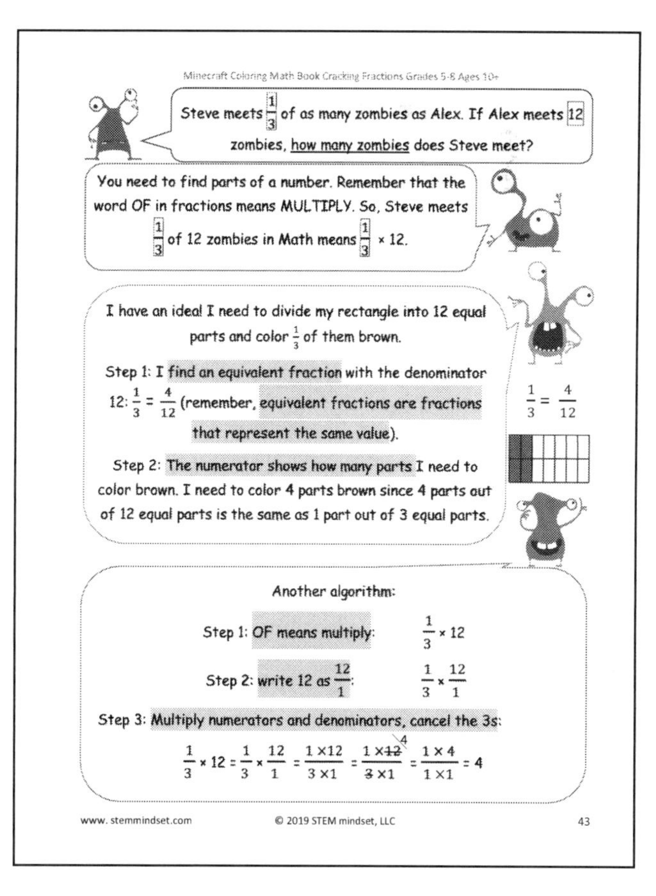

Steve meets $\frac{1}{3}$ of as many zombies as Alex. If Alex meets 12 zombies, how many zombies does Steve meet?

You need to find parts of a number. Remember that the word OF in fractions means MULTIPLY. So, Steve meets $\frac{1}{3}$ of 12 zombies in Math means $\frac{1}{3} \times 12$.

I have an idea! I need to divide my rectangle into 12 equal parts and color $\frac{1}{3}$ of them brown.

Step 1: I find an equivalent fraction with the denominator 12: $\frac{1}{3} = \frac{4}{12}$ (remember, equivalent fractions are fractions that represent the same value).

Step 2: The numerator shows how many parts I need to color brown. I need to color 4 parts brown since 4 parts out of 12 equal parts is the same as 1 part out of 3 equal parts.

$\frac{1}{3} = \frac{4}{12}$

Another algorithm:

Step 1: OF means multiply: $\frac{1}{3} \times 12$

Step 2: write 12 as $\frac{12}{1}$: $\frac{1}{3} \times \frac{12}{1}$

Step 3: Multiply numerators and denominators, cancel the 3s:

$\frac{1}{3} \times 12 = \frac{1}{3} \times \frac{12}{1} = \frac{1 \times 12}{3 \times 1} = \frac{1 \times \cancel{12}^{4}}{\cancel{3} \times 1} = \frac{1 \times 4}{1 \times 1} = 4$

Aha... I have another idea! All the zombies are divided equally into 3 equal groups. And I meet only $\frac{1}{3}$ of them.

The question is: If there are 12 zombies altogether, how many zombies are in one of the 3 equal parts?

Step 1: find how many zombies are in one of the three equal parts: $12 \div 3 = 4$

(4 zombies are in each of the three equal groups)

Step 2: find how many zombies are in one group: $1 \times 4 = 4$

(4 zombies in 1 group).

I wonder, if I meet $\frac{2}{3}$ of 12 zombies, how many zombies do I meet?

That's the question I like. My favorite algorithm:

$\frac{2}{3} \times 12 = \frac{2}{3} \times \frac{12}{1} = \frac{2 \times 12}{3 \times 1} = \frac{2 \times \cancel{12}}{\cancel{3} \times 1} = \frac{2 \times 4}{1 \times 1} = 8$ (zombies)

I like to solve it slower: $\frac{2}{3} \times 12$

Step 1: $12 \div 3 = 4$ (find $\frac{1}{3}$ out of 12)

(4 zombies in each of the three equal groups)

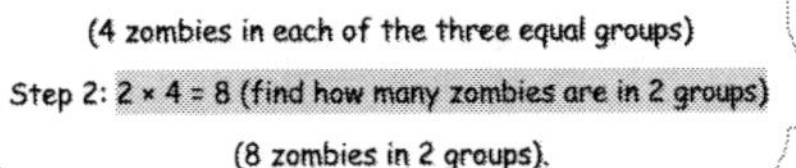

Step 2: $2 \times 4 = 8$ (find how many zombies are in 2 groups)

(8 zombies in 2 groups).

1. Multiply and simplify if possible. Color the correct number of images.

Find $\frac{1}{5}$ of 15 pigs

$\frac{1}{5} \times 15 = \frac{1}{5} \times \frac{15}{1} = \frac{1 \times 15}{5 \times 1} = 3$

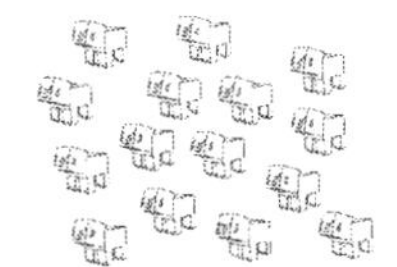

Find $\frac{3}{4}$ of 16 pigs

Step 1: $16 \div 4 = 4$ (find $\frac{1}{4}$ out of 16)

Step 2: $3 \times 4 = 12$

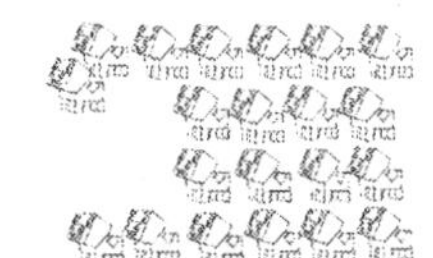

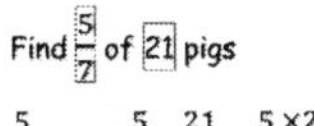

Find $\frac{5}{7}$ of 21 pigs

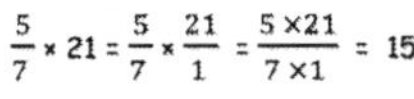

$\frac{5}{7} \times 21 = \frac{5}{7} \times \frac{21}{1} = \frac{5 \times 21}{7 \times 1} = 15$

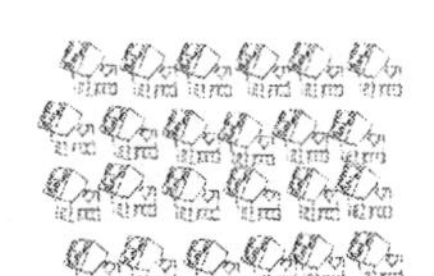

Find $\frac{5}{8}$ of 24 pigs

Step 1: $24 \div 8 = 3$ (find $\frac{1}{8}$ out of 24)

Step 2: $5 \times 3 = 15$

1. Reduce each fraction to simplest form, write the missing numbers, and compare fractions using "<," ">," or "=." Color the fraction of each shape. Cross out the wrong word.

$\frac{14}{16} = \frac{7}{8} > \frac{14}{20} = \frac{7}{10}$

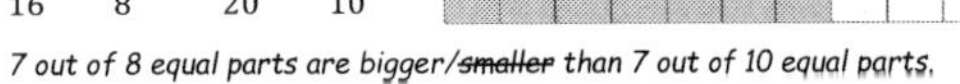

7 out of 8 equal parts are bigger/~~smaller~~ than 7 out of 10 equal parts.

$\frac{16}{18} = \frac{8}{9} > \frac{16}{22} = \frac{8}{11}$

8 out of 9 equal parts are bigger/~~smaller~~ than 8 out of 11 equal parts.

$\frac{20}{24} = \frac{5}{6} > \frac{10}{14} = \frac{5}{7}$

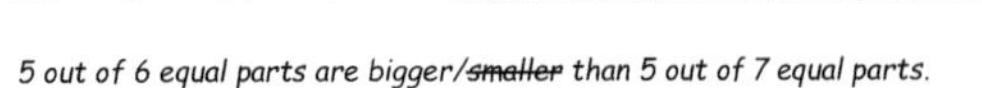

5 out of 6 equal parts are bigger/~~smaller~~ than 5 out of 7 equal parts.

2. Find and circle the area of a field to plant potatoes that was drawn by a villager. Hint: Figures are not drawn to scale.

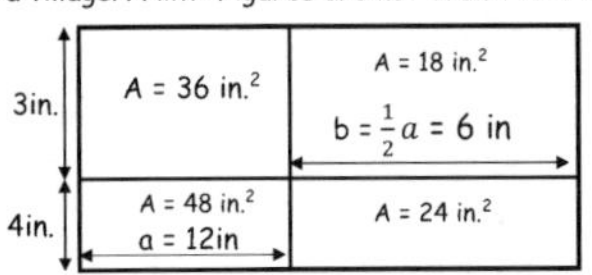

a) 162 b) 136 c) 126

1. Write in the missing numbers on the Minecraft factor tree and cross out the common prime factors.

A prime number is a counting number that has exactly two different factors

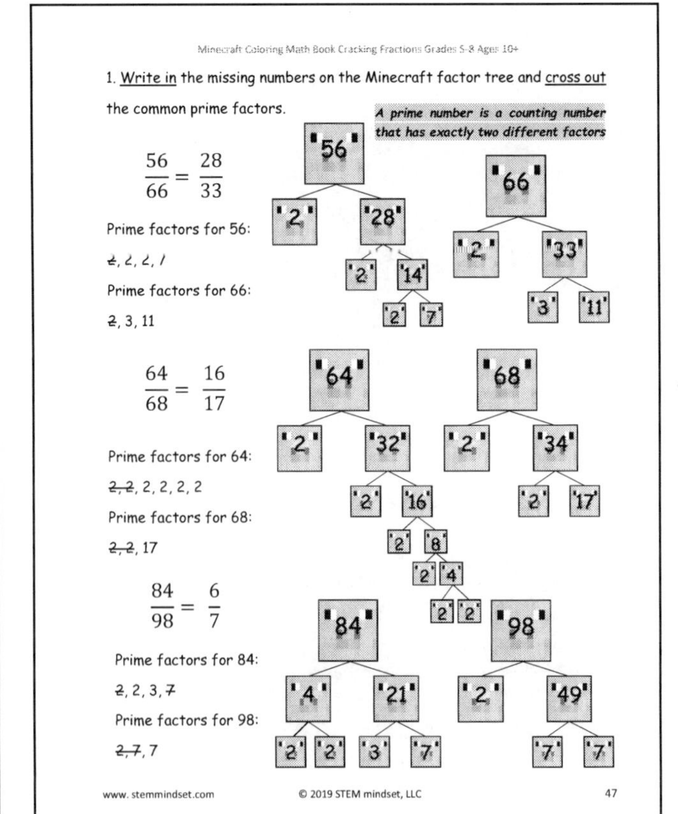

$\frac{56}{66} = \frac{28}{33}$

Prime factors for 56:

~~2~~, 2, 2, 7

Prime factors for 66:

~~2~~, 3, 11

$\frac{64}{68} = \frac{16}{17}$

Prime factors for 64:

~~2, 2~~, 2, 2, 2, 2

Prime factors for 68:

~~2, 2~~, 17

$\frac{84}{98} = \frac{6}{7}$

Prime factors for 84:

~~2~~, 2, 3, ~~7~~

Prime factors for 98:

~~2, 7~~, 7

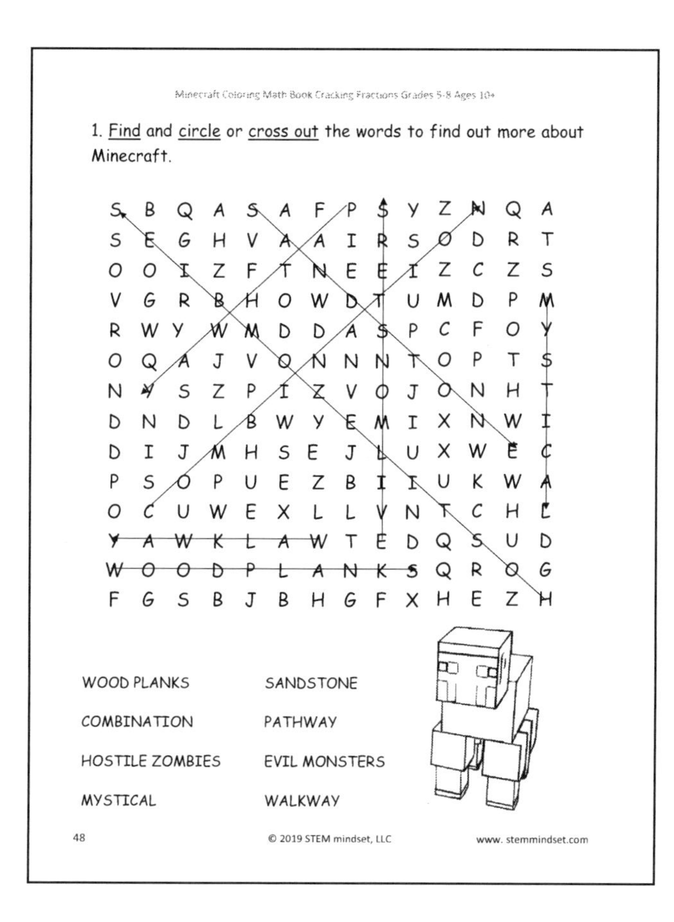

1. Find and circle or cross out the words to find out more about Minecraft.

S B Q A S A F P S Y Z N Q A
S E G H V A A I R S O D R T
O O I Z F T N E E I Z C Z S
V G R B H O W D T U M D P M
R W Y W M D D A S P C F O Y
O Q A J V Q N N N T O P T S
N Y S Z P I Z V O J O N H T
D N D L B W Y E M I X N W I
D I J M H S E J L U X W E C
P S O P U E Z B I I U K W A
O C U W E X L L V N T C H L
Y A W K L A W T E D Q S U D
W O O D P L A N K S Q R O G
F G S B J B H G F X H E Z H

WOOD PLANKS	SANDSTONE
COMBINATION	PATHWAY
HOSTILE ZOMBIES	EVIL MONSTERS
MYSTICAL	WALKWAY

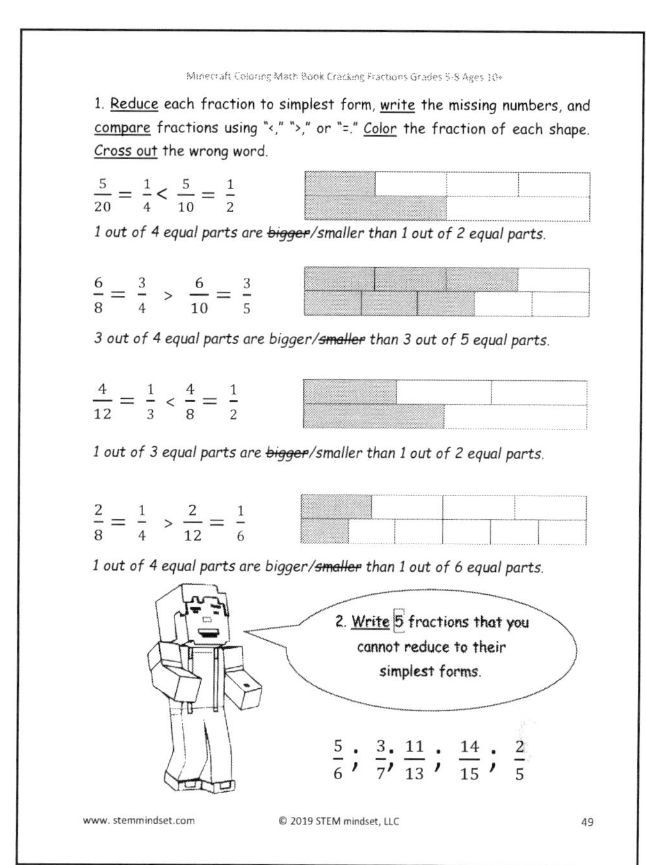

1. Reduce each fraction to simplest form, write the missing numbers, and compare fractions using "<," ">," or "=." Color the fraction of each shape. Cross out the wrong word.

$\frac{5}{20} = \frac{1}{4} < \frac{5}{10} = \frac{1}{2}$

1 out of 4 equal parts are ~~bigger~~/smaller than 1 out of 2 equal parts.

$\frac{6}{8} = \frac{3}{4} > \frac{6}{10} = \frac{3}{5}$

3 out of 4 equal parts are bigger/~~smaller~~ than 3 out of 5 equal parts.

$\frac{4}{12} = \frac{1}{3} < \frac{4}{8} = \frac{1}{2}$

1 out of 3 equal parts are ~~bigger~~/smaller than 1 out of 2 equal parts.

$\frac{2}{8} = \frac{1}{4} > \frac{2}{12} = \frac{1}{6}$

1 out of 4 equal parts are bigger/~~smaller~~ than 1 out of 6 equal parts.

2. Write 5 fractions that you cannot reduce to their simplest forms.

$\frac{5}{6}$; $\frac{3}{7}$; $\frac{11}{13}$; $\frac{14}{15}$; $\frac{2}{5}$

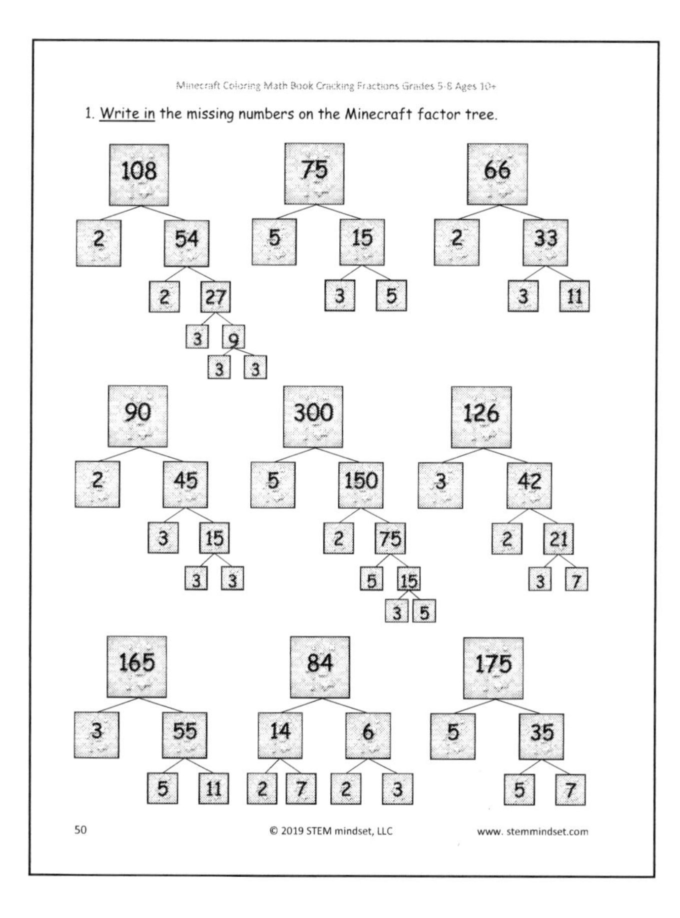

1. Write in the missing numbers on the Minecraft factor tree.

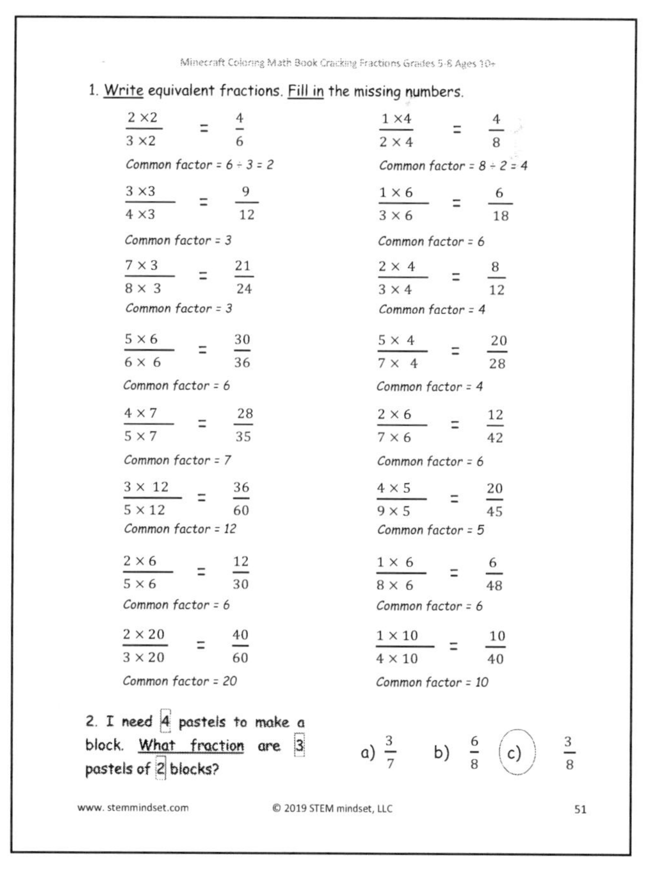

1. Write equivalent fractions. Fill in the missing numbers.

$\frac{2\times2}{3\times2} = \frac{4}{6}$ *Common factor = 6 ÷ 3 = 2*

$\frac{1\times4}{2\times4} = \frac{4}{8}$ *Common factor = 8 ÷ 2 = 4*

$\frac{3\times3}{4\times3} = \frac{9}{12}$ *Common factor = 3*

$\frac{1\times6}{3\times6} = \frac{6}{18}$ *Common factor = 6*

$\frac{7\times3}{8\times3} = \frac{21}{24}$ *Common factor = 3*

$\frac{2\times4}{3\times4} = \frac{8}{12}$ *Common factor = 4*

$\frac{5\times6}{6\times6} = \frac{30}{36}$ *Common factor = 6*

$\frac{5\times4}{7\times4} = \frac{20}{28}$ *Common factor = 4*

$\frac{4\times7}{5\times7} = \frac{28}{35}$ *Common factor = 7*

$\frac{2\times6}{7\times6} = \frac{12}{42}$ *Common factor = 6*

$\frac{3\times12}{5\times12} = \frac{36}{60}$ *Common factor = 12*

$\frac{4\times5}{9\times5} = \frac{20}{45}$ *Common factor = 5*

$\frac{2\times6}{5\times6} = \frac{12}{30}$ *Common factor = 6*

$\frac{1\times6}{8\times6} = \frac{6}{48}$ *Common factor = 6*

$\frac{2\times20}{3\times20} = \frac{40}{60}$ *Common factor = 20*

$\frac{1\times10}{4\times10} = \frac{10}{40}$ *Common factor = 10*

2. I need 4 pastels to make a block. What fraction are 3 pastels of 2 blocks?

a) $\frac{3}{7}$ b) $\frac{6}{8}$ c) $\frac{3}{8}$

1. Multiply and simplify if possible.

Find and color $\frac{2}{3}$ of 21 blocks.

$\frac{2}{3} \times 21 = \frac{2}{3} \times \frac{21}{1} = \frac{2 \times 21}{3 \times 1} = 14$

Find and color $\frac{6}{11}$ of 33 blocks.

Step 1: 33 ÷ 11 = 3 (find $\frac{1}{11}$ out of 33)

Step 2: 6 × 3 = 18

Find and color $\frac{7}{9}$ of 27 blocks.

$\frac{7}{9} \times 27 = \frac{7}{9} \times \frac{27}{1} = \frac{7 \times 27}{9 \times 1} = 21$

Find and color $\frac{4}{7}$ of 28 blocks.

Step 1: 28 ÷ 7 = 4 (find $\frac{1}{7}$ out of 28)

Step 2: 4 × 4 = 16

Minecraft Coloring Math Book Cracking Fractions Grades 5-8 Ages 10+

1. Reduce each fraction to simplest form, write the missing numbers, and compare fractions using "<," ">," or "=." Color the fraction of each shape. Cross out the wrong word.

$\frac{36}{48} = \frac{3}{4} > \frac{36}{60} = \frac{3}{5}$

3 out of 4 equal parts are bigger/~~smaller~~ than 3 out of 5 equal parts.

$\frac{45}{70} = \frac{9}{14} > \frac{45}{80} = \frac{9}{16}$

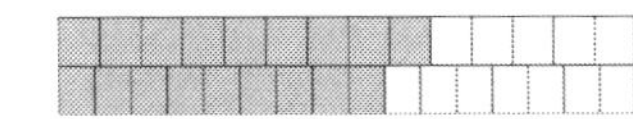

9 out of 14 equal parts are bigger/~~smaller~~ than 9 out of 16 equal parts.

$\frac{40}{64} = \frac{5}{8} > \frac{60}{108} = \frac{5}{9}$

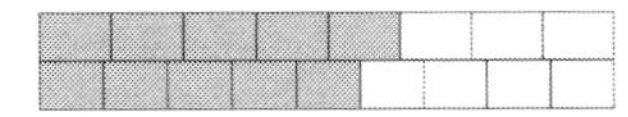

5 out of 8 equal parts are bigger/~~smaller~~ than 5 out of 9 equal parts.

$\frac{28}{49} = \frac{4}{7} > \frac{28}{63} = \frac{4}{9}$

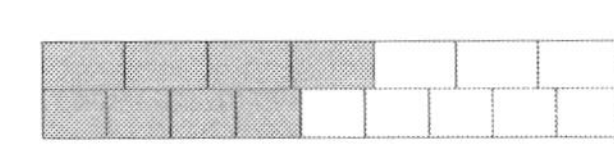

4 out of 7 equal parts are bigger/~~smaller~~ than 4 out of 9 equal parts.

2. Find d and circle the correct answer. Hint: Figures are not drawn to scale.

a) d = 9   (b) d = 8)   c) d = 7

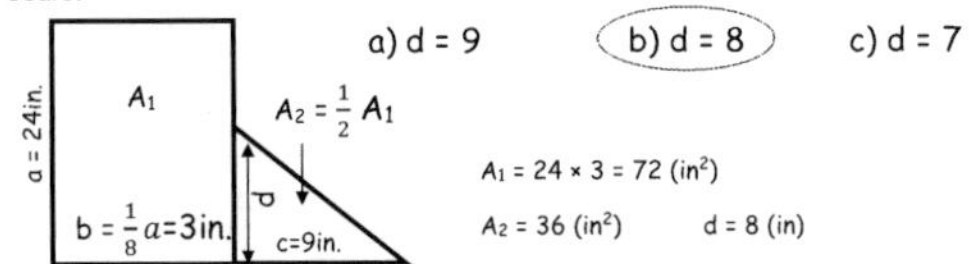

Minecraft Coloring Math Book Cracking Fractions Grades 5-8 Ages 10+

1. Multiply and simplify if possible. Circle the right answer (letter).

Find and color $\frac{6}{11}$ of 22 blocks.

$\frac{6}{11} \times 22 = \frac{6}{11} \times \frac{22}{1} = \frac{6 \times 22}{11 \times 1} = 12$

a) 10
(b) 12)
c) 15

Find and color $\frac{7}{10}$ of 30 blocks.

Step 1: 30 ÷ 10 = 3 (find $\frac{1}{10}$ out of 30)

Step 2: 7 × 3 = 21

(a) 21)
b) 35
c) 20

Find and color $\frac{3}{8}$ of 32 blocks.

$\frac{3}{8} \times 32 = \frac{3}{8} \times \frac{32}{1} = \frac{3 \times 32}{8 \times 1} = 12$

a) 16
b) 10
(c) 12)

Find and color $\frac{5}{7}$ of 35 blocks.

Step 1: 35 ÷ 7 = 5 (find $\frac{1}{7}$ out of 35)

Step 2: 5 × 5 = 25

a) 20
b) 35
(c) 25)

Minecraft Coloring Math Book Cracking Fractions Grades 5-8 Ages 10+

1. Reduce each fraction to simplest form, write the missing numbers, and compare fractions using "<," ">," or "=." Color the fraction of each shape. Cross out the wrong word.

$\frac{42}{54} = \frac{7}{9} < \frac{42}{48} = \frac{7}{8}$

7 out of 9 equal parts are ~~bigger~~/smaller than 7 out of 8 equal parts.

$\frac{24}{40} = \frac{3}{5} > \frac{24}{64} = \frac{5}{8}$

3 out of 5 equal parts are bigger/~~smaller~~ than 3 out of 8 equal parts.

$\frac{20}{24} = \frac{5}{6} > \frac{40}{64} = \frac{5}{8}$

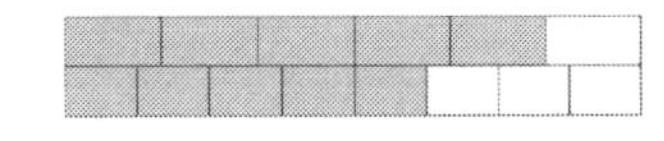

5 out of 6 equal parts are bigger/~~smaller~~ than 5 out of 8 equal parts.

2. Find and circle the area of a field to plant potatoes that was drawn by a villager. Hint: Figures are not drawn to scale.

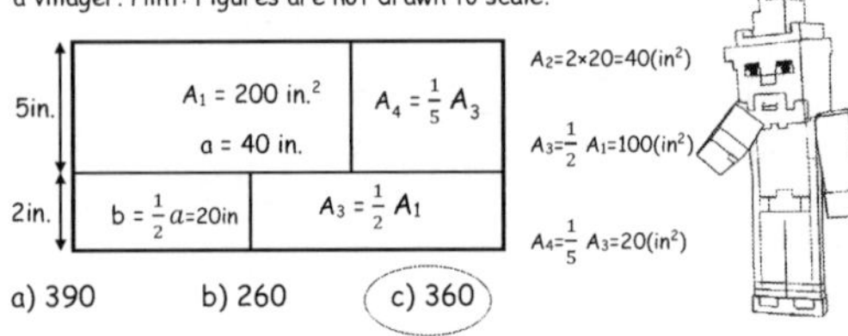

a) 390   b) 260   (c) 360)

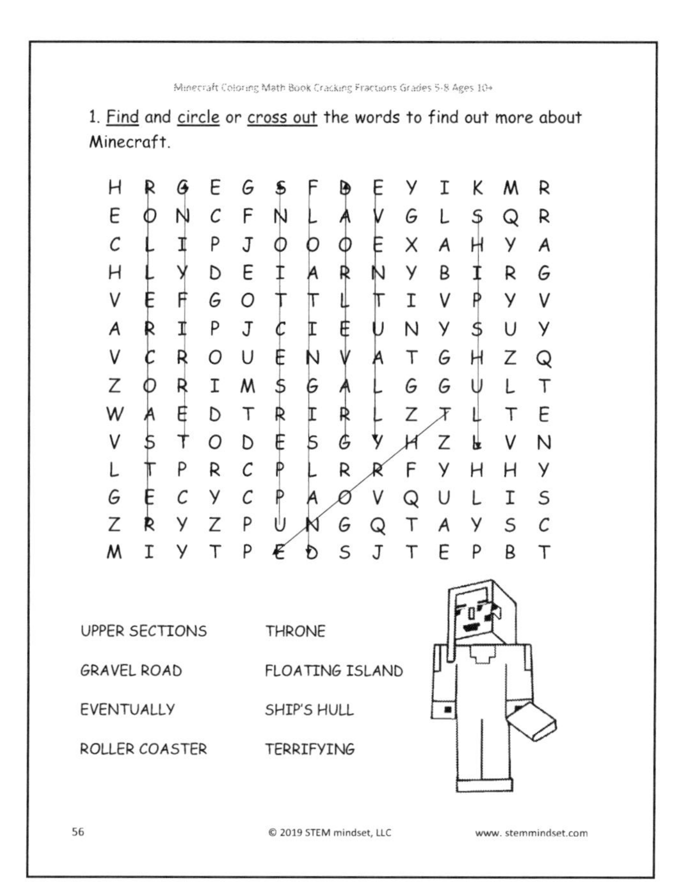

Minecraft Coloring Math Book Cracking Fractions Grades 5-8 Ages 10+

1. Find and circle or cross out the words to find out more about Minecraft.

H	R	G	E	G	S	F	D	E	Y	I	K	M	R
E	O	N	C	F	N	L	A	V	G	L	S	Q	R
C	L	I	P	J	O	O	O	E	X	A	H	Y	A
H	L	Y	D	E	I	A	R	N	Y	B	I	R	G
V	E	F	G	O	T	T	L	T	I	V	P	Y	V
A	R	I	P	J	C	I	E	U	N	Y	S	U	Y
V	C	R	O	U	E	N	V	A	T	G	H	Z	Q
Z	O	R	I	M	S	G	A	L	G	G	U	L	T
W	A	E	D	T	R	I	R	L	Z	T	L	T	E
V	S	T	O	D	E	S	G	Y	H	Z	L	V	N
L	T	P	R	C	P	L	R	R	F	Y	H	H	Y
G	E	C	Y	C	P	A	O	V	Q	U	L	I	S
Z	R	Y	Z	P	U	N	G	Q	T	A	Y	S	C
M	I	Y	T	P	E	D	S	J	T	E	P	B	T

UPPER SECTIONS | THRONE
GRAVEL ROAD | FLOATING ISLAND
EVENTUALLY | SHIP'S HULL
ROLLER COASTER | TERRIFYING

56 © 2019 STEM mindset, LLC www. stemmindset.com

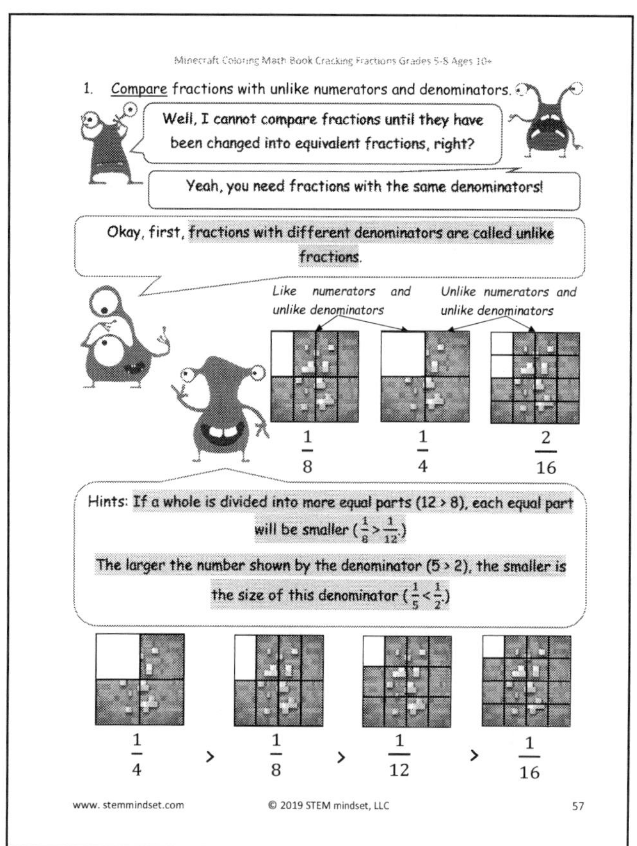

Minecraft Coloring Math Book Cracking Fractions Grades 5-8 Ages 10+

1. Compare fractions with unlike numerators and denominators.

Well, I cannot compare fractions until they have been changed into equivalent fractions, right?

Yeah, you need fractions with the same denominators!

Okay, first, fractions with different denominators are called unlike fractions.

Like numerators and unlike denominators — Unlike numerators and unlike denominators

$\frac{1}{8}$ $\frac{1}{4}$ $\frac{2}{16}$

Hints: If a whole is divided into more equal parts (12 > 8), each equal part will be smaller ($\frac{1}{8} > \frac{1}{12}$)

The larger the number shown by the denominator (5 > 2), the smaller is the size of this denominator ($\frac{1}{5} < \frac{1}{2}$)

$\frac{1}{4} > \frac{1}{8} > \frac{1}{12} > \frac{1}{16}$

www. stemmindset.com © 2019 STEM mindset, LLC 57

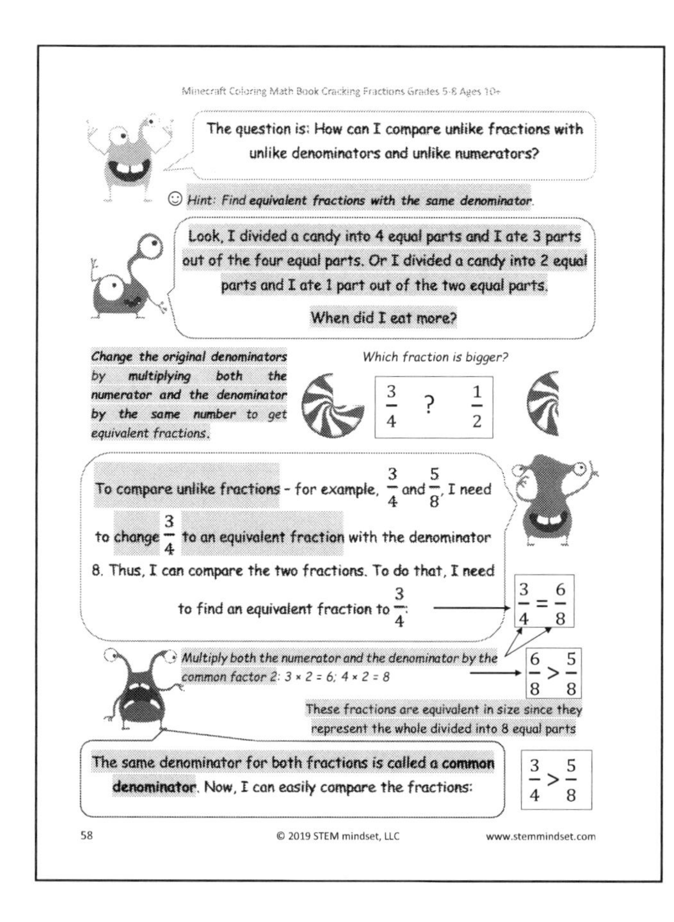

Minecraft Coloring Math Book Cracking Fractions Grades 5-8 Ages 10+

The question is: How can I compare unlike fractions with unlike denominators and unlike numerators?

Hint: Find equivalent fractions with the same denominator.

Look, I divided a candy into 4 equal parts and I ate 3 parts out of the four equal parts. Or I divided a candy into 2 equal parts and I ate 1 part out of the two equal parts.

When did I eat more?

Change the original denominators by multiplying both the numerator and the denominator by the same number to get equivalent fractions.

Which fraction is bigger?

$\frac{3}{4}\ ?\ \frac{1}{2}$

To compare unlike fractions - for example, $\frac{3}{4}$ and $\frac{5}{8}$, I need to change $\frac{3}{4}$ to an equivalent fraction with the denominator 8. Thus, I can compare the two fractions. To do that, I need to find an equivalent fraction to $\frac{3}{4}$: $\frac{3}{4} = \frac{6}{8}$

Multiply both the numerator and the denominator by the common factor 2: 3 × 2 = 6; 4 × 2 = 8 $\frac{6}{8} > \frac{5}{8}$

These fractions are equivalent in size since they represent the whole divided into 8 equal parts

The same denominator for both fractions is called a common denominator. Now, I can easily compare the fractions: $\frac{3}{4} > \frac{5}{8}$

58 © 2019 STEM mindset, LLC www.stemmindset.com

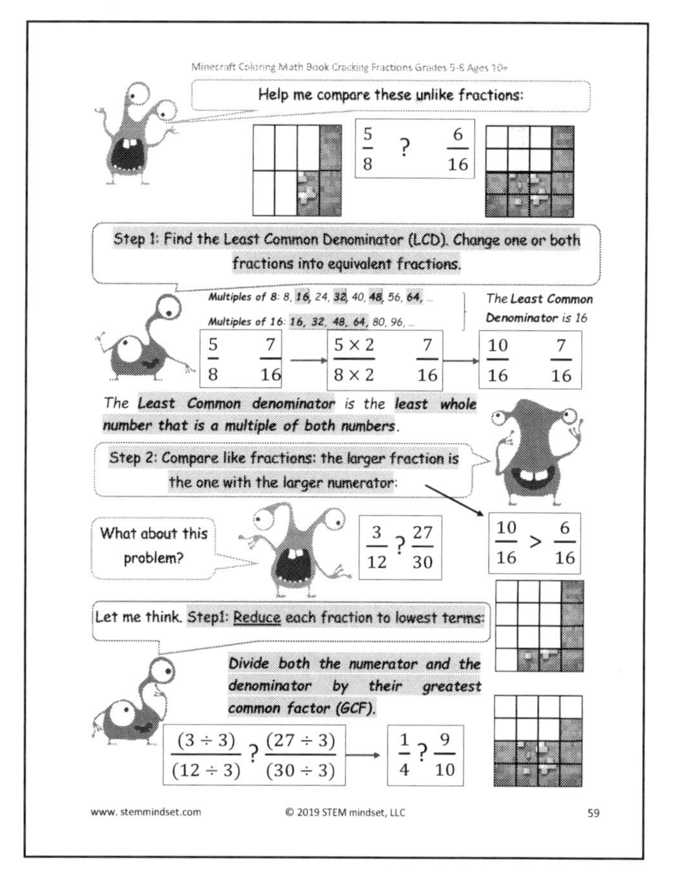

Minecraft Coloring Math Book Cracking Fractions Grades 5-8 Ages 10+

Help me compare these unlike fractions:

$\frac{5}{8}\ ?\ \frac{6}{16}$

Step 1: Find the Least Common Denominator (LCD). Change one or both fractions into equivalent fractions.

Multiples of 8: 8, 16, 24, 32, 40, 48, 56, 64, ...

Multiples of 16: 16, 32, 48, 64, 80, 96, ...

The Least Common Denominator is 16

$\frac{5}{8}\ \frac{7}{16} \rightarrow \frac{5 \times 2}{8 \times 2}\ \frac{7}{16} \rightarrow \frac{10}{16}\ \frac{7}{16}$

*The **Least Common denominator** is the **least whole number that is a multiple of both numbers.***

Step 2: Compare like fractions: the larger fraction is the one with the larger numerator: $\frac{10}{16} > \frac{6}{16}$

What about this problem? $\frac{3}{12}\ ?\ \frac{27}{30}$

Let me think. Step1: Reduce each fraction to lowest terms:

Divide both the numerator and the denominator by their greatest common factor (GCF).

$\frac{(3 \div 3)}{(12 \div 3)}\ ?\ \frac{(27 \div 3)}{(30 \div 3)} \rightarrow \frac{1}{4}\ ?\ \frac{9}{10}$

www. stemmindset.com © 2019 STEM mindset, LLC 59

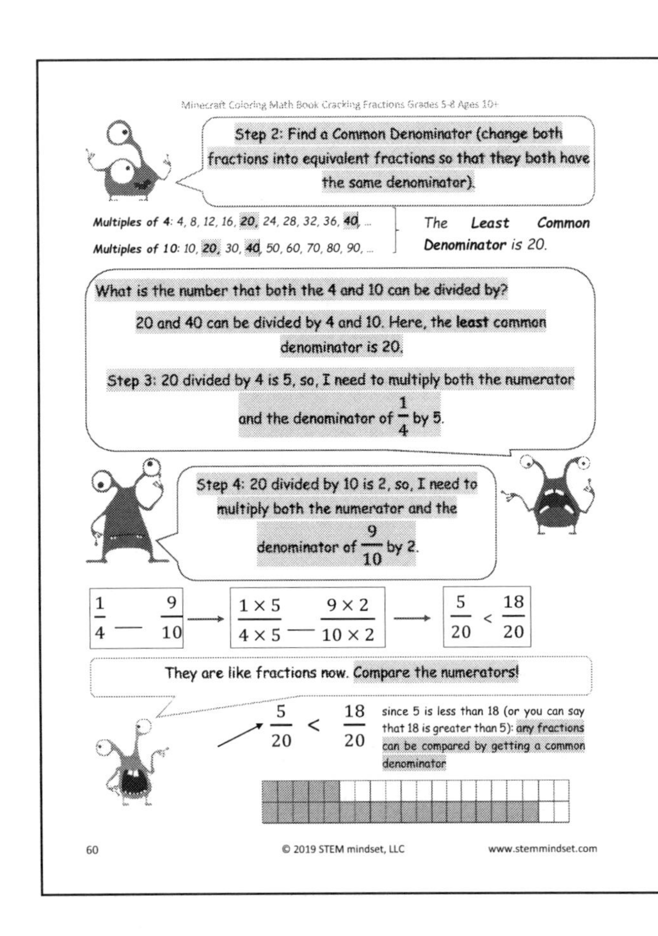

Step 2: Find a Common Denominator (change both fractions into equivalent fractions so that they both have the same denominator).

Multiples of 4: 4, 8, 12, 16, 20, 24, 28, 32, 36, 40, ...
Multiples of 10: 10, 20, 30, 40, 50, 60, 70, 80, 90, ...

*The **Least** Common Denominator is 20.*

What is the number that both the 4 and 10 can be divided by?

20 and 40 can be divided by 4 and 10. Here, the **least** common denominator is 20.

Step 3: 20 divided by 4 is 5, so, I need to multiply both the numerator and the denominator of $\frac{1}{4}$ by 5.

Step 4: 20 divided by 10 is 2, so, I need to multiply both the numerator and the denominator of $\frac{9}{10}$ by 2.

$\frac{1}{4}$ — $\frac{9}{10}$ → $\frac{1\times5}{4\times5}$ — $\frac{9\times2}{10\times2}$ → $\frac{5}{20} < \frac{18}{20}$

They are like fractions now. Compare the numerators!

$\frac{5}{20} < \frac{18}{20}$ since 5 is less than 18 (or you can say that 18 is greater than 5): any fractions can be compared by getting a common denominator

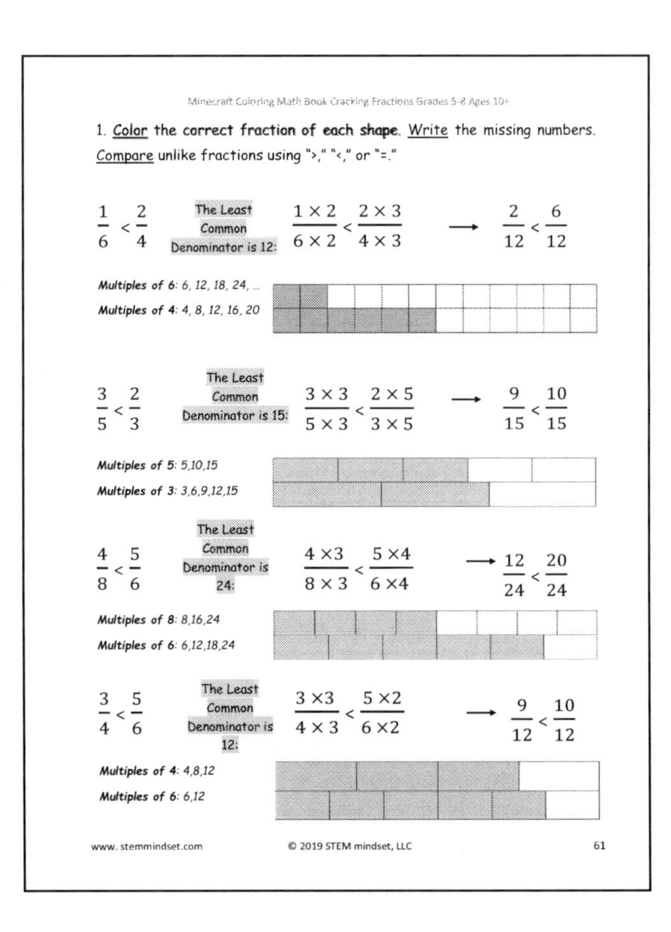

1. Color the correct fraction of each shape. Write the missing numbers. Compare unlike fractions using ">," "<," or "="

$\frac{1}{6} < \frac{2}{4}$ The Least Common Denominator is 12: $\frac{1\times2}{6\times2} < \frac{2\times3}{4\times3}$ → $\frac{2}{12} < \frac{6}{12}$

Multiples of 6: 6, 12, 18, 24, ...
Multiples of 4: 4, 8, 12, 16, 20

$\frac{3}{5} < \frac{2}{3}$ The Least Common Denominator is 15: $\frac{3\times3}{5\times3} < \frac{2\times5}{3\times5}$ → $\frac{9}{15} < \frac{10}{15}$

Multiples of 5: 5,10,15
Multiples of 3: 3,6,9,12,15

$\frac{4}{8} < \frac{5}{6}$ The Least Common Denominator is 24: $\frac{4\times3}{8\times3} < \frac{5\times4}{6\times4}$ → $\frac{12}{24} < \frac{20}{24}$

Multiples of 8: 8,16,24
Multiples of 6: 6,12,18,24

$\frac{3}{4} < \frac{5}{6}$ The Least Common Denominator is 12: $\frac{3\times3}{4\times3} < \frac{5\times2}{6\times2}$ → $\frac{9}{12} < \frac{10}{12}$

Multiples of 4: 4,8,12
Multiples of 6: 6,12

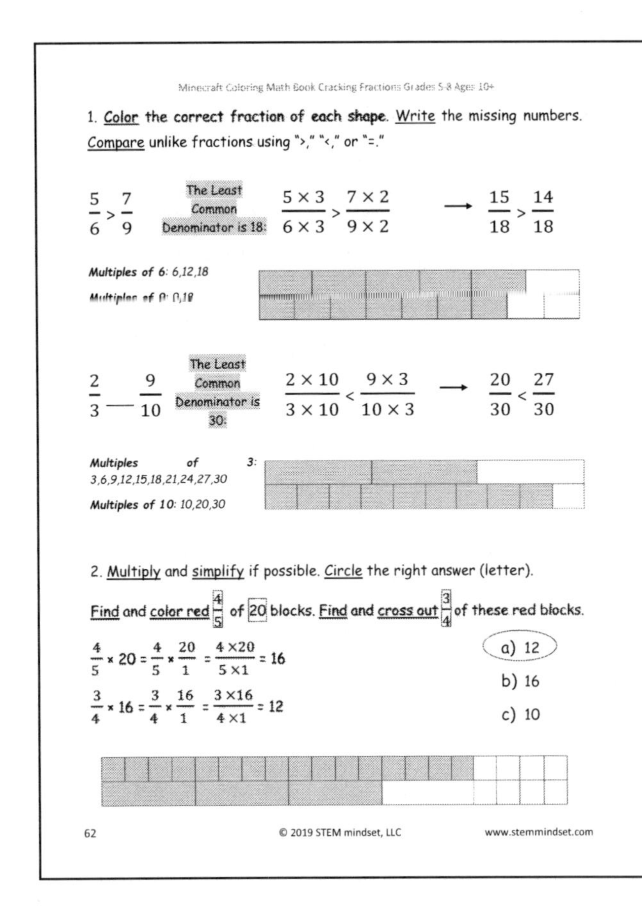

1. Color the correct fraction of each shape. Write the missing numbers. Compare unlike fractions using ">," "<," or "="

$\frac{5}{6} > \frac{7}{9}$ The Least Common Denominator is 18: $\frac{5\times3}{6\times3} > \frac{7\times2}{9\times2}$ → $\frac{15}{18} > \frac{14}{18}$

Multiples of 6: 6,12,18
Multiples of 9: 9,18

$\frac{2}{3}$ — $\frac{9}{10}$ The Least Common Denominator is 30: $\frac{2\times10}{3\times10} < \frac{9\times3}{10\times3}$ → $\frac{20}{30} < \frac{27}{30}$

Multiples of 3: 3,6,9,12,15,18,21,24,27,30
Multiples of 10: 10,20,30

2. Multiply and simplify if possible. Circle the right answer (letter).

Find and color red $\frac{4}{5}$ of 20 blocks. Find and cross out $\frac{3}{4}$ of these red blocks.

$\frac{4}{5} \times 20 = \frac{4}{5} \times \frac{20}{1} = \frac{4\times20}{5\times1} = 16$

$\frac{3}{4} \times 16 = \frac{3}{4} \times \frac{16}{1} = \frac{3\times16}{4\times1} = 12$

a) 12
b) 16
c) 10

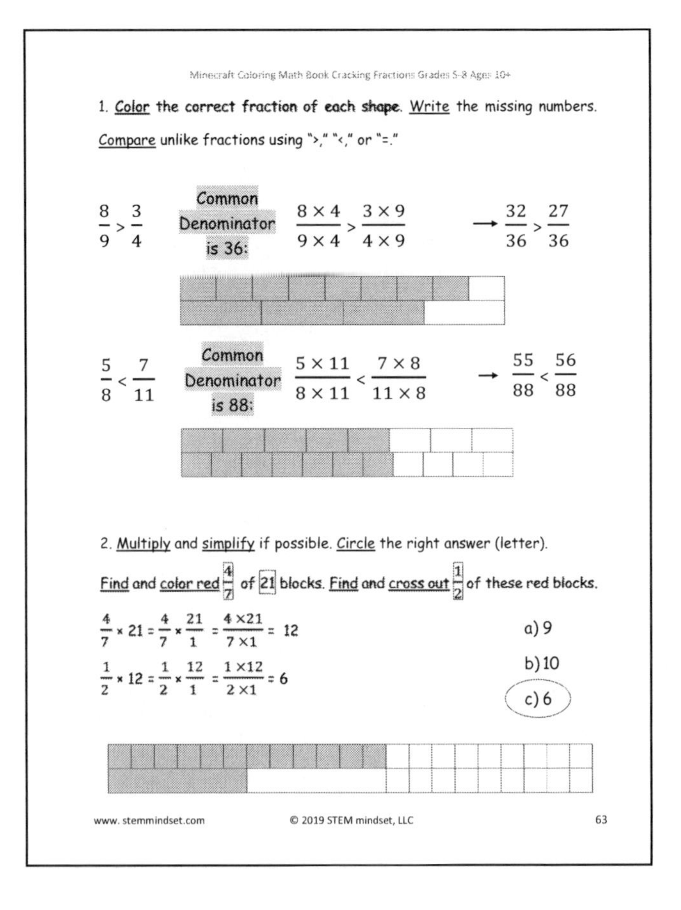

1. Color the correct fraction of each shape. Write the missing numbers. Compare unlike fractions using ">," "<," or "="

$\frac{8}{9} > \frac{3}{4}$ Common Denominator is 36: $\frac{8\times4}{9\times4} > \frac{3\times9}{4\times9}$ → $\frac{32}{36} > \frac{27}{36}$

$\frac{5}{8} < \frac{7}{11}$ Common Denominator is 88: $\frac{5\times11}{8\times11} < \frac{7\times8}{11\times8}$ → $\frac{55}{88} < \frac{56}{88}$

2. Multiply and simplify if possible. Circle the right answer (letter).

Find and color red $\frac{4}{7}$ of 21 blocks. Find and cross out $\frac{1}{2}$ of these red blocks.

$\frac{4}{7} \times 21 = \frac{4}{7} \times \frac{21}{1} = \frac{4\times21}{7\times1} = 12$

$\frac{1}{2} \times 12 = \frac{1}{2} \times \frac{12}{1} = \frac{1\times12}{2\times1} = 6$

a) 9
b) 10
c) 6

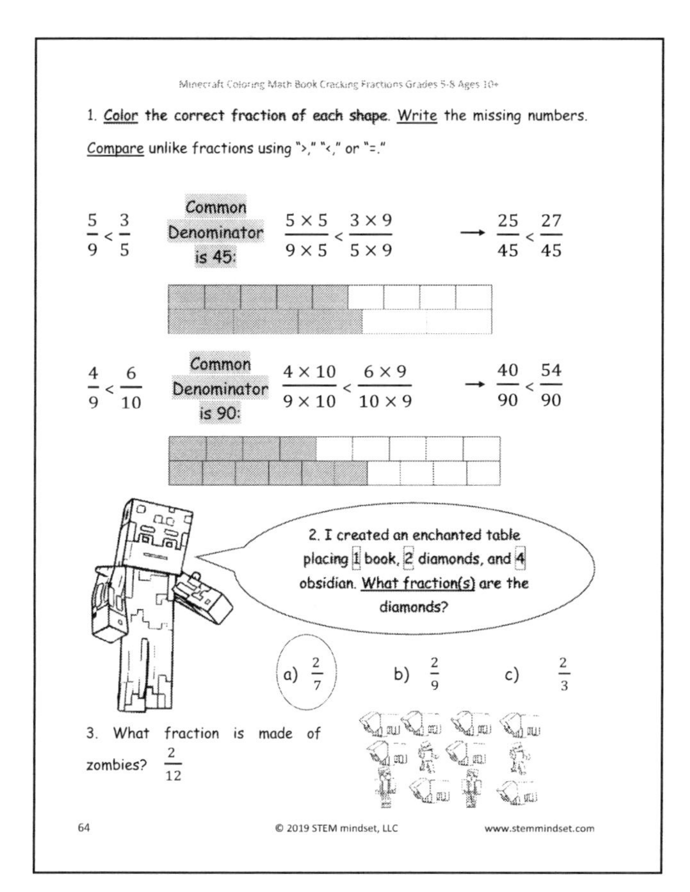
Minecraft Coloring Math Book Cracking Fractions Grades 5-8 Ages 10+

1. Color the correct fraction of each shape. Write the missing numbers. Compare unlike fractions using ">," "<," or "=."

$\frac{5}{9} < \frac{3}{5}$ Common Denominator is 45: $\frac{5 \times 5}{9 \times 5} < \frac{3 \times 9}{5 \times 9} \rightarrow \frac{25}{45} < \frac{27}{45}$

$\frac{4}{9} < \frac{6}{10}$ Common Denominator is 90: $\frac{4 \times 10}{9 \times 10} < \frac{6 \times 9}{10 \times 9} \rightarrow \frac{40}{90} < \frac{54}{90}$

2. I created an enchanted table placing 1 book, 2 diamonds, and 4 obsidian. What fraction(s) are the diamonds?

a) $\frac{2}{7}$ b) $\frac{2}{9}$ c) $\frac{2}{3}$

3. What fraction is made of zombies? $\frac{2}{12}$

64 © 2019 STEM mindset, LLC www.stemmindset.com

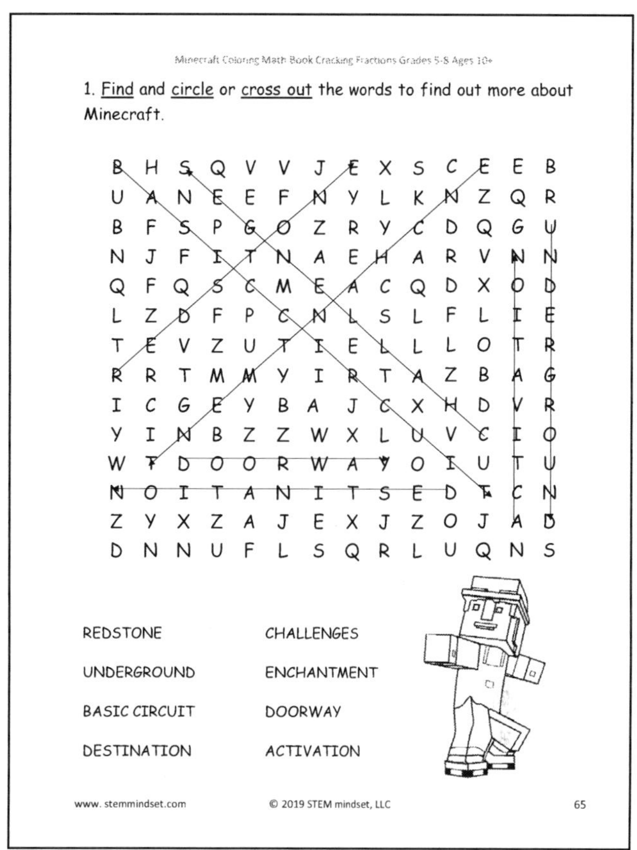
Minecraft Coloring Math Book Cracking Fractions Grades 5-8 Ages 10+

1. Find and circle or cross out the words to find out more about Minecraft.

B H S Q V V J E X S C E E B
U A N E E F N Y L K N Z Q R
B F S P G O Z R Y C D Q G U
N J F I T N A E H A R V N N
Q F Q S C M E A C Q D X O D
L Z D F P C N L S L F L I E
T E V Z U T I E L L L O T R
R R T M M Y I R T A Z B A G
I C G E Y B A J C X H D V R
Y I N B Z Z W X L U V C I O
W T D O O R W A Y O I U T U
N O I T A N I T S E D T C N
Z Y X Z A J E X J Z O J A D
D N N U F L S Q R L U Q N S

REDSTONE CHALLENGES
UNDERGROUND ENCHANTMENT
BASIC CIRCUIT DOORWAY
DESTINATION ACTIVATION

www. stemmindset.com © 2019 STEM mindset, LLC 65

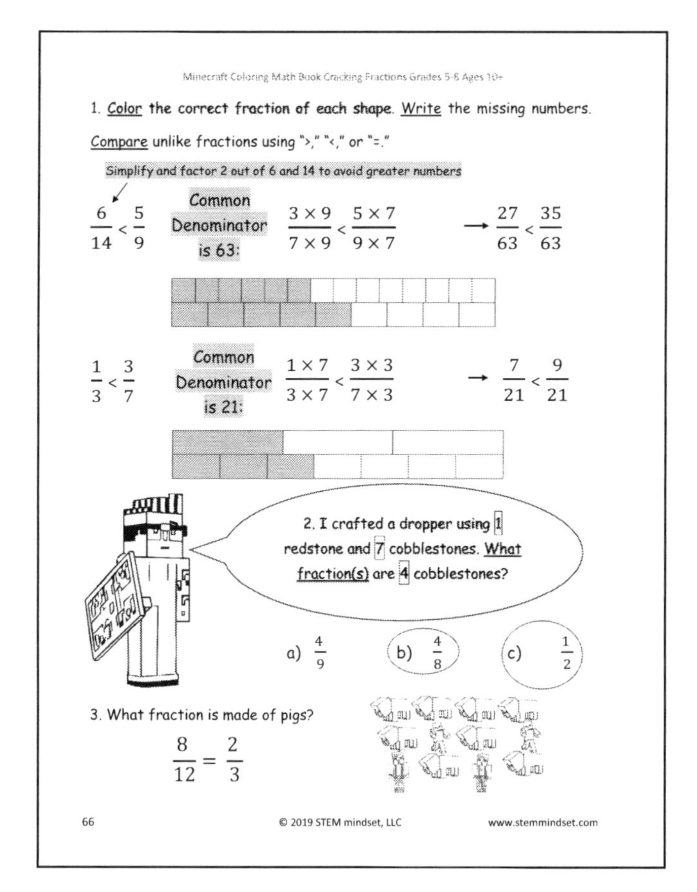
Minecraft Coloring Math Book Cracking Fractions Grades 5-8 Ages 10+

1. Color the correct fraction of each shape. Write the missing numbers. Compare unlike fractions using ">," "<," or "=."

Simplify and factor 2 out of 6 and 14 to avoid greater numbers

$\frac{6}{14} < \frac{5}{9}$ Common Denominator is 63: $\frac{3 \times 9}{7 \times 9} < \frac{5 \times 7}{9 \times 7} \rightarrow \frac{27}{63} < \frac{35}{63}$

$\frac{1}{3} < \frac{3}{7}$ Common Denominator is 21: $\frac{1 \times 7}{3 \times 7} < \frac{3 \times 3}{7 \times 3} \rightarrow \frac{7}{21} < \frac{9}{21}$

2. I crafted a dropper using 1 redstone and 7 cobblestones. What fraction(s) are 4 cobblestones?

a) $\frac{4}{9}$ b) $\frac{4}{8}$ c) $\frac{1}{2}$

3. What fraction is made of pigs?

$\frac{8}{12} = \frac{2}{3}$

66 © 2019 STEM mindset, LLC www.stemmindset.com

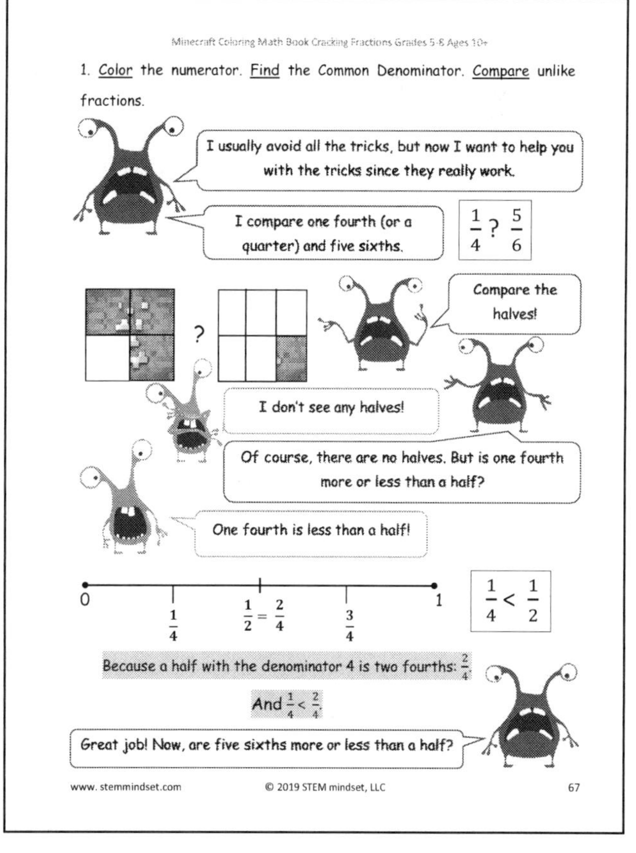
Minecraft Coloring Math Book Cracking Fractions Grades 5-8 Ages 10+

1. Color the numerator. Find the Common Denominator. Compare unlike fractions.

I usually avoid all the tricks, but now I want to help you with the tricks since they really work.
I compare one fourth (or a quarter) and five sixths.
$\frac{1}{4} ? \frac{5}{6}$
Compare the halves!
?
I don't see any halves!
Of course, there are no halves. But is one fourth more or less than a half?
One fourth is less than a half!
0 $\frac{1}{4}$ $\frac{1}{2} = \frac{2}{4}$ $\frac{3}{4}$ 1
$\frac{1}{4} < \frac{1}{2}$

Because a half with the denominator 4 is two fourths: $\frac{2}{4}$.

And $\frac{1}{4} < \frac{2}{4}$.

Great job! Now, are five sixths more or less than a half?

www. stemmindset.com © 2019 STEM mindset, LLC 67

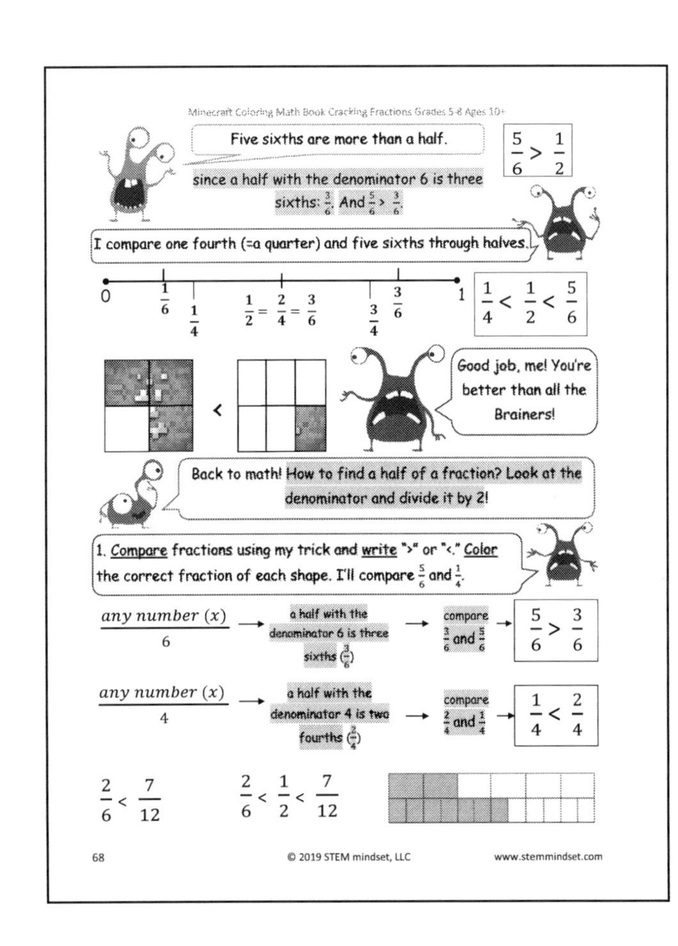

Five sixths are more than a half.

$\frac{5}{6} > \frac{1}{2}$

since a half with the denominator 6 is three sixths: $\frac{3}{6}$. And $\frac{5}{6} > \frac{3}{6}$.

I compare one fourth (=a quarter) and five sixths through halves.

0, $\frac{1}{6}$, $\frac{1}{4}$, $\frac{1}{2} = \frac{2}{4} = \frac{3}{6}$, $\frac{3}{4}$, $\frac{3}{6}$, 1

$\frac{1}{4} < \frac{1}{2} < \frac{5}{6}$

<

Good job, me! You're better than all the Brainers!

Back to math! How to find a half of a fraction? Look at the denominator and divide it by 2!

1. Compare fractions using my trick and write ">" or "<." Color the correct fraction of each shape. I'll compare $\frac{5}{6}$ and $\frac{1}{4}$.

$\frac{any\ number\ (x)}{6}$ → a half with the denominator 6 is three sixths ($\frac{3}{6}$) → compare $\frac{3}{6}$ and $\frac{5}{6}$ → $\frac{5}{6} > \frac{3}{6}$

$\frac{any\ number\ (x)}{4}$ → a half with the denominator 4 is two fourths ($\frac{2}{4}$) → compare $\frac{2}{4}$ and $\frac{1}{4}$ → $\frac{1}{4} < \frac{2}{4}$

$\frac{2}{6} < \frac{7}{12}$ $\qquad$ $\frac{2}{6} < \frac{1}{2} < \frac{7}{12}$

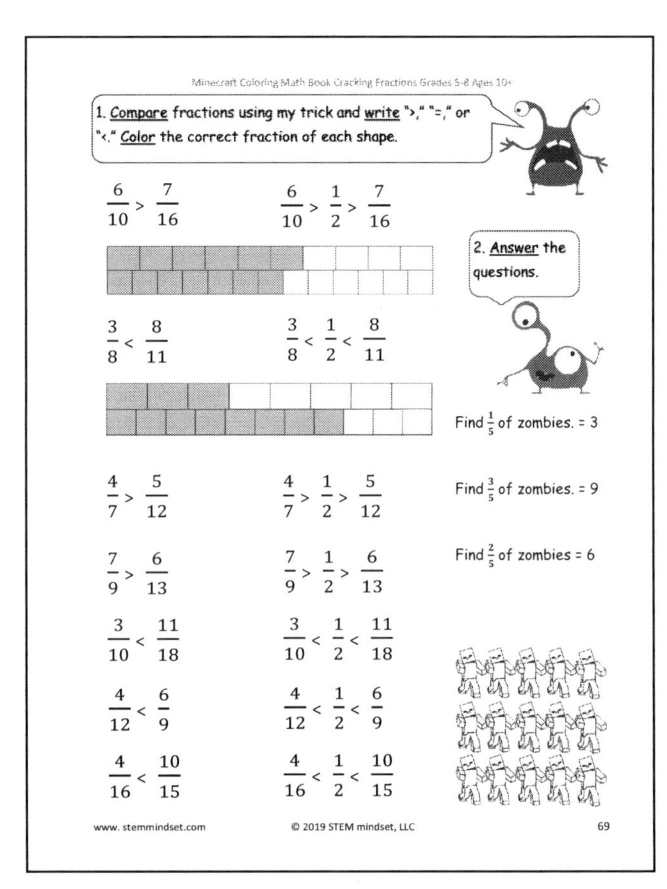

1. Compare fractions using my trick and write ">," "=," or "<." Color the correct fraction of each shape.

$\frac{6}{10} > \frac{7}{16}$ $\qquad$ $\frac{6}{10} > \frac{1}{2} > \frac{7}{16}$

$\frac{3}{8} < \frac{8}{11}$ $\qquad$ $\frac{3}{8} < \frac{1}{2} < \frac{8}{11}$

$\frac{4}{7} > \frac{5}{12}$ $\qquad$ $\frac{4}{7} > \frac{1}{2} > \frac{5}{12}$

$\frac{7}{9} > \frac{6}{13}$ $\qquad$ $\frac{7}{9} > \frac{1}{2} > \frac{6}{13}$

$\frac{3}{10} < \frac{11}{18}$ $\qquad$ $\frac{3}{10} < \frac{1}{2} < \frac{11}{18}$

$\frac{4}{12} < \frac{6}{9}$ $\qquad$ $\frac{4}{12} < \frac{1}{2} < \frac{6}{9}$

$\frac{4}{16} < \frac{10}{15}$ $\qquad$ $\frac{4}{16} < \frac{1}{2} < \frac{10}{15}$

2. Answer the questions.

Find $\frac{1}{5}$ of zombies. = 3

Find $\frac{3}{5}$ of zombies. = 9

Find $\frac{2}{5}$ of zombies = 6

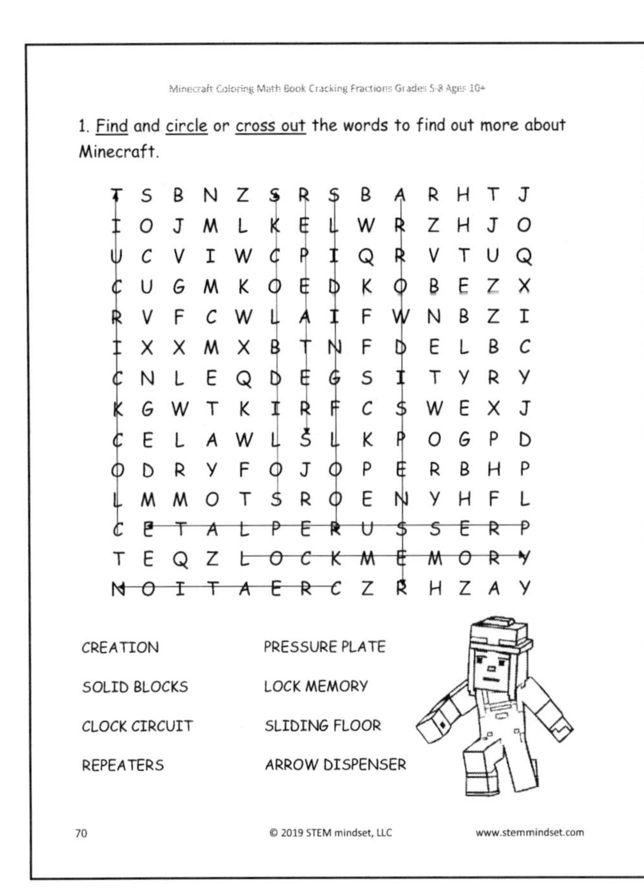

1. Find and circle or cross out the words to find out more about Minecraft.

T	S	B	N	Z	S	R	S	B	A	R	H	T	J
I	O	J	M	L	K	E	L	W	R	Z	H	J	O
U	C	V	I	W	C	P	I	Q	R	V	T	U	Q
C	U	G	M	K	O	E	D	K	O	B	E	Z	X
R	V	F	C	W	L	A	I	F	W	N	B	Z	I
I	X	X	M	X	B	T	N	F	D	E	L	B	C
C	N	L	E	Q	D	E	G	S	I	T	Y	R	Y
K	G	W	T	K	I	R	F	C	S	W	E	X	J
C	E	L	A	W	L	S	L	K	P	O	G	P	D
O	D	R	Y	F	O	J	O	P	E	R	B	H	P
L	M	M	O	T	S	R	O	E	N	Y	H	F	L
C	E	T	A	L	P	E	R	U	S	S	E	R	P
T	E	Q	Z	L	O	C	K	M	E	M	O	R	Y
N	O	I	T	A	E	R	C	Z	R	H	Z	A	Y

CREATION $\qquad$ PRESSURE PLATE

SOLID BLOCKS $\qquad$ LOCK MEMORY

CLOCK CIRCUIT $\qquad$ SLIDING FLOOR

REPEATERS $\qquad$ ARROW DISPENSER

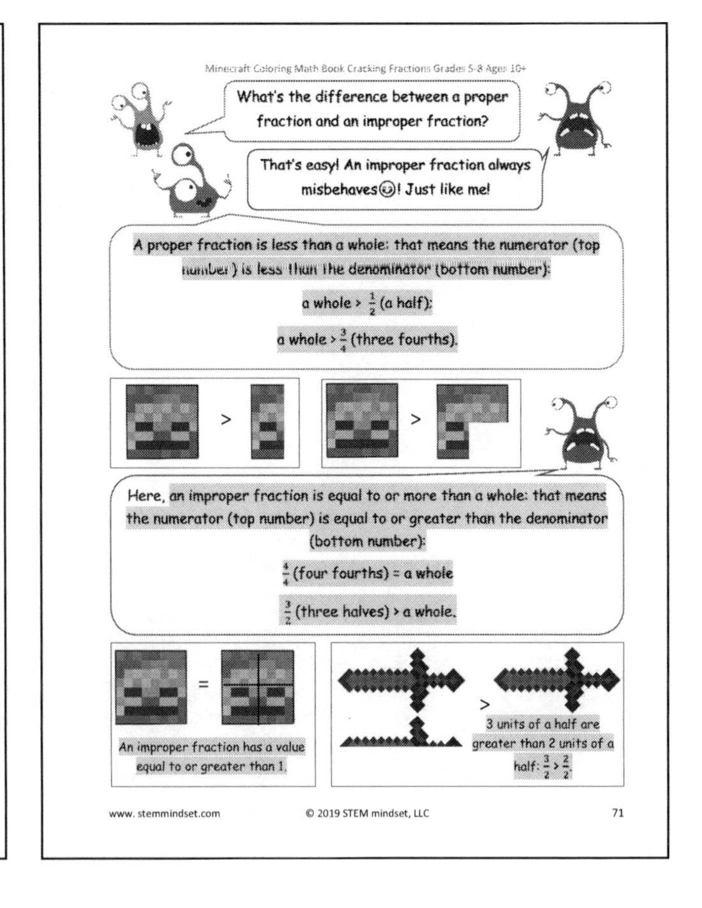

What's the difference between a proper fraction and an improper fraction?
That's easy! An improper fraction always misbehaves☺! Just like me!

A proper fraction is less than a whole: that means the numerator (top number) is less than the denominator (bottom number):

a whole > $\frac{1}{2}$ (a half);

a whole > $\frac{3}{4}$ (three fourths).

\> $\qquad$ >

Here, an improper fraction is equal to or more than a whole: that means the numerator (top number) is equal to or greater than the denominator (bottom number):

$\frac{4}{4}$ (four fourths) = a whole

$\frac{3}{2}$ (three halves) > a whole.

=

An improper fraction has a value equal to or greater than 1.

\>

3 units of a half are greater than 2 units of a half: $\frac{3}{2} > \frac{2}{2}$.

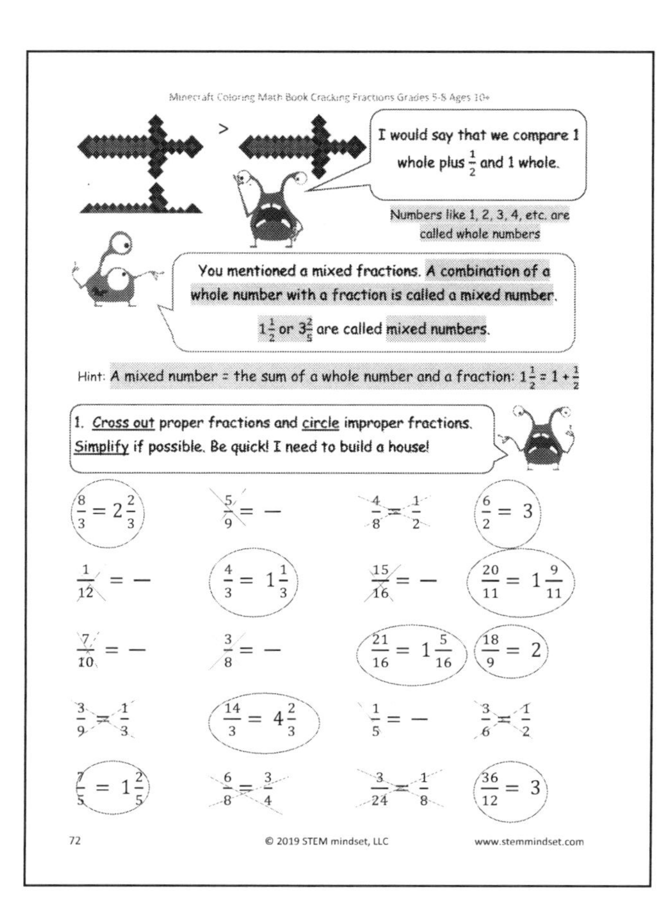

$>$

I would say that we compare 1 whole plus $\frac{1}{2}$ and 1 whole.

Numbers like 1, 2, 3, 4, etc. are called whole numbers

You mentioned a mixed fractions. A combination of a whole number with a fraction is called a mixed number.

$1\frac{1}{2}$ or $3\frac{2}{5}$ are called mixed numbers.

Hint: A mixed number = the sum of a whole number and a fraction: $1\frac{1}{2} = 1 + \frac{1}{2}$

1. Cross out proper fractions and circle improper fractions. Simplify if possible. Be quick! I need to build a house!

$\frac{8}{3} = 2\frac{2}{3}$ $\frac{5}{9} = —$ $\frac{4}{8} = \frac{1}{2}$ $\frac{6}{2} = 3$

$\frac{1}{12} = —$ $\frac{4}{3} = 1\frac{1}{3}$ $\frac{15}{16} = —$ $\frac{20}{11} = 1\frac{9}{11}$

$\frac{7}{10} = —$ $\frac{3}{8} = —$ $\frac{21}{16} = 1\frac{5}{16}$ $\frac{18}{9} = 2$

$\frac{3}{9} = \frac{1}{3}$ $\frac{14}{3} = 4\frac{2}{3}$ $\frac{1}{5} = —$ $\frac{3}{6} = \frac{1}{2}$

$\frac{7}{5} = 1\frac{2}{5}$ $\frac{6}{8} = \frac{3}{4}$ $\frac{3}{24} = \frac{1}{8}$ $\frac{36}{12} = 3$

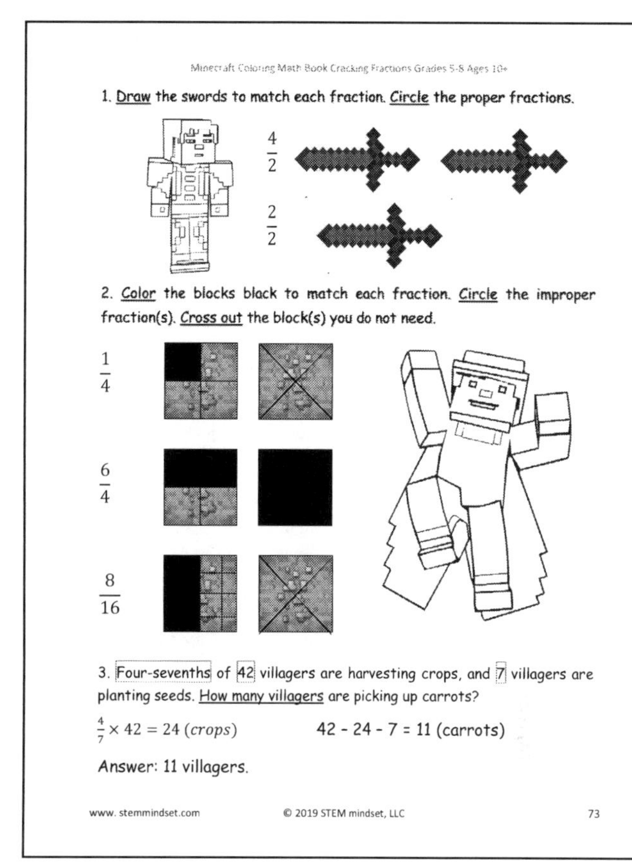

1. Draw the swords to match each fraction. Circle the proper fractions.

$\frac{4}{2}$

$\frac{2}{2}$

2. Color the blocks black to match each fraction. Circle the improper fraction(s). Cross out the block(s) you do not need.

$\frac{1}{4}$

$\frac{6}{4}$

$\frac{8}{16}$

3. Four-sevenths of 42 villagers are harvesting crops, and 7 villagers are planting seeds. How many villagers are picking up carrots?

$\frac{4}{7} \times 42 = 24$ (crops) 42 - 24 - 7 = 11 (carrots)

Answer: 11 villagers.

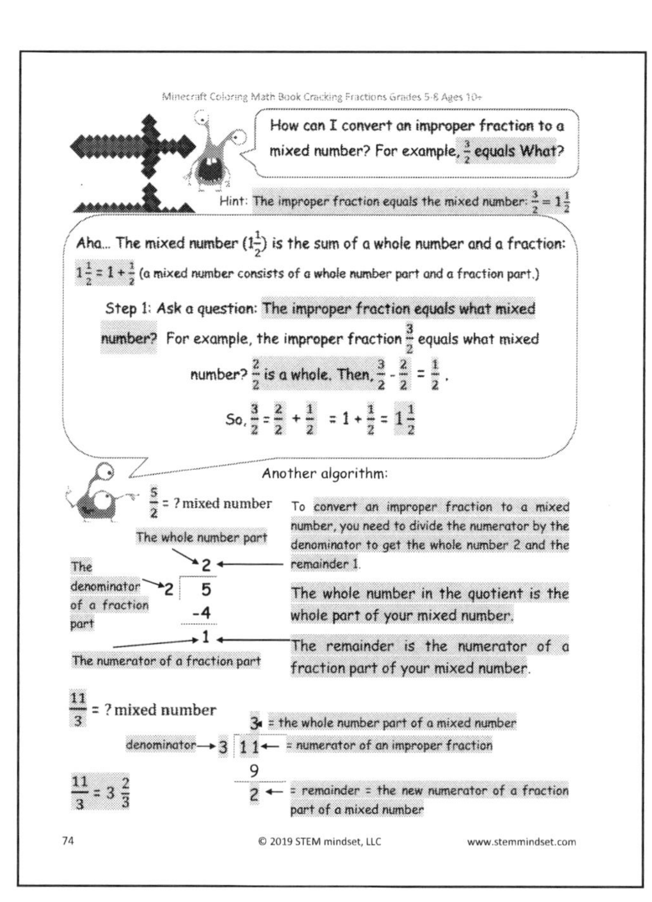

How can I convert an improper fraction to a mixed number? For example, $\frac{3}{2}$ equals What?

Hint: The improper fraction equals the mixed number: $\frac{3}{2} = 1\frac{1}{2}$

Aha... The mixed number ($1\frac{1}{2}$) is the sum of a whole number and a fraction:

$1\frac{1}{2} = 1 + \frac{1}{2}$ (a mixed number consists of a whole number part and a fraction part.)

Step 1: Ask a question: The improper fraction equals what mixed number? For example, the improper fraction $\frac{3}{2}$ equals what mixed number? $\frac{2}{2}$ is a whole. Then, $\frac{3}{2} - \frac{2}{2} = \frac{1}{2}$.

So, $\frac{3}{2} = \frac{2}{2} + \frac{1}{2} = 1 + \frac{1}{2} = 1\frac{1}{2}$

Another algorithm:

$\frac{5}{2}$ = ? mixed number

The whole number part

The denominator of a fraction part

The numerator of a fraction part

2 ⟌ 5 → 2, −4, 1

To convert an improper fraction to a mixed number, you need to divide the numerator by the denominator to get the whole number 2 and the remainder 1.

The whole number in the quotient is the whole part of your mixed number.

The remainder is the numerator of a fraction part of your mixed number.

$\frac{11}{3}$ = ? mixed number

3 = the whole number part of a mixed number

denominator → 3 ⟌ 11 ← = numerator of an improper fraction

9

2 ← = remainder = the new numerator of a fraction part of a mixed number

$\frac{11}{3} = 3\frac{2}{3}$

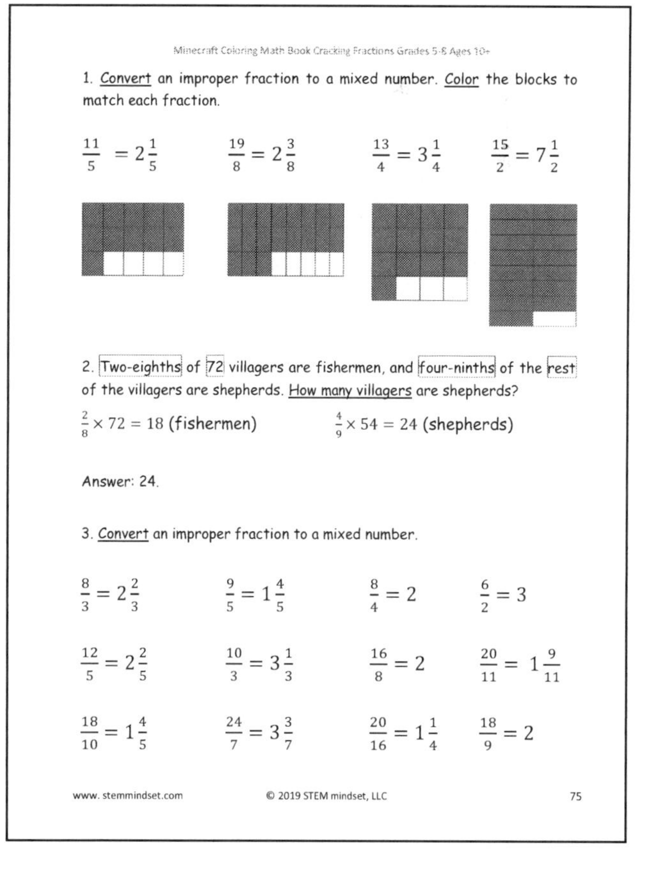

1. Convert an improper fraction to a mixed number. Color the blocks to match each fraction.

$\frac{11}{5} = 2\frac{1}{5}$ $\frac{19}{8} = 2\frac{3}{8}$ $\frac{13}{4} = 3\frac{1}{4}$ $\frac{15}{2} = 7\frac{1}{2}$

2. Two-eighths of 72 villagers are fishermen, and four-ninths of the rest of the villagers are shepherds. How many villagers are shepherds?

$\frac{2}{8} \times 72 = 18$ (fishermen) $\frac{4}{9} \times 54 = 24$ (shepherds)

Answer: 24.

3. Convert an improper fraction to a mixed number.

$\frac{8}{3} = 2\frac{2}{3}$ $\frac{9}{5} = 1\frac{4}{5}$ $\frac{8}{4} = 2$ $\frac{6}{2} = 3$

$\frac{12}{5} = 2\frac{2}{5}$ $\frac{10}{3} = 3\frac{1}{3}$ $\frac{16}{8} = 2$ $\frac{20}{11} = 1\frac{9}{11}$

$\frac{18}{10} = 1\frac{4}{5}$ $\frac{24}{7} = 3\frac{3}{7}$ $\frac{20}{16} = 1\frac{1}{4}$ $\frac{18}{9} = 2$

1. Convert an improper fraction to a mixed number. Simplify.

$\frac{9}{3} = 3$ $\frac{14}{3} = 4\frac{2}{3}$ $\frac{5}{2} = 2\frac{1}{2}$ $\frac{7}{6} = 1\frac{1}{6}$

$\frac{7}{5} = 1\frac{2}{5}$ $\frac{8}{5} = 1\frac{3}{5}$ $\frac{24}{3} = 8$ $\frac{36}{12} = 3$

$\frac{22}{3} = 7\frac{1}{3}$ $\frac{34}{6} = 5\frac{2}{3}$ $\frac{11}{2} = 5\frac{1}{2}$ $\frac{28}{5} = 5\frac{3}{5}$

$\frac{42}{9} = 4\frac{2}{3}$ $\frac{25}{4} = 6\frac{1}{4}$ $\frac{63}{8} = 7\frac{7}{8}$ $\frac{36}{10} = 2\frac{3}{5}$

$\frac{31}{7} = 4\frac{3}{7}$ $\frac{46}{5} = 9\frac{1}{5}$ $\frac{25}{12} = 2\frac{1}{12}$ $\frac{41}{15} = 2\frac{11}{15}$

2. Two-sevenths of 35 spiders are cave spiders. How many more or less cave spiders than regular spiders are there?

$\frac{2}{7} \times 35 = 10\ (cave)$ 35 - 10 = 25 (regular)

Answer: 15 less.

3. I have run away from five-eighths of the 32 zombies I had met in my game. My brother run away from four-ninths of the 27 zombies he had met in his game. Who was luckier and run away from the greatest number of zombies?

$\frac{5}{8} \times 32 = 20\ (I)$ $\frac{4}{9} \times 27 = 12\ (Brother)$

Answer: I was luckier.

1. Find and circle or cross out the words to find out more about Minecraft.

E	O	L	M	H	W	G	K	P	H	S	C
C	Z	U	E	P	O	J	O	C	C	V	N
E	N	V	D	G	T	R	R	U	G	Q	O
I	U	F	I	Q	P	O	S	P	D	C	I
P	V	N	E	V	P	W	G	E	H	J	T
R	V	B	V	R	B	S	A	W	S	M	C
E	F	L	A	B	Y	R	I	N	T	H	E
T	O	E	L	G	N	I	N	W	A	P	S
N	R	I	F	E	O	U	P	U	P	M	S
E	F	D	O	M	A	O	V	D	W	D	K
C	X	F	P	I	L	L	A	R	S	G	T
B	R	F	V	A	J	X	X	U	K	W	Z

HORSES
REAR PORCH
LABYRINTH
SECTION
CENTERPIECE
MEDIEVAL
PILLARS
SPAWNING

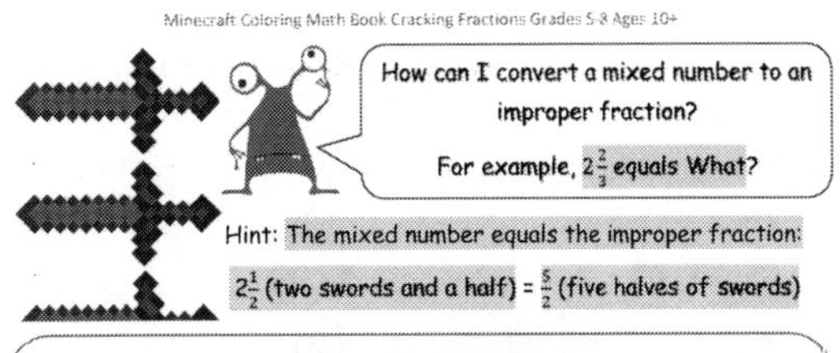

How can I convert a mixed number to an improper fraction?

For example, $2\frac{2}{3}$ equals What?

Hint: The mixed number equals the improper fraction: $2\frac{1}{2}$ (two swords and a half) = $\frac{5}{2}$ (five halves of swords)

Correct. Let's think!

Step 1: The whole number part equals what fraction? The whole number part of 1 equals: $1 = \frac{2}{2}$. The whole number part of 2 equals: $2 = \frac{4}{2}$.

Step 2: Add the two fractions: $\frac{4}{2} + \frac{1}{2} = \frac{4+1}{2} = \frac{5}{2}$

Another algorithm: $2\frac{1}{2} = ?$

Step 1: Multiply the whole number part by the denominator (in our problem we multiply 2 by 2). → 2 × 2 = 4 (the product is the numerator of the whole number part)

Step 2: Add the product to the numerator of a fraction part (in our problem it's 1). → 4 + 1 = 5 (we added the numerator of the whole number part and the numerator of the fraction part that equals the numerator of the improper fraction)

$$2\frac{1}{2} = \frac{5}{2}$$

1. Convert a mixed number to an improper fraction.

$1\frac{3}{5} = \frac{8}{5}$ $2\frac{5}{8} = \frac{21}{8}$ $3\frac{1}{4} = \frac{13}{8}$

1) 1 × 5 = 5   1) 2 × 8 = 16   1) 3 × 4 = 12

2) 5 + 3 = 8   2) 16 + 5 = 21   2) 12 + 1 = 13

2. Three-tenths of 50 blooming flowers were blue orchid. If sunflower were one-tenths more than blue orchid, and the rest of the flowers were divided by peonies and lilacs, and roses respectively, how many lilacs were blooming?

$\frac{3}{10} \times 50 = 15\ (BO)$ $\frac{4}{10} \times 50 = 20\ (S)$

Answer: $\frac{1}{10} \times 50 = 5\ (lilacs)$

3. Convert a mixed number to an improper fraction. Write the missing numbers.

$4\frac{2}{5} = \frac{(4\times5)+2}{5} = \frac{22}{5}$ $7\frac{1}{2} = \frac{(7\times2)+1}{2} = \frac{15}{2}$

$2\frac{1}{4} = \frac{(2\times4)+1}{4} = \frac{9}{4}$ $6\frac{2}{3} = \frac{(6\times3)+2}{3} = \frac{20}{3}$

$3\frac{1}{2} = \frac{(3\times2)+1}{2} = \frac{7}{2}$ $3\frac{5}{8} = \frac{(3\times8)+5}{8} = \frac{29}{8}$

1. Convert a mixed number to an improper fraction. Write the missing numbers.

$7\frac{3}{4} = \frac{(7\times4)+3}{4} = \frac{31}{4}$ $9\frac{4}{5} = \frac{(9\times5)+4}{5} = \frac{49}{5}$

$3\frac{3}{8} = \frac{(3\times8)+3}{8} = \frac{27}{8}$ $5\frac{5}{8} = \frac{(5\times8)+5}{8} = \frac{45}{8}$

$6\frac{1}{10} = \frac{(6\times10)+1}{10} = \frac{61}{10}$ $3\frac{2}{9} = \frac{(3\times9)+2}{9} = \frac{29}{9}$

$10\frac{3}{5} = \frac{(10\times5)+3}{5} = \frac{53}{5}$ $7\frac{9}{11} = \frac{(7\times11)+9}{11} = \frac{86}{11}$

2. I met some horses of two colors. Three-sixths of the horses were white. 24 horses were black. How many horses did I meet?

$\frac{3}{6}\times x = white$ $\frac{3}{6} = 24 = black$ $\frac{6}{6} = white + black = 48$

Answer: 48

3. The villager planted some melon and pumpkin seeds. Two-fifths of them were melon seeds. If he planted 45 pumpkin seeds, how many melon and pumpkin seeds did he have at first?

$\frac{3}{5}\times x = 45 \rightarrow \frac{1}{5} = 15 \rightarrow \frac{5}{5} = 75$

Answer: 75 seeds.

1. Convert an improper fraction to a mixed number. Circle the right answer.

$\frac{9}{2} = 4\frac{1}{2}$ $\frac{25}{8} = 3\frac{1}{8}$ $\frac{31}{4} = 7\frac{3}{4}$ $\frac{37}{5} = 7\frac{2}{5}$

a) $5\frac{2}{2}$ a) $2\frac{5}{8}$ a) $5\frac{3}{4}$ a) $7\frac{2}{5}$

b) $1\frac{3}{2}$ b) $3\frac{1}{8}$ b) $8\frac{1}{4}$ b) $6\frac{7}{5}$

c) $4\frac{1}{2}$ c) $5\frac{1}{8}$ c) $4\frac{3}{4}$ c) $5\frac{21}{5}$

d) $4\frac{2}{2}$ d) $3\frac{2}{8}$ d) $7\frac{3}{4}$ d) $4\frac{4}{5}$

2. I am an improper fraction. The sum of my numerator and denominator is 14, their difference is 2^3 (2 × 2 × 2). What fraction am I?

a) $\frac{8}{6}$ b) $\frac{15}{1}$ c) $\frac{6}{8}$ d) $\frac{11}{3}$

3. True or false? Explain. Circle the right answer.

To convert an improper fraction to a mixed number, you need to divide the denominator by the numerator.

To convert an improper fraction to a mixed number, you need to divide the numerator by the denominator.

a) TRUE b) FALSE

1. Convert a mixed number to an improper fraction. Circle the right answer.

$2\frac{1}{4} = \frac{9}{4}$ $7\frac{3}{8} = \frac{59}{8}$ $5\frac{7}{12} = \frac{67}{12}$ $3\frac{2}{5} = \frac{17}{5}$

a) $\frac{8}{4}$ a) $\frac{60}{8}$ a) $\frac{66}{12}$ a) $\frac{17}{5}$

b) $\frac{3}{4}$ b) $\frac{56}{8}$ b) $\frac{70}{12}$ b) $\frac{10}{5}$

c) $\frac{9}{4}$ c) $\frac{58}{8}$ c) $\frac{67}{12}$ c) $\frac{21}{5}$

d) $\frac{10}{4}$ d) $\frac{59}{8}$ d) $\frac{77}{12}$ d) $\frac{16}{5}$

2. I am an improper fraction. The product of my numerator and denominator is 27, their quotient is 3 What fraction am I?

a) $\frac{27}{9}$ b) $\frac{15}{14}$ c) $\frac{12}{4}$ d) $\frac{9}{3}$

3. True or false? Explain. Circle the right answer.

The smaller the denominator the greater the fraction.

You need to compare the numerators and the denominators.

b) TRUE b) FALSE

1. Find and circle or cross out the words to find out more about Minecraft.

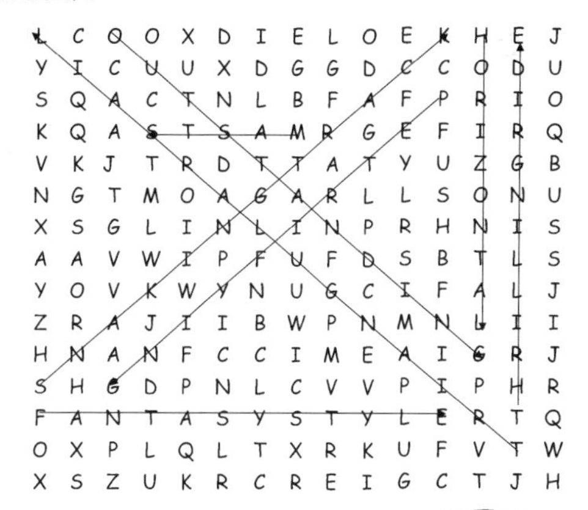

SNAKING TRACK MASTS

OUTSTANDING FANTASY-STYLE

HORIZONTAL TRIANGULAR SAIL

PETRIFYING THRILLING RIDE

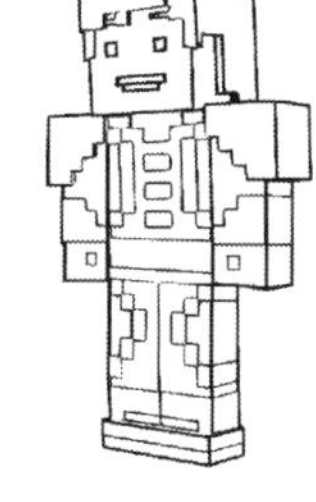

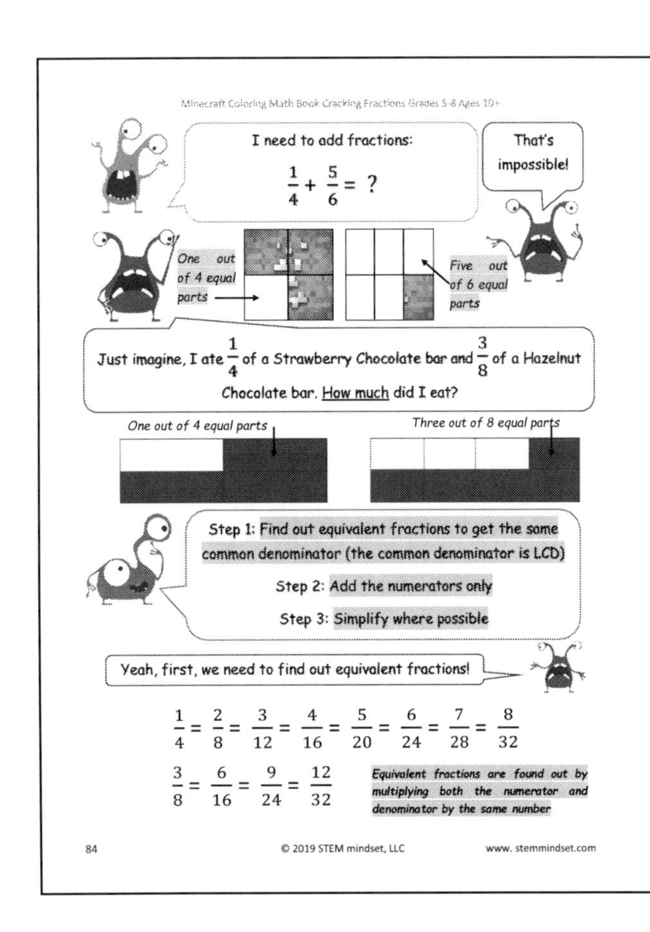

Minecraft Coloring Math Book Cracking Fractions Grades 5-8 Ages 10+

I need to add fractions:

$\frac{1}{4} + \frac{5}{6} = ?$

That's impossible!

One out of 4 equal parts

Five out of 6 equal parts

Just imagine, I ate $\frac{1}{4}$ of a Strawberry Chocolate bar and $\frac{3}{8}$ of a Hazelnut Chocolate bar. <u>How much</u> did I eat?

One out of 4 equal parts

Three out of 8 equal parts

Step 1: Find out equivalent fractions to get the same common denominator (the common denominator is LCD)

Step 2: Add the numerators only

Step 3: Simplify where possible

Yeah, first, we need to find out equivalent fractions!

$$\frac{1}{4} = \frac{2}{8} = \frac{3}{12} = \frac{4}{16} = \frac{5}{20} = \frac{6}{24} = \frac{7}{28} = \frac{8}{32}$$

$$\frac{3}{8} = \frac{6}{16} = \frac{9}{24} = \frac{12}{32}$$

Equivalent fractions are found out by multiplying both the numerator and denominator by the same number

84 © 2019 STEM mindset, LLC www. stemmindset.com

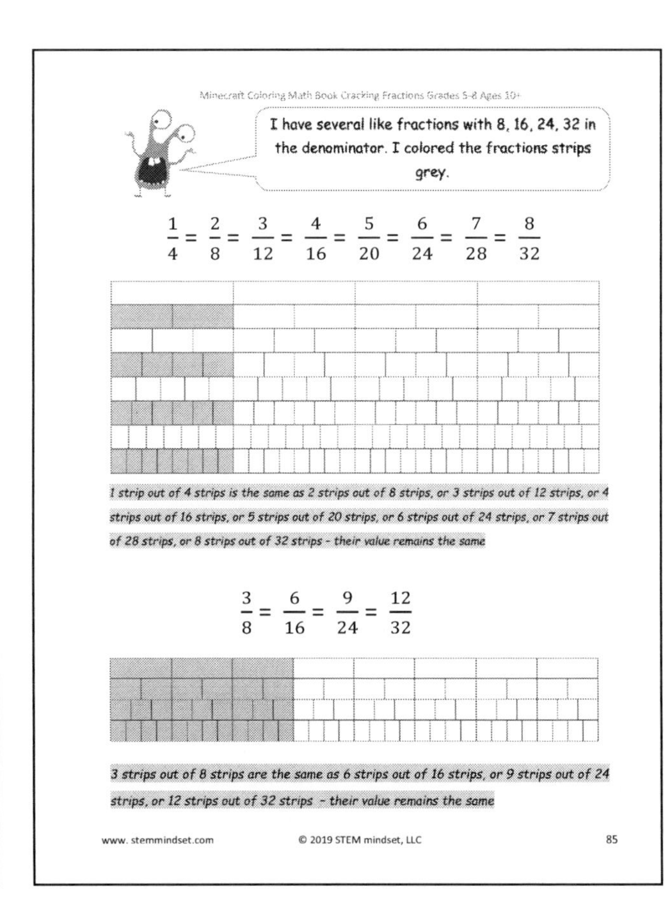

Minecraft Coloring Math Book Cracking Fractions Grades 5-8 Ages 10+

I have several like fractions with 8, 16, 24, 32 in the denominator. I colored the fractions strips grey.

$$\frac{1}{4} = \frac{2}{8} = \frac{3}{12} = \frac{4}{16} = \frac{5}{20} = \frac{6}{24} = \frac{7}{28} = \frac{8}{32}$$

1 strip out of 4 strips is the same as 2 strips out of 8 strips, or 3 strips out of 12 strips, or 4 strips out of 16 strips, or 5 strips out of 20 strips, or 6 strips out of 24 strips, or 7 strips out of 28 strips, or 8 strips out of 32 strips - their value remains the same

$$\frac{3}{8} = \frac{6}{16} = \frac{9}{24} = \frac{12}{32}$$

3 strips out of 8 strips are the same as 6 strips out of 16 strips, or 9 strips out of 24 strips, or 12 strips out of 32 strips - their value remains the same

www. stemmindset.com © 2019 STEM mindset, LLC 85

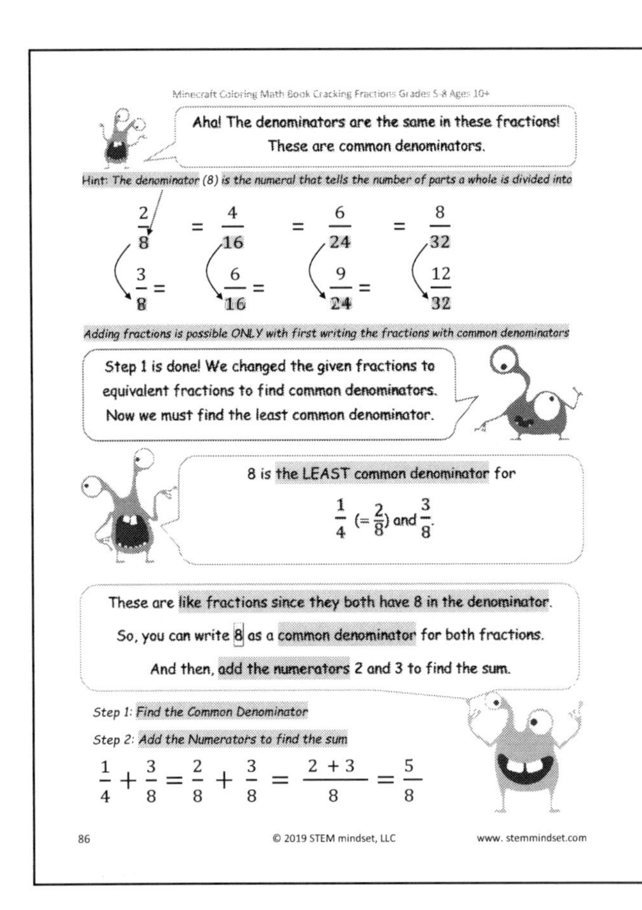

Minecraft Coloring Math Book Cracking Fractions Grades 5-8 Ages 10+

Aha! The denominators are the same in these fractions! These are common denominators.

Hint: The denominator (8) is the numeral that tells the number of parts a whole is divided into

$$\frac{2}{8} = \frac{4}{16} = \frac{6}{24} = \frac{8}{32}$$

$$\frac{3}{8} = \quad \frac{6}{16} = \quad \frac{9}{24} = \quad \frac{12}{32}$$

Adding fractions is possible ONLY with first writing the fractions with common denominators

Step 1 is done! We changed the given fractions to equivalent fractions to find common denominators. Now we must find the least common denominator.

8 is the LEAST common denominator for

$\frac{1}{4}$ $(=\frac{2}{8})$ and $\frac{3}{8}$.

These are like fractions since they both have 8 in the denominator.

So, you can write 8 as a common denominator for both fractions.

And then, add the numerators 2 and 3 to find the sum.

Step 1: Find the Common Denominator

Step 2: Add the Numerators to find the sum

$$\frac{1}{4} + \frac{3}{8} = \frac{2}{8} + \frac{3}{8} = \frac{2+3}{8} = \frac{5}{8}$$

86 © 2019 STEM mindset, LLC www. stemmindset.com

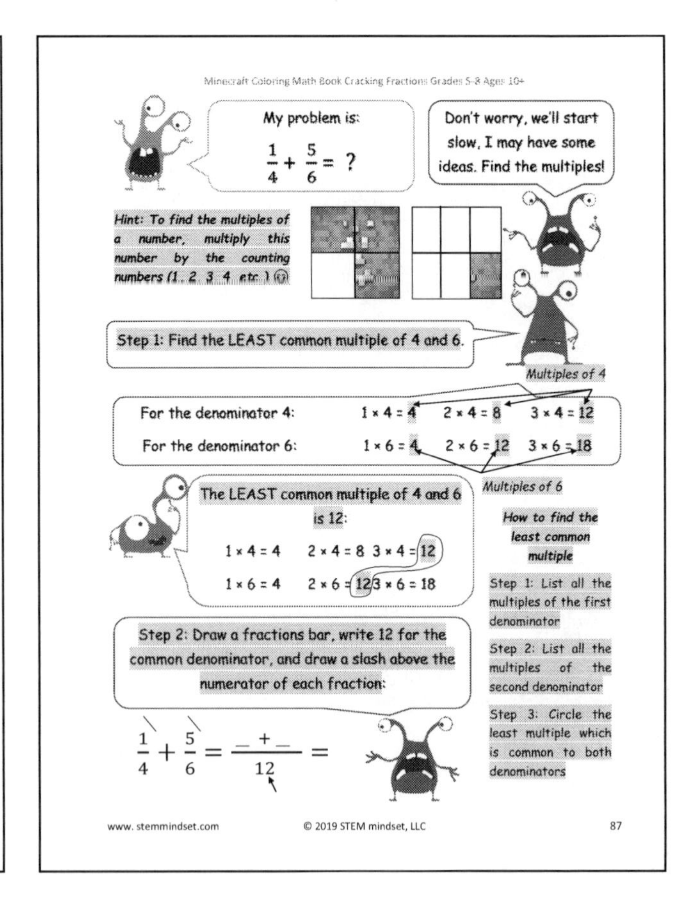

Minecraft Coloring Math Book Cracking Fractions Grades 5-8 Ages 10+

My problem is:

$\frac{1}{4} + \frac{5}{6} = ?$

Don't worry, we'll start slow, I may have some ideas. Find the multiples!

Hint: To find the multiples of a number, multiply this number by the counting numbers (1, 2, 3, 4, etc.)

Step 1: Find the LEAST common multiple of 4 and 6.

Multiples of 4

For the denominator 4:	1 × 4 = 4	2 × 4 = 8	3 × 4 = 12
For the denominator 6:	1 × 6 = 4	2 × 6 = 12	3 × 6 = 18

Multiples of 6

The LEAST common multiple of 4 and 6 is 12:

1 × 4 = 4 2 × 4 = 8 3 × 4 = 12

1 × 6 = 4 2 × 6 = 12 3 × 6 = 18

How to find the least common multiple

Step 1: List all the multiples of the first denominator

Step 2: List all the multiples of the second denominator

Step 3: Circle the least multiple which is common to both denominators

Step 2: Draw a fractions bar, write 12 for the common denominator, and draw a slash above the numerator of each fraction:

$$\frac{1}{4} + \frac{5}{6} = \frac{\_ + \_}{12} =$$

www. stemmindset.com © 2019 STEM mindset, LLC 87

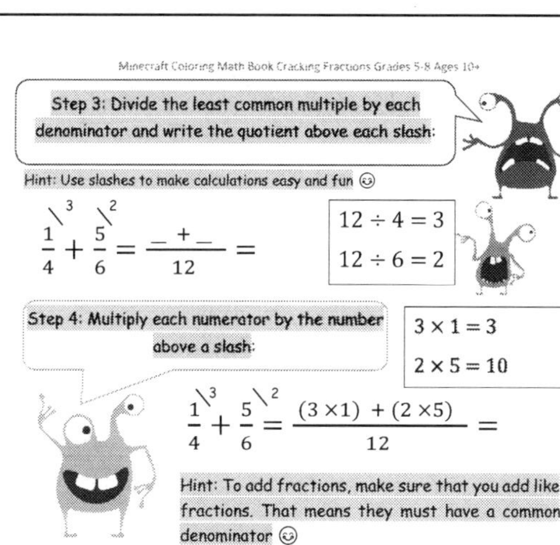

Step 3: Divide the least common multiple by each denominator and write the quotient above each slash:

Hint: Use slashes to make calculations easy and fun ☺

$$\frac{1}{4}+\frac{5}{6}=\frac{\_+\_}{12}=$$

$12 \div 4 = 3$

$12 \div 6 = 2$

Step 4: Multiply each numerator by the number above a slash:

$3 \times 1 = 3$

$2 \times 5 = 10$

$$\frac{1}{4}+\frac{5}{6}=\frac{(3\times1)+(2\times5)}{12}=$$

Hint: To add fractions, make sure that you add like fractions. That means they must have a common denominator ☺

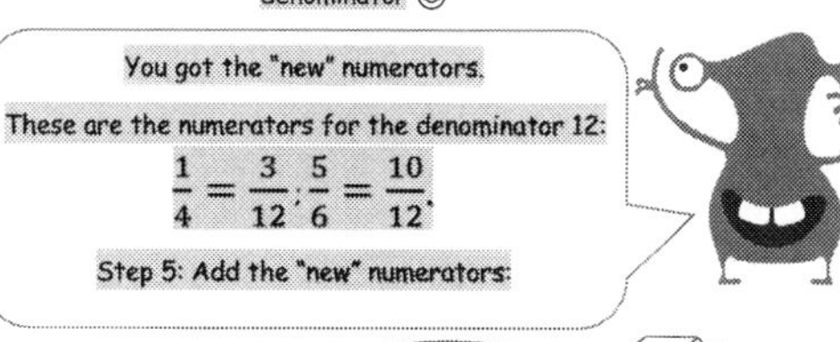

You got the "new" numerators.

These are the numerators for the denominator 12:

$$\frac{1}{4}=\frac{3}{12};\ \frac{5}{6}=\frac{10}{12}.$$

Step 5: Add the "new" numerators:

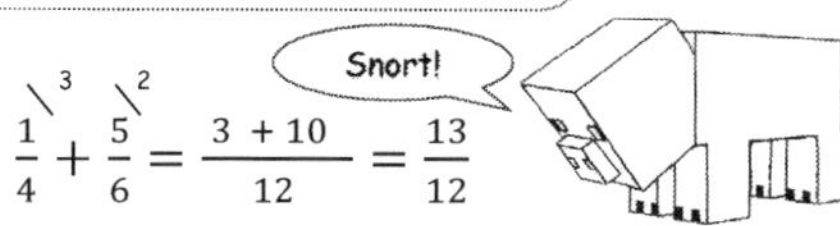

$$\frac{1}{4}+\frac{5}{6}=\frac{3+10}{12}=\frac{13}{12}$$

1. Add fractions. Write the missing numbers. Color the fraction of each shape. Simplify if possible.

$\frac{1}{4}+\frac{1}{3}=?$

$\frac{3}{4}+\frac{2}{5}=?$

Step 1: Find the least common denominator

$\frac{1}{4}=\frac{2}{8}=\frac{3}{12}$

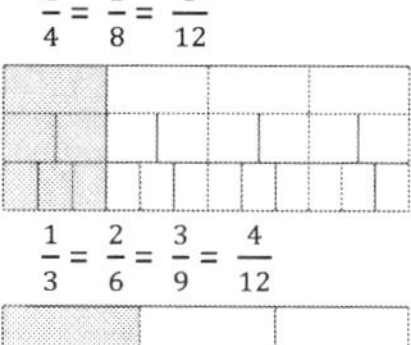

$\frac{1}{3}=\frac{2}{6}=\frac{3}{9}=\frac{4}{12}$

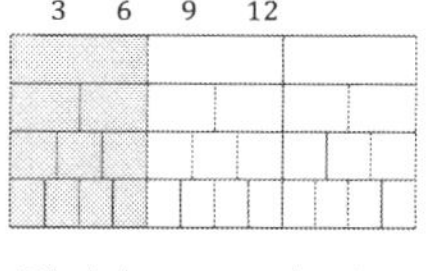

12 is the least common denominator

$\frac{3}{4}=\frac{6}{8}=\frac{9}{12}=\frac{12}{16}=\frac{15}{20}$

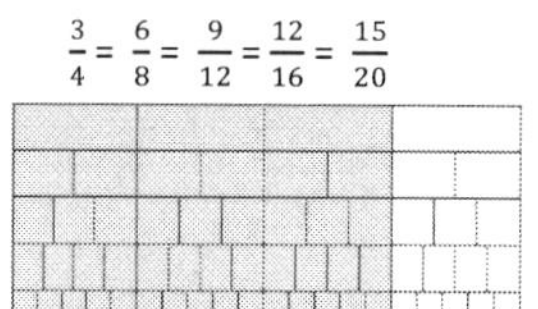

$\frac{2}{5}=\frac{4}{10}=\frac{6}{15}=\frac{8}{20}$

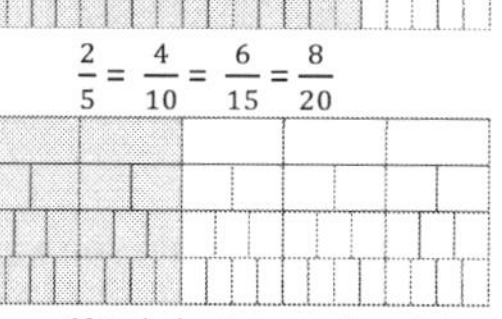

20 is the least common denominator

Step 2: Add the Numerators to find the sum

$\frac{1}{4}+\frac{1}{3}=\frac{3+4}{12}=\frac{7}{12}=\_\ \_$

$\frac{3}{4}+\frac{2}{5}=\frac{15+8}{20}=\frac{23}{20}=1\frac{3}{20}$

2. True or false? Explain. Circle the right answer.

The greater the numerator in like fractions the smaller the fraction.

The greater the numerator in like fractions the greater the fraction.

a) TRUE b) FALSE

1. Add fractions. Write the missing numbers. Color the fraction of each shape. Simplify if possible.

$\frac{2}{3}+\frac{5}{12}=?$

$\frac{5}{6}+\frac{3}{4}=?$

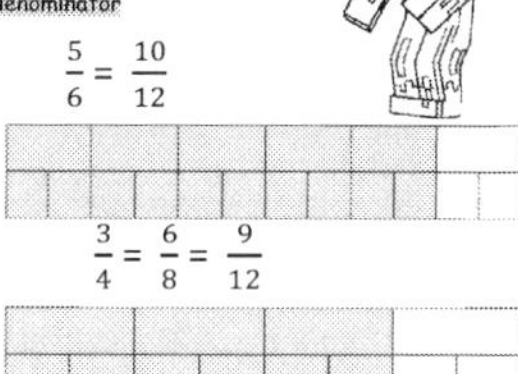

Step 1: Find the least common denominator

$\frac{2}{3}=\frac{4}{6}=\frac{6}{9}=\frac{8}{12}$

$\frac{5}{12}$

12 is the least common denominator

$\frac{5}{6}=\frac{10}{12}$

$\frac{3}{4}=\frac{6}{8}=\frac{9}{12}$

12 is the least common denominator

Step 2: Add the Numerators to find the sum

$\frac{2}{3}+\frac{5}{12}=\frac{8+5}{12}=\frac{13}{12}=1\frac{1}{12}$

$\frac{5}{6}+\frac{3}{4}=\frac{10+9}{12}=\frac{19}{12}=1\frac{7}{12}$

2. True or false? Explain. Circle the right answer.

The greater the denominator in unlike fractions the greater the fraction.

You can compare only fractions with the like denominators.

a) TRUE b) FALSE

1. Add fractions. Write the missing numbers. Simplify if possible.

$\frac{2}{6}+\frac{1}{2}=\frac{4+6}{12}=\frac{10}{12}=\frac{5}{6}$

$\frac{2}{4}+\frac{4}{5}=\frac{10+16}{20}=\frac{26}{20}=1\frac{3}{10}$

$\frac{1}{3}+\frac{1}{4}=\frac{4+3}{12}=\frac{7}{12}$

$\frac{2}{3}+\frac{1}{6}=\frac{4+1}{6}=\frac{5}{6}$

$\frac{4}{6}+\frac{2}{4}=\frac{8+6}{12}=\frac{14}{12}=1\frac{1}{6}$

$\frac{4}{9}+\frac{6}{18}=\frac{8+6}{18}=\frac{14}{18}=\frac{7}{9}$

$\frac{5}{14}+\frac{3}{7}=\frac{5+6}{14}=\frac{11}{14}$

$\frac{1}{6}+\frac{3}{8}=\frac{4+9}{24}=\frac{13}{24}$

$\frac{3}{4}+\frac{7}{16}=\frac{12+7}{16}=\frac{19}{16}=1\frac{3}{16}$

$\frac{4}{5}+\frac{5}{6}=\frac{24+25}{30}=\frac{49}{30}$

$\frac{7}{12}+\frac{7}{8}=\frac{14+21}{24}=\frac{35}{24}=1\frac{11}{24}$

$\frac{3}{10}+\frac{4}{6}=\frac{18+40}{60}=\frac{58}{60}=\frac{29}{30}$

Hint: Find the least common denominator by multiplying the greatest given denominator by 2, then, by 3, and so on until you get a number that is exactly divided by the given denominators.

2. True or false? Explain. Circle the right answer.

The greater the denominator in unlike fractions the greater the fraction.

Any fractions can be compared by getting a common denominator.

a) TRUE b) FALSE

1. Add fractions. Write the missing numbers. Simplify if possible.

$\frac{9}{16}+\frac{11}{12}=\frac{27+44}{48}=\frac{71}{48}=1\frac{23}{48}$ $\frac{9}{18}+\frac{1}{4}=\frac{18+9}{36}=\frac{27}{36}=\frac{3}{4}$

$\frac{7}{20}+\frac{5}{8}=\frac{14+25}{40}=\frac{39}{40}$ $\frac{13}{18}+\frac{7}{10}=\frac{65+63}{90}=\frac{128}{90}=1\frac{19}{45}$

$\frac{8}{25}+\frac{4}{15}=\frac{24+20}{75}=\frac{44}{75}$ $\frac{10}{28}+\frac{3}{4}=\frac{10+21}{28}=\frac{31}{28}=1\frac{3}{28}$

$\frac{9}{20}+\frac{11}{14}+\frac{2}{7}=\frac{63+110+40}{140}=\frac{213}{140}=1\frac{73}{140}$

$\frac{3}{7}+\frac{9}{14}+\frac{13}{21}=\frac{18+27+26}{42}=\frac{71}{42}=1\frac{29}{42}$

$\frac{1}{2}+\frac{7}{13}+\frac{5}{26}=\frac{13+14+5}{26}=\frac{32}{26}=1\frac{6}{13}$

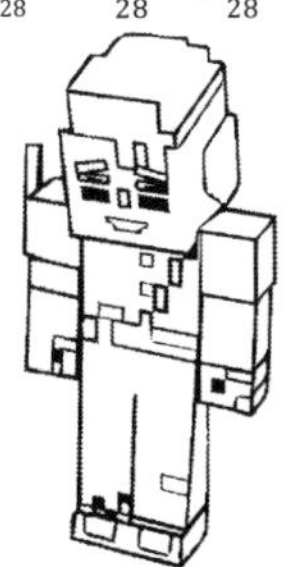

2. True or false? Explain. Circle the right answer.

The least common denominator is the least factor for the numerator.

The least common denominator is the least multiple of two numbers.

a) TRUE b) FALSE

1. Add fractions. Write the missing numbers. Simplify if possible.

$\frac{2}{13}+\frac{5}{39}=\frac{6+5}{39}=\frac{11}{39}$ $\frac{2}{5}+\frac{3}{10}=\frac{4+3}{10}=\frac{7}{10}$

$\frac{5}{8}+\frac{5}{6}=\frac{15+20}{24}=\frac{35}{24}=1\frac{11}{24}$ $\frac{5}{6}+\frac{7}{15}=\frac{25+14}{30}=\frac{39}{30}=1\frac{3}{10}$

$\frac{7}{18}+\frac{8}{15}=\frac{35+48}{90}=\frac{83}{90}$ $\frac{5}{18}+\frac{3}{12}=\frac{30+27}{108}=\frac{57}{108}=\frac{19}{36}$

A common denominator may be found by multiplying together given denominators!

$\frac{11}{15}+\frac{7}{9}=1\frac{23}{45}$ $\frac{9}{13}+\frac{1}{4}=\frac{49}{52}$

$\frac{13}{16}+\frac{7}{8}=1\frac{11}{16}$ $\frac{3}{21}+\frac{1}{6}=\frac{13}{42}$

$\frac{5}{12}+\frac{3}{8}=\frac{19}{24}$ $\frac{7}{16}+\frac{5}{7}=1\frac{17}{112}$

2. True or false? Explain. Circle the right answer.

The mixed number is the sum of a proper fraction and an improper fraction.

a) TRUE b) FALSE

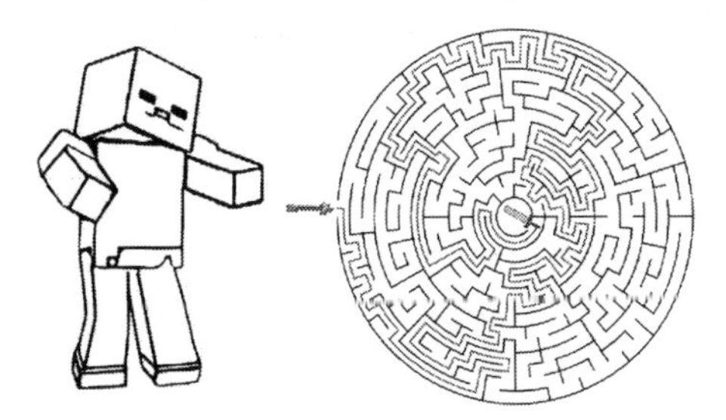

1. Find and circle or cross out the words to find out more about Minecraft.

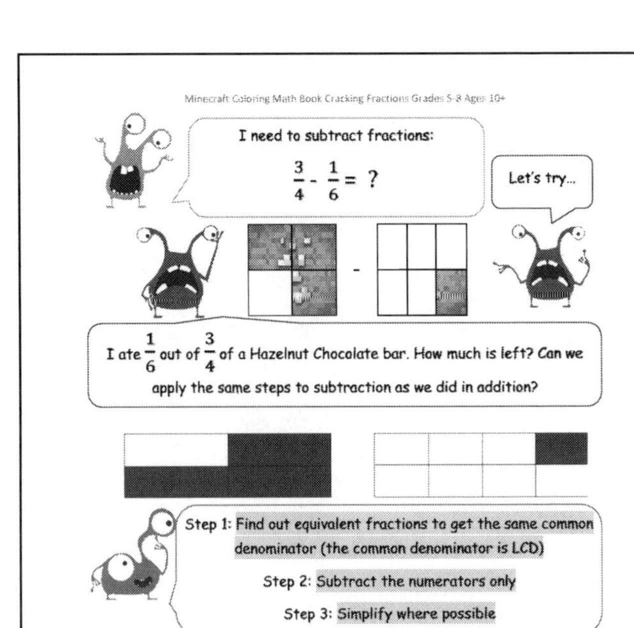

I need to subtract fractions:
3/4 - 1/6 = ?
Let's try...
-
I ate 1/6 out of 3/4 of a Hazelnut Chocolate bar. How much is left? Can we apply the same steps to subtraction as we did in addition?
Step 1: Find out equivalent fractions to get the same common denominator (the common denominator is LCD)
Step 2: Subtract the numerators only
Step 3: Simplify where possible

Find out equivalent fractions!

$$\frac{3}{4}=\frac{6}{8}=\frac{9}{12}=\frac{12}{16}=\frac{15}{20}=\frac{18}{24}=\frac{21}{28}=\frac{24}{32}=\frac{27}{36}$$

$$\frac{1}{6}=\frac{2}{12}=\frac{3}{18}=\frac{4}{24}=\frac{5}{30}=\frac{6}{36}$$

Equivalent fractions are found out by multiplying both the numerator and denominator by the same number

I have several like fractions with 12, 24, 36 in the denominator. I colored the fractions strips grey.

$$\frac{3}{4} = \frac{6}{8} = \frac{9}{12} = \frac{12}{16} = \frac{15}{20} = \frac{18}{24} = \frac{21}{28} = \frac{24}{32} = \frac{27}{36}$$

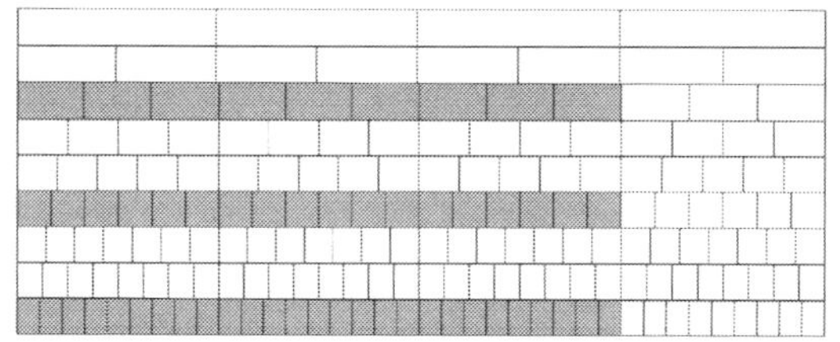

3 strips out of 4 strips are the same as 6 strips out of 8 strips, or 9 strips out of 12 strips, or 12 strips out of 16 strips, or 15 strips out of 20 strips, or 18 strips out of 24 strips, or 21 strips out of 28 strips, or 24 strips out of 32 strips, or 27 strips out of 36 strips - their value remains the same

$$\frac{1}{6} = \frac{2}{12} = \frac{3}{18} = \frac{4}{24} = \frac{5}{30} = \frac{6}{36}$$

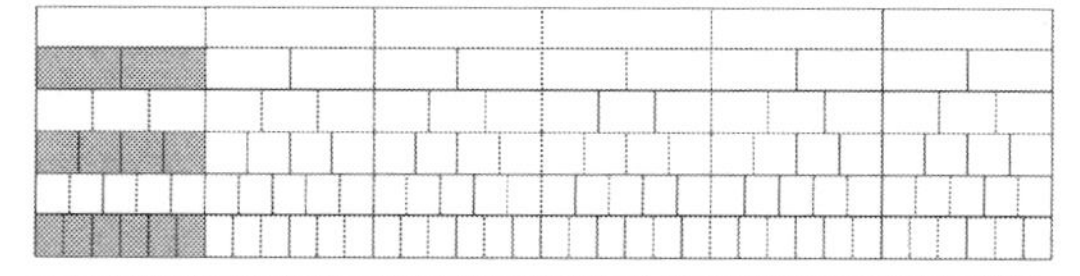

1 strip out of 6 strips is the same as 2 strips out of 12 strips, or 3 strips out of 18 strips, or 4 strips out of 24 strips, or 5 strips out of 30 strips, or 6 strips out of 36 strips - their value remains the same

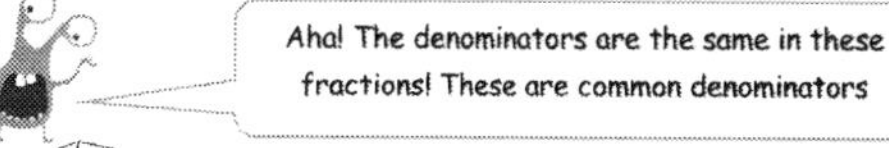

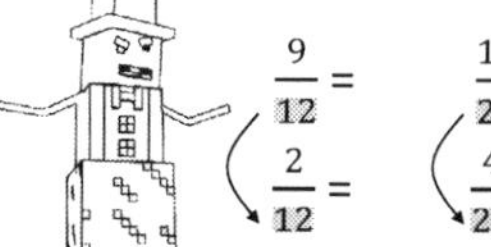

$\frac{9}{12} = \frac{2}{12}$ $\frac{18}{24} = \frac{4}{24}$ $\frac{27}{36} = \frac{6}{36}$

Hint: You have a list of equivalent fractions ☺

Step 1 is done! We changed the given fractions to equivalent fractions to find common denominators. Now we must find the least common denominator.

12 is the LEAST common denominator for $\frac{3}{4}$ $(= \frac{9}{12})$ and $\frac{1}{6}$ $(= \frac{2}{12})$.

Hint: The least common denominator is the denominator which is common to both denominators of the given fractions.

These are like fractions since they both have 12 in the denominator. So, you can write 12 as a common denominator for both fractions.

And then, subtract the numerators 9 and 2.

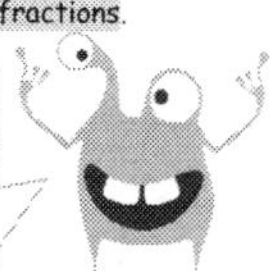

$$\frac{3}{4} - \frac{1}{6} = \frac{9}{12} - \frac{2}{12} = \frac{9-2}{12} = \frac{7}{12}$$

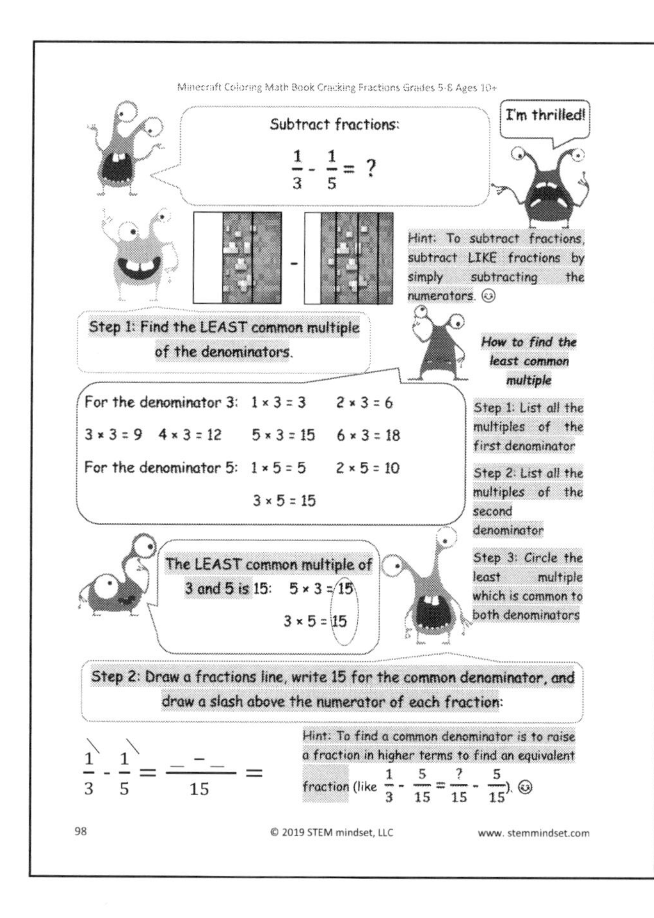

Subtract fractions:

$$\frac{1}{3} - \frac{1}{5} = ?$$

I'm thrilled!

Hint: To subtract fractions, subtract LIKE fractions by simply subtracting the numerators. ☺

Step 1: Find the LEAST common multiple of the denominators.

How to find the least common multiple

For the denominator 3: 1 × 3 = 3 2 × 3 = 6

3 × 3 = 9 4 × 3 = 12 5 × 3 = 15 6 × 3 = 18

For the denominator 5: 1 × 5 = 5 2 × 5 = 10

3 × 5 = 15

Step 1: List all the multiples of the first denominator

Step 2: List all the multiples of the second denominator

Step 3: Circle the least multiple which is common to both denominators

The LEAST common multiple of 3 and 5 is 15: 5 × 3 = 15

3 × 5 = 15

Step 2: Draw a fractions line, write 15 for the common denominator, and draw a slash above the numerator of each fraction:

$$\frac{1}{3} - \frac{1}{5} = \frac{\_ - \_}{15} =$$

Hint: To find a common denominator is to raise a fraction in higher terms to find an equivalent fraction (like $\frac{1}{3} - \frac{5}{15} = \frac{?}{15} - \frac{5}{15}$). ☺

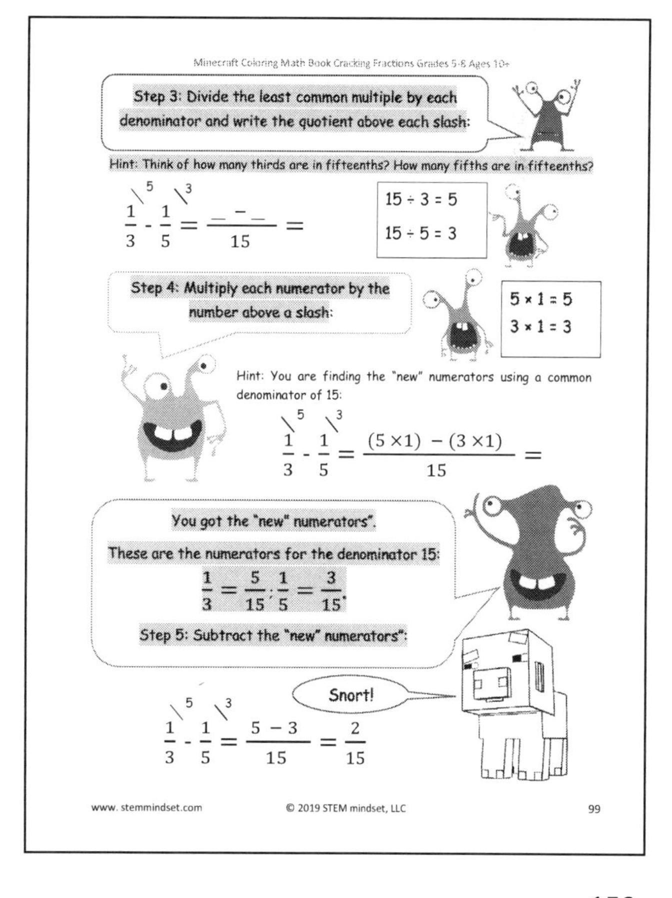

Step 3: Divide the least common multiple by each denominator and write the quotient above each slash:

Hint: Think of how many thirds are in fifteenths? How many fifths are in fifteenths?

$$\overset{5}{\frac{1}{3}} - \overset{3}{\frac{1}{5}} = \frac{\_ - \_}{15} =$$

15 ÷ 3 = 5

15 ÷ 5 = 3

Step 4: Multiply each numerator by the number above a slash:

5 × 1 = 5

3 × 1 = 3

Hint: You are finding the "new" numerators using a common denominator of 15:

$$\overset{5}{\frac{1}{3}} - \overset{3}{\frac{1}{5}} = \frac{(5 \times 1) - (3 \times 1)}{15} =$$

You got the "new" numerators.

These are the numerators for the denominator 15: $\frac{1}{3} = \frac{5}{15}; \frac{1}{5} = \frac{3}{15}$.

Step 5: Subtract the "new" numerators:

Snort!

$$\overset{5}{\frac{1}{3}} - \overset{3}{\frac{1}{5}} = \frac{5-3}{15} = \frac{2}{15}$$

1. Subtract fractions. Write the missing numbers. Color the fraction of each shape. Simplify if possible.

$\frac{3}{4} - \frac{1}{3} = ?$

Step 1: Find the least common denominator

$\frac{3}{4} = \frac{6}{8} = \frac{9}{12}$

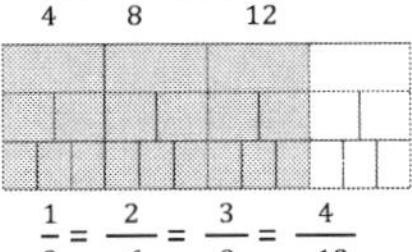

$\frac{1}{3} = \frac{2}{6} = \frac{3}{9} = \frac{4}{12}$

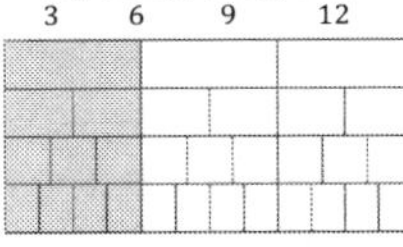

12 is the least common denominator

$\frac{3}{4} - \frac{1}{5} = ?$

$\frac{3}{4} = \frac{6}{8} = \frac{9}{12} = \frac{12}{16} = \frac{15}{20}$

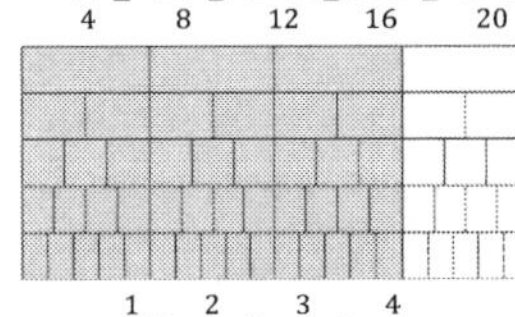

$\frac{1}{5} = \frac{2}{10} = \frac{3}{15} = \frac{4}{20}$

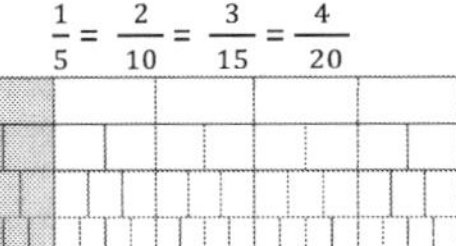

20 is the least common denominator

Step 2: Add the Numerators to find the sum

$\frac{3}{4} - \frac{1}{3} = \frac{9-4}{12} = \frac{5}{12}$

$\frac{3}{4} - \frac{2}{5} = \frac{15-8}{20} = \frac{7}{20}$

2. I had 54 bones in the beginning of the game. I used $\frac{3}{6}$ of the bones to tame one wolf and $\frac{2}{9}$ of the leftover bones to tame another wolf. How many bones did I use?

$\frac{3}{6} \times 54 = 27$ $\frac{2}{9} \times 27 = 6$

Answer: 27 + 6 = 33 (bones).

1. Subtract fractions. Write the missing numbers. Color the fraction of each shape. Simplify if possible.

$\frac{2}{3} - \frac{7}{12} = ?$

Step 1: Find the least common denominator

$\frac{2}{3} = \frac{4}{6} = \frac{6}{9} = \frac{8}{12}$

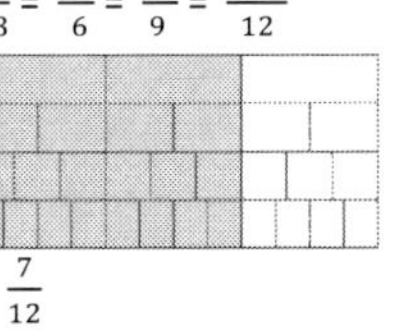

$\frac{7}{12}$

12 is the least common denominator

$\frac{1}{4} - \frac{1}{6} = ?$

$\frac{1}{4} = \frac{2}{8} = \frac{3}{12}$

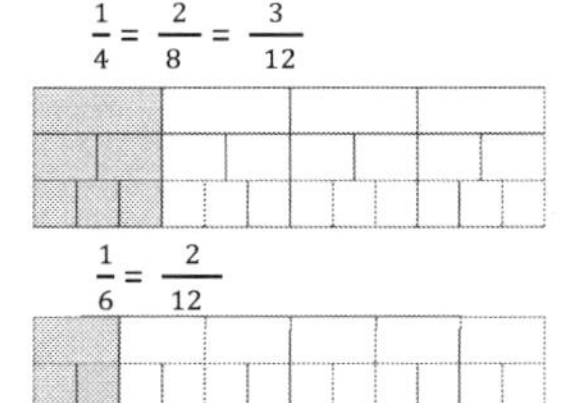

$\frac{1}{6} = \frac{2}{12}$

12 is the least common denominator

Step 2: Add the Numerators to find the sum

$\frac{2}{3} - \frac{7}{12} = \frac{8-7}{12} = \frac{1}{12}$

$\frac{1}{4} - \frac{1}{6} = \frac{3-2}{12} = \frac{1}{12}$

2. I encountered 48 creepers in the game. I destroyed $\frac{1}{3}$ of the creepers with my iron sword and $\frac{5}{8}$ of the leftover creepers with my bow and arrow. How many creepers did I destroy?

$\frac{1}{3} \times 48 = 16$ $\frac{5}{8} \times 32 = 20$

Answer: 36 creepers.

1. Subtract fractions. Write the missing numbers. Simplify if possible.

$\frac{3}{8} - \frac{1}{6} = \frac{9-4}{24} = \frac{5}{24}$

$\frac{7}{9} - \frac{1}{3} = \frac{7-3}{9} = \frac{4}{9}$

$\frac{5}{8} - \frac{1}{4} = \frac{5-2}{8} = \frac{3}{8}$

$\frac{11}{15} - \frac{2}{5} = \frac{11-6}{15} = \frac{5}{15} = \frac{1}{3}$

$\frac{3}{10} - \frac{1}{4} = \frac{6-5}{20} = \frac{1}{20}$

$\frac{11}{16} - \frac{3}{8} = \frac{11-6}{16} = \frac{5}{16}$

$\frac{8}{12} - \frac{2}{6} = \frac{8-4}{12} = \frac{4}{12} = \frac{1}{3}$

$\frac{7}{8} - \frac{3}{6} = \frac{21-12}{24} = \frac{9}{24} = \frac{3}{8}$

$\frac{13}{18} - \frac{2}{3} = \frac{13-12}{18} = \frac{1}{18}$

$\frac{7}{15} - \frac{1}{3} = \frac{7-5}{15} = \frac{2}{15}$

$\frac{5}{7} - \frac{9}{14} = \frac{10-9}{14} = \frac{1}{14}$

$\frac{9}{10} - \frac{5}{6} = \frac{27-25}{30} = \frac{2}{30} = \frac{1}{15}$

2. The villager used $\frac{1}{3}$ of his water bucket to water the carrots and $\frac{7}{24}$ of his water bucket to water the bushes, and $\frac{5}{16}$ of his water bucket to water the flowers. How much water (—) was left?

$\frac{1}{3} + \frac{7}{24} + \frac{5}{16} = \frac{16+14+15}{48} = \frac{45}{48} = \frac{15}{16}$

Answer: $\frac{1}{16}$

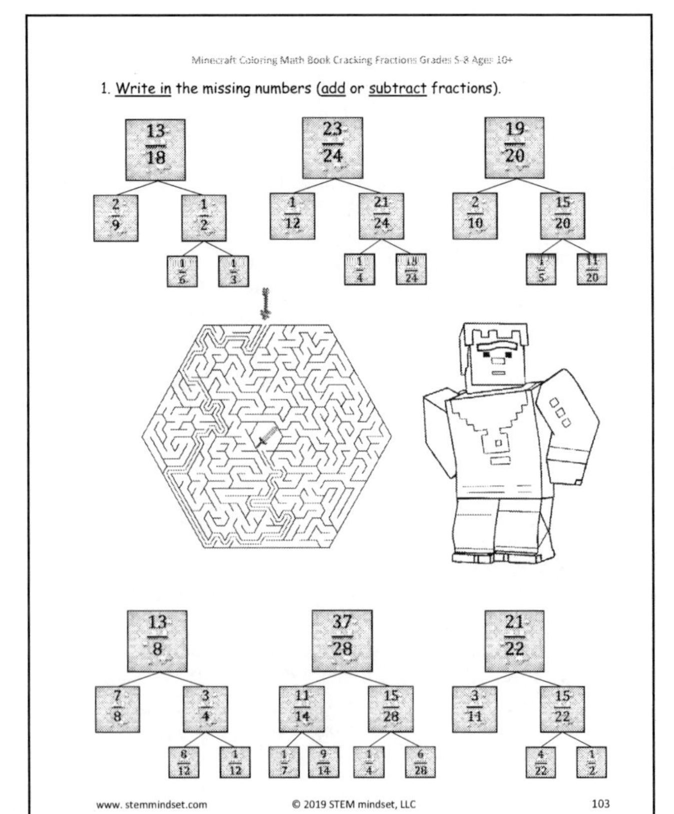

1. Subtract fractions. Write the missing numbers. Simplify if possible.

$\frac{1}{2} - \frac{1}{6} = \frac{3-1}{6} = \frac{2}{6} = \frac{1}{3}$

$\frac{13}{30} - \frac{2}{5} = \frac{13-12}{30} = \frac{1}{30}$

$\frac{19}{24} - \frac{5}{8} = \frac{19-15}{24} = \frac{4}{24} = \frac{1}{6}$

$\frac{15}{28} - \frac{2}{4} = \frac{15-14}{28} = \frac{1}{28}$

$\frac{3}{4} - \frac{3}{12} = \frac{9-3}{12} = \frac{6}{12} = \frac{1}{2}$

$\frac{5}{7} - \frac{4}{9} = \frac{45-28}{63} = \frac{17}{63}$

$\frac{4}{11} - \frac{2}{22} = \frac{8-2}{22} = \frac{6}{22} = \frac{3}{11}$

$\frac{7}{10} - \frac{3}{40} = \frac{28-3}{40} = \frac{25}{40} = \frac{5}{8}$

$\frac{15}{20} - \frac{2}{5} = \frac{15-8}{20} = \frac{7}{20}$

$\frac{5}{6} - \frac{3}{8} = \frac{20-9}{24} = \frac{11}{24}$

$\frac{10}{15} - \frac{6}{12} = \frac{40-30}{60} = \frac{10}{60} = \frac{1}{6}$

$\frac{12}{36} - \frac{8}{48} = \frac{48-24}{144} = \frac{24}{144} = \frac{1}{6}$

2. The villager had some trees. He chopped down $\frac{2}{9}$ of the trees and he still had 21 trees left. How many trees did the villager have in the beginning?

$\frac{7}{9} \times x = 21 \quad \rightarrow \quad \frac{1}{9} = 3 \quad \rightarrow \quad \frac{9}{9} = 27$

Answer: 27 trees.

1. Subtract fractions. Write the missing numbers. Simplify if possible.

$\frac{5}{6} - \frac{3}{24} = \frac{17}{24}$

$\frac{7}{9} - \frac{1}{3} = \frac{4}{9}$

$\frac{5}{9} - \frac{1}{27} = \frac{14}{27}$

$\frac{7}{16} - \frac{5}{48} = \frac{1}{3}$

$\frac{23}{24} - \frac{1}{3} - \frac{1}{4} = \frac{23-8-6}{24} = \frac{3}{8}$

$\frac{5}{6} - \frac{1}{5} - \frac{1}{15} = \frac{25-6-2}{30} = \frac{17}{30}$

$\frac{6}{7} - \frac{2}{5} - \frac{5}{70} = \frac{60-28-5}{70} = \frac{27}{70}$

$\frac{11}{12} - \frac{5}{48} - \frac{7}{24} = \frac{44-5-14}{48} = \frac{25}{48}$

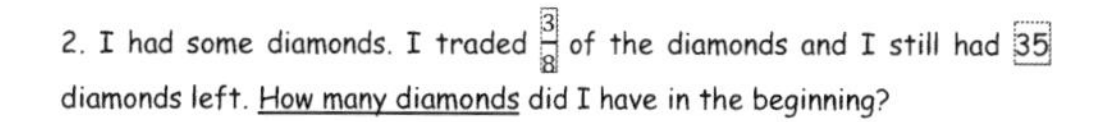

2. I had some diamonds. I traded $\frac{3}{8}$ of the diamonds and I still had 35 diamonds left. How many diamonds did I have in the beginning?

$\frac{5}{8} \times x = 35 \quad \rightarrow \quad \frac{1}{8} = 7 \quad \rightarrow \quad \frac{8}{8} = 56$

Answer: 56 diamonds.

1. Find and circle or cross out the words to find out more about Minecraft.

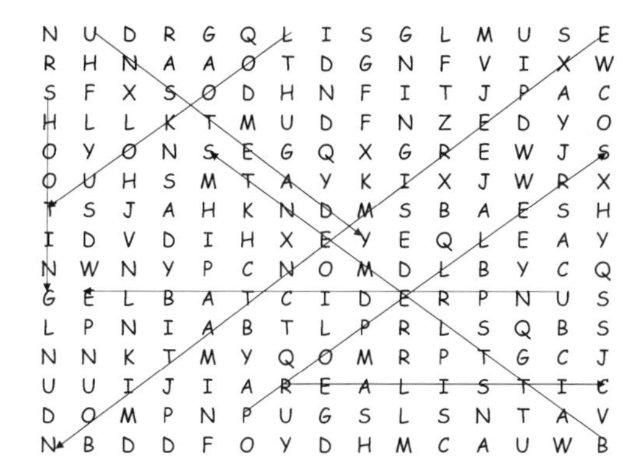

LOOKOUT
BATTLEMENTS
PROPELLERS
REDESIGNING
EXPERIMENTATION
REALISTIC
UNSTEADY
SHOOTING
UNPREDICTABLE

1. Write in the missing numbers (add or subtract).

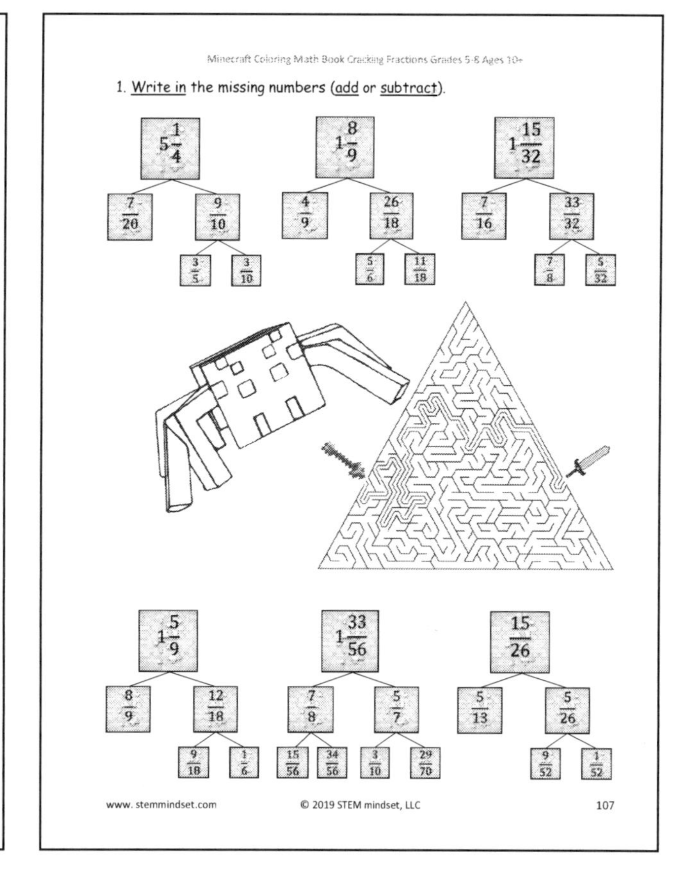

1. Subtract fractions. Write the missing numbers. Simplify if possible.

$1 - \frac{1}{6} = \frac{6}{6} - \frac{1}{6} = \frac{6-1}{6} = \frac{5}{6}$ $\quad 1 - \frac{1}{3} = \frac{3}{3} - \frac{1}{3} = \frac{3-1}{3} = \frac{2}{3}$

Hint: To subtract a fraction from a whole number: change a whole number to a fraction with the same denominator as that in the fraction to be subtracted

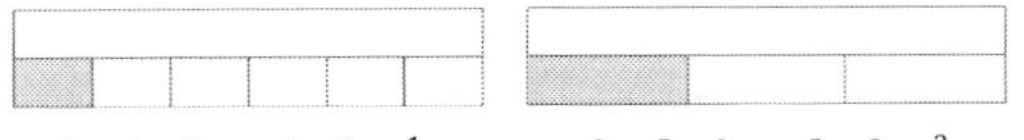

$1 - \frac{3}{4} = \frac{4}{4} - \frac{3}{4} = \frac{4-3}{4} = \frac{1}{4}$ $\quad 1 - \frac{2}{5} = \frac{5}{5} - \frac{2}{5} = \frac{5-2}{5} = \frac{3}{5}$

Hint: Rewrite a whole number to a fraction so that both fractions have a common denominator:

$1 - \frac{5}{7} = \frac{7}{7} - \frac{5}{7} = \frac{7-5}{7} = \frac{2}{7}$ $\quad 1 - \frac{3}{8} = \frac{8}{8} - \frac{3}{8} = \frac{8-3}{8} = \frac{5}{8}$

Hint: First, simplify if possible! Then, subtract!

$2 - \frac{2}{6} = \frac{12}{6} - \frac{2}{6} = \frac{12-2}{6} = \frac{10}{6}$ $\quad 3 - \frac{3}{6} = \frac{18}{6} - \frac{3}{6} = \frac{18-3}{6} = \frac{15}{6}$

$4 - \frac{8}{12} = \frac{12}{3} - \frac{2}{3} = \frac{12-2}{3} = \frac{10}{3}$ $\quad 5 - \frac{9}{15} = 5 - \frac{3}{5} = \frac{25-3}{5} = \frac{22}{5}$

1. Subtract fractions. Write the missing numbers. Simplify if possible.

Hint: To subtract a fraction from a whole number: rewrite a whole number to a mixed number: a whole number and a fraction with the same denominator (like $2\frac{9}{9}$).

$3 - \frac{4}{9} = 2\frac{9}{9} - \frac{4}{9} = 2\frac{9-4}{9} = 2\frac{5}{9}$ $\quad 6 - \frac{8}{12} = 5\frac{12}{12} - \frac{8}{12} = 5\frac{12-8}{12} = 5\frac{1}{3}$

$4 - \frac{2}{3} = 3\frac{3}{3} - \frac{2}{3} = 3\frac{3-2}{3} = 3\frac{1}{3}$ $\quad 5 - \frac{2}{5} = 4\frac{5}{5} - \frac{2}{5} = 4\frac{5-2}{5} = 4\frac{3}{5}$

$6 - \frac{9}{14} = 5\frac{14}{14} - \frac{9}{14} = 5\frac{14-9}{14} = 5\frac{5}{14}$ $\quad 7 - \frac{5}{6} = 6\frac{6}{6} - \frac{5}{6} = 6\frac{6-5}{6} = 6\frac{1}{6}$

2. Multiply and simplify if possible. Circle the right answer (letter).

Find $\frac{10}{18}$ of 54 arrows.

$\frac{10}{18} \times 54 = \frac{10 \times 54}{18 \times 1} = 30$

a) 50
b) 44
c) 30

Find $\frac{2}{3}$ of 72 donkeys.

Step 1: 72 ÷ 3 = 24

Step 2: 2 × 24 = 48

a) 48
b) 52
c) 60

Find $\frac{3}{4}$ of 56 wood planks.

Step 1: 56 ÷ 4 = 14

Step 2: 3 × 14 = 42

d) 32
e) 40
f) 42

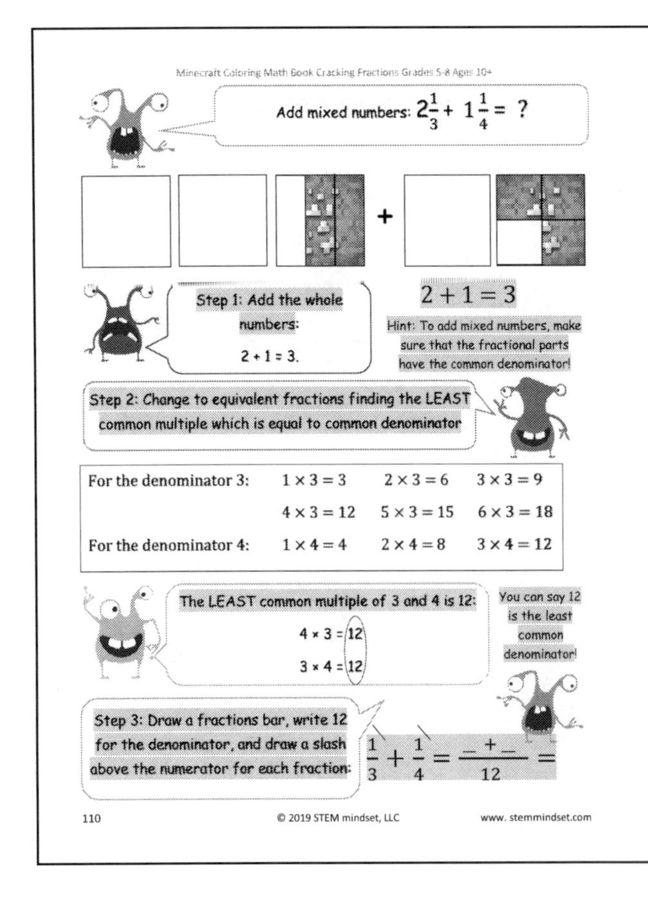
Minecraft Coloring Math Book Cracking Fractions Grades 5-8 Ages 10+
Add mixed numbers: $2\frac{1}{3} + 1\frac{1}{4} = ?$
+
Step 1: Add the whole numbers:
2 + 1 = 3.
$2 + 1 = 3$
Hint: To add mixed numbers, make sure that the fractional parts have the common denominator!
Step 2: Change to equivalent fractions finding the LEAST common multiple which is equal to common denominator
For the denominator 3: 1 × 3 = 3 2 × 3 = 6 3 × 3 = 9
4 × 3 = 12 5 × 3 = 15 6 × 3 = 18
For the denominator 4: 1 × 4 = 4 2 × 4 = 8 3 × 4 = 12
The LEAST common multiple of 3 and 4 is 12:
4 × 3 = 12
3 × 4 = 12
You can say 12 is the least common denominator!
Step 3: Draw a fractions bar, write 12 for the denominator, and draw a slash above the numerator for each fraction:
$\frac{1}{3} + \frac{1}{4} = \frac{_ + _}{12} =$
110 © 2019 STEM mindset, LLC www. stemmindset.com

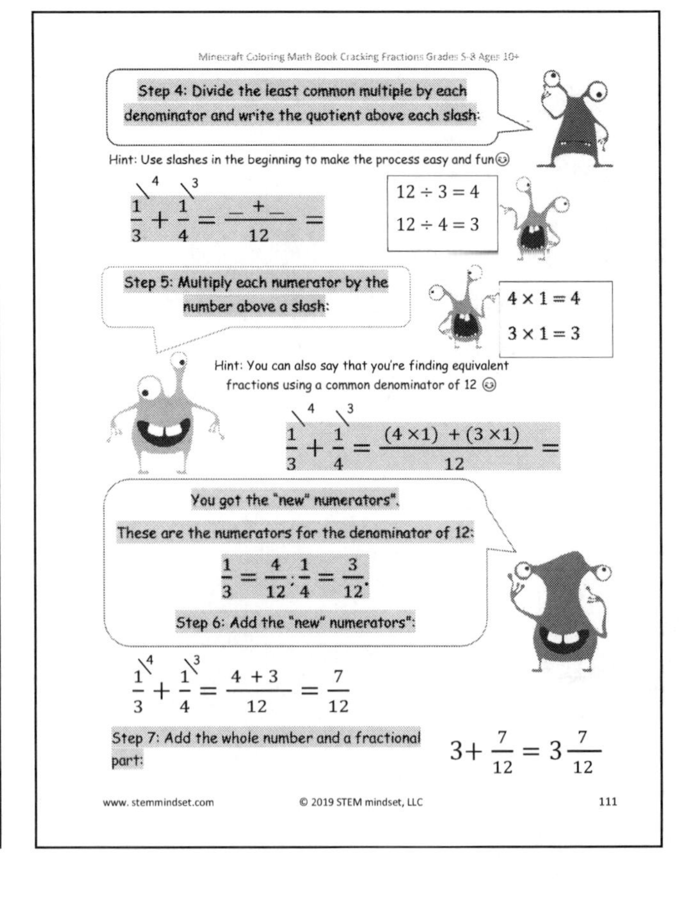
Minecraft Coloring Math Book Cracking Fractions Grades 5-8 Ages 10+
Step 4: Divide the least common multiple by each denominator and write the quotient above each slash:
Hint: Use slashes in the beginning to make the process easy and fun☺
$\frac{1}{3} + \frac{1}{4} = \frac{_ + _}{12} =$
12 ÷ 3 = 4
12 ÷ 4 = 3
Step 5: Multiply each numerator by the number above a slash:
4 × 1 = 4
3 × 1 = 3
Hint: You can also say that you're finding equivalent fractions using a common denominator of 12 ☺
$\frac{1}{3} + \frac{1}{4} = \frac{(4 \times 1) + (3 \times 1)}{12} =$
You got the "new" numerators".
These are the numerators for the denominator of 12:
$\frac{1}{3} = \frac{4}{12}, \frac{1}{4} = \frac{3}{12}.$
Step 6: Add the "new" numerators":
$\frac{1}{3} + \frac{1}{4} = \frac{4+3}{12} = \frac{7}{12}$
Step 7: Add the whole number and a fractional part:
$3 + \frac{7}{12} = 3\frac{7}{12}$
www. stemmindset.com © 2019 STEM mindset, LLC 111

1. Add fractions. Write the missing numbers. Simplify if possible.

$2\frac{3}{8}+4\frac{5}{9}=6\frac{67}{72}$

1) Add whole numbers — 2) Add fractions

1) 2 + 4 = 6 — 2) $\frac{3}{8}+\frac{5}{9}=\frac{27+40}{72}=\frac{67}{72}$

$3\frac{3}{4}+5\frac{3}{5}=9\frac{7}{20}$

1) 3 + 5 = 8 — 2) $\frac{3}{4}+\frac{3}{5}=\frac{15+12}{20}=\frac{27}{20}=1\frac{7}{20}$

$6\frac{3}{10}+4\frac{5}{6}=11\frac{2}{15}$

1) 6 + 4 = 10 — 2) $\frac{3}{10}+\frac{5}{6}=\frac{9+25}{30}=\frac{34}{30}=1\frac{2}{15}$

2. Change the mixed number to an improper faction. Circle the correct answer.

$3\frac{7}{13}=\frac{46}{13}$ — $8\frac{5}{9}=\frac{77}{9}$

a) $\frac{39}{13}$ b) $\frac{23}{13}$ c) $\frac{46}{13}$ (circled) — a) $\frac{72}{9}$ b) $\frac{75}{9}$ c) $\frac{77}{9}$ (circled)

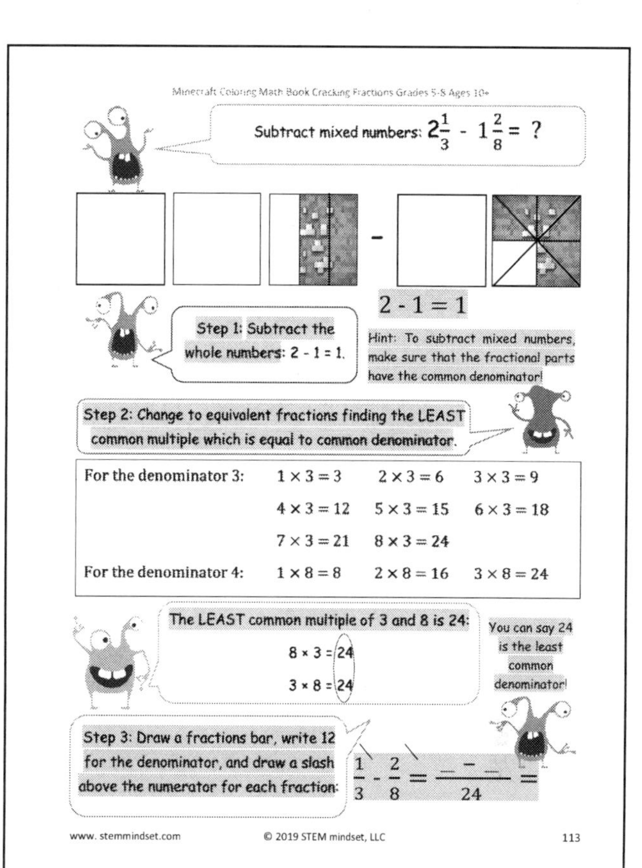

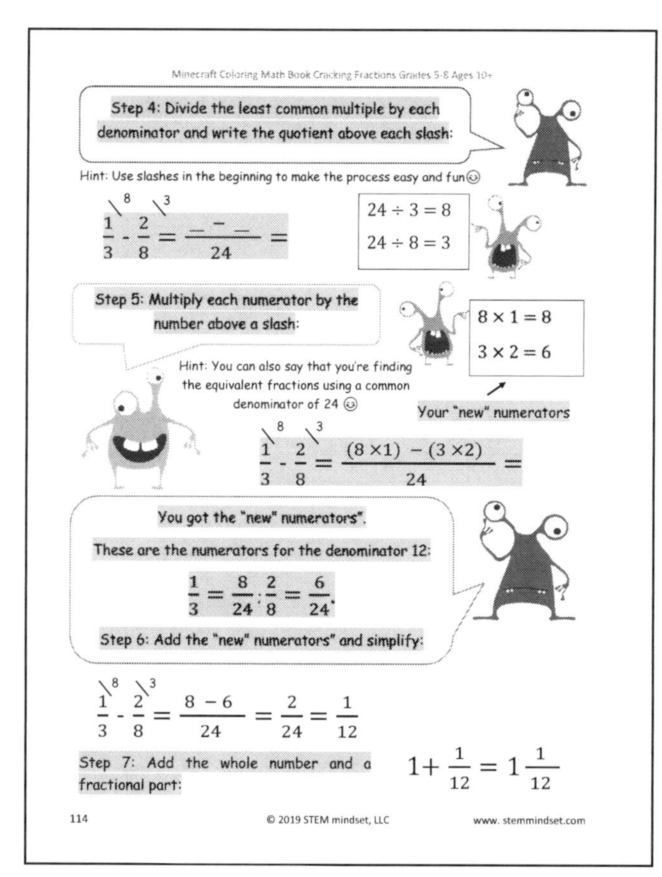

1. Subtract fractions. Write the missing numbers. Simplify if possible.

$3\frac{5}{8}-1\frac{5}{6}=1\frac{19}{24}$

1) Subtract whole numbers — 2) Subtract fractions

To subtract a greater fraction from a smaller fraction: Step 1: borrow one from a whole number; Step 2: rewrite the borrowed whole number to a fraction: $3\frac{1}{3}-1\frac{3}{8}=1\frac{23}{24}$

Step 1: 3-1=~~2~~ 1 — Step 2: $\frac{1}{3}-\frac{3}{8}=\frac{8-9}{24}=\frac{8+24-9}{24}=\frac{23}{24}$

1) 3 - 1 = ~~2~~ 1 — 2) $\frac{5}{8}-\frac{5}{6}=\frac{15+24-20}{24}=\frac{19}{24}$

$8\frac{7}{10}-2\frac{3}{4}=5\frac{19}{20}$

1) 8 - 2 = ~~6~~ 5 — 2) $\frac{7}{10}-\frac{3}{4}=\frac{14+20-15}{20}=\frac{19}{20}$

$9\frac{1}{15}-3\frac{7}{9}=5\frac{13}{45}$

1) 9 - 3 = ~~6~~ 5 — 2) $\frac{1}{15}-\frac{7}{9}=\frac{3+45-35}{45}=\frac{13}{45}$

2. Change the improper faction to a mixed number. Circle the correct answer.

$\frac{32}{5}=6\frac{2}{5}$ — $\frac{49}{11}=4\frac{5}{11}$

a) $3\frac{2}{5}$ b) $6\frac{1}{5}$ c) $6\frac{2}{5}$ (circled) — a) $11\frac{5}{11}$ b) $1\frac{38}{11}$ c) $4\frac{5}{11}$ (circled)

1. Multiply fractions. Write the missing numbers. Simplify if possible.

Step 1: Change the mixed number to an improper fraction

Step 2: Multiply the numerators and denominators

$$3\times 2\frac{3}{15} = \frac{3}{1}\times\frac{33}{15} = \frac{3}{1}\times\frac{11}{5} = \frac{11+11+11}{5} = 6\frac{3}{5}$$

Hint: Simplify before multiplying where you can! (cancel 3s from 33 and 15)

Hint: Change the mixed number to improper fractions!

$$6\times 3\frac{5}{12} = \frac{6}{1}\times\frac{41}{12} = \frac{6\times 41}{1\times 12} = \frac{41}{2} = 20\frac{1}{2}$$

Hint: Cancel or cross out the common factors of the numerator and the denominator to simplify where it's possible ☺

$$5\times 2\frac{7}{20} = \frac{5}{1}\times\frac{47}{20} = \frac{5\times 47}{1\times 20} = \frac{47}{4} = 11\frac{3}{4}$$

$$4\times 5\frac{5}{16} = \frac{4}{1}\times\frac{85}{16} = \frac{4\times 85}{1\times 16} = \frac{85}{4} = 21\frac{1}{4}$$

2. Multiply. Express your answer in simplest form.

$\frac{9}{16}\times\frac{20}{36}$

a) $\frac{15}{36}$ 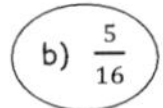 b) $\frac{5}{16}$ c) $\frac{180}{432}$ d) $2\frac{2}{5}$

Minecraft Coloring Math Book Cracking Fractions Grades 5-8 Ages 10+

1. Find and circle or cross out the words to find out more about Minecraft.

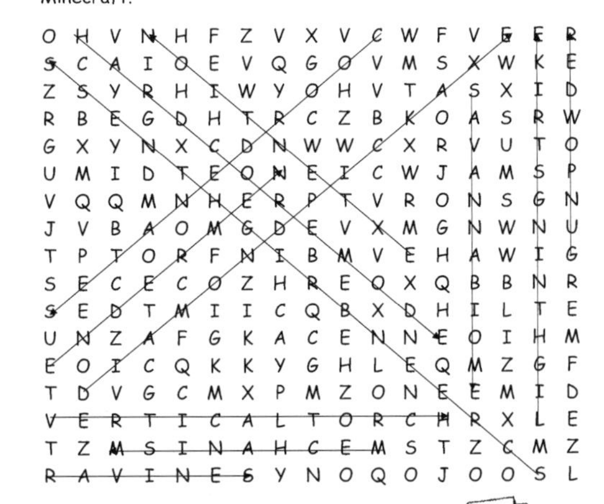

MECHANISM	SCREEN BRIGHTNESS
VERTICAL TORCH	DIAMOND PICKAXE
EXTENTION	GUNPOWDER
HARDCORE MODE	RAVINES
ENDERMEN	SAVANNA BIOME
COORDINATES	LIGHTNING STRIKE

Minecraft Coloring Math Book Cracking Fractions Grades 5-8 Ages 10+

1. Multiply fractions. Write the missing numbers. Simplify if possible.

Step 1: Change the mixed number to an improper fraction

Step 2: Multiply the numerators and denominators

$$4\frac{1}{5}\times 4\frac{2}{7} = \frac{21}{5}\times\frac{30}{7} = \frac{21\times 30}{5\times 7} = \frac{18}{1} = 18$$

Hint: Factor the numerator and the denominator by the same number to reduce the fraction to lowest terms! ☺

$$2\frac{1}{4}\times 3\frac{4}{6} = \frac{9}{4}\times\frac{22}{6} = \frac{9\times 22}{4\times 6} = \frac{33}{4} = 8\frac{1}{4}$$

$$3\frac{1}{2}\times 2\frac{2}{7} = \frac{7}{2}\times\frac{16}{7} = \frac{7\times 16}{2\times 7} = 8$$

$$9\frac{5}{8}\times 3\frac{7}{11} = \frac{77}{8}\times\frac{40}{11} = \frac{77\times 40}{8\times 11} = 35$$

2. Multiply. Express your answer in simplest form.

$\frac{5}{12}\times\frac{36}{55}$

a) $\frac{3}{10}$ b) $\frac{36}{60}$ c) $\frac{3}{11}$ d) $1\frac{2}{5}$

Minecraft Coloring Math Book Cracking Fractions Grades 5-8 Ages 10+

1. Write in the missing numbers (practice adding and subtracting mixed numbers).

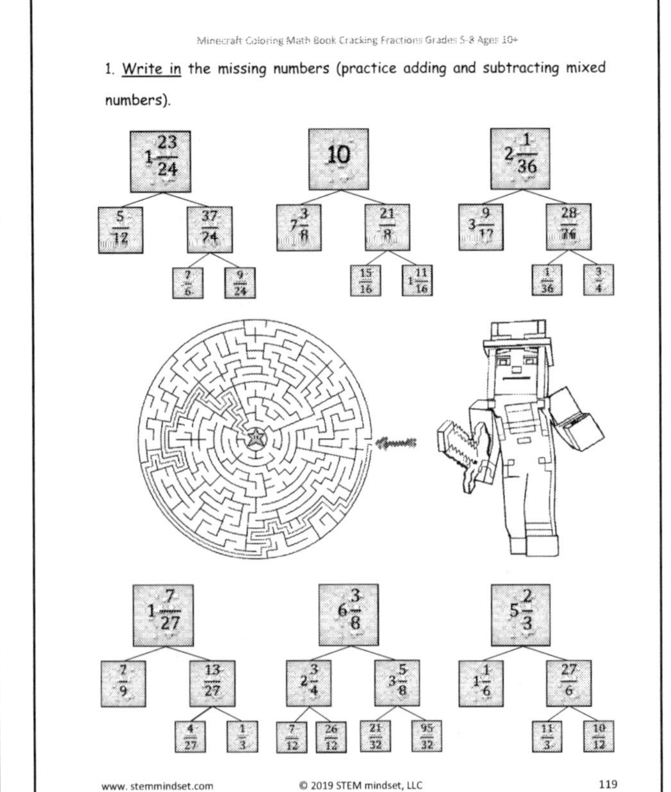

1. Divide fractions. Write the missing numbers. Simplify if possible.

Divide fractions! That's impossible! I cannot divide 3 blocks by, say, $\frac{5}{9}$!

$3 \div \frac{5}{9}$ = no answer, no solution, let's skip it.

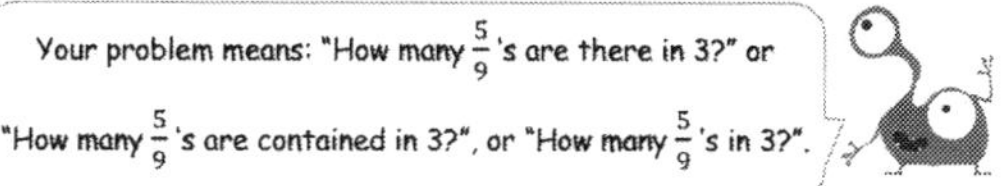

Your problem means: "How many $\frac{5}{9}$'s are there in 3?" or "How many $\frac{5}{9}$'s are contained in 3?", or "How many $\frac{5}{9}$'s in 3?".

Step 1: Change 3 to $\frac{1}{9}$'s. Use a common denominator of 9. $3 = \frac{27}{9}$

Step 2: Divide the numerators, then, divide the denominators, then, divide the numerator by the denominator.

$$\text{So, } 3 \div \frac{5}{9} = \frac{27}{9} \div \frac{5}{9} = \frac{(27 \div 5)}{(9 \div 9)} = \frac{(27 \div 5)}{1} = \frac{27}{5} = 5\frac{2}{5}$$

5 =The whole number
Denominator 5 | 27 =Numerator
25
2 = Remainder

! Dividend = Numerator
Divisor = Denominator
Quotient = The Whole Number
Remainder = Numerator of a fraction part

$$5 \div \frac{2}{7} = \frac{35}{7} \div \frac{2}{7} = \frac{(35 \div 2)}{(7 \div 7)} = \frac{35}{2} = 17\frac{1}{2}$$

$$4 \div \frac{2}{3} = \frac{12}{3} \div \frac{2}{3} = \frac{(12 \div 2)}{(3 \div 3)} = \frac{12}{2} = 6$$

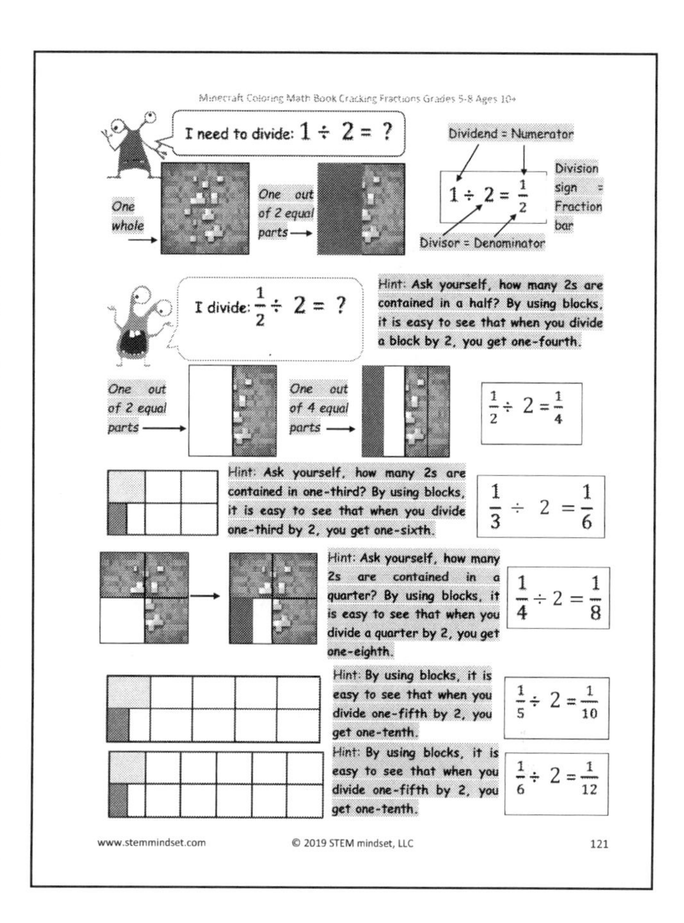

I need to divide: $1 \div 2 = ?$

One whole → One out of 2 equal parts →

Dividend = Numerator

$1 \div 2 = \frac{1}{2}$

Division sign = Fraction bar

Divisor = Denominator

I divide: $\frac{1}{2} \div 2 = ?$

Hint: Ask yourself, how many 2s are contained in a half? By using blocks, it is easy to see that when you divide a block by 2, you get one-fourth.

One out of 2 equal parts → One out of 4 equal parts →

$\frac{1}{2} \div 2 = \frac{1}{4}$

Hint: Ask yourself, how many 2s are contained in one-third? By using blocks, it is easy to see that when you divide one-third by 2, you get one-sixth.

$\frac{1}{3} \div 2 = \frac{1}{6}$

Hint: Ask yourself, how many 2s are contained in a quarter? By using blocks, it is easy to see that when you divide a quarter by 2, you get one-eighth.

$\frac{1}{4} \div 2 = \frac{1}{8}$

Hint: By using blocks, it is easy to see that when you divide one-fifth by 2, you get one-tenth.

$\frac{1}{5} \div 2 = \frac{1}{10}$

Hint: By using blocks, it is easy to see that when you divide one-fifth by 2, you get one-tenth.

$\frac{1}{6} \div 2 = \frac{1}{12}$

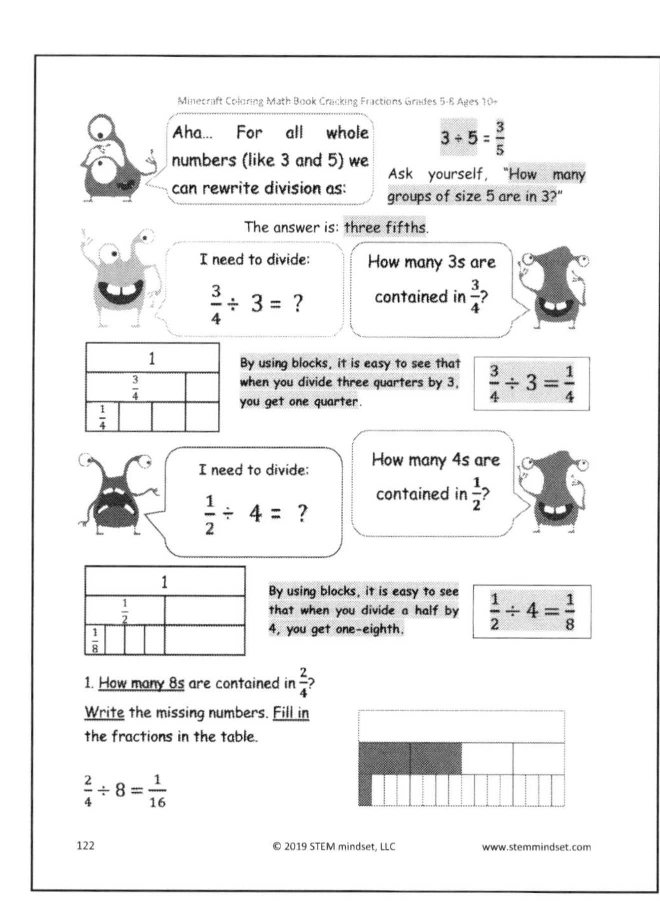

Aha... For all whole numbers (like 3 and 5) we can rewrite division as:

$3 \div 5 = \frac{3}{5}$

Ask yourself, "How many groups of size 5 are in 3?"

The answer is: three fifths.

I need to divide: $\frac{3}{4} \div 3 = ?$

How many 3s are contained in $\frac{3}{4}$?

1
$\frac{3}{4}$
$\frac{1}{4}$

By using blocks, it is easy to see that when you divide three quarters by 3, you get one quarter.

$\frac{3}{4} \div 3 = \frac{1}{4}$

I need to divide: $\frac{1}{2} \div 4 = ?$

How many 4s are contained in $\frac{1}{2}$?

1
$\frac{1}{2}$
$\frac{1}{8}$

By using blocks, it is easy to see that when you divide a half by 4, you get one-eighth.

$\frac{1}{2} \div 4 = \frac{1}{8}$

1. How many 8s are contained in $\frac{2}{4}$?

Write the missing numbers. Fill in the fractions in the table.

$\frac{2}{4} \div 8 = \frac{1}{16}$

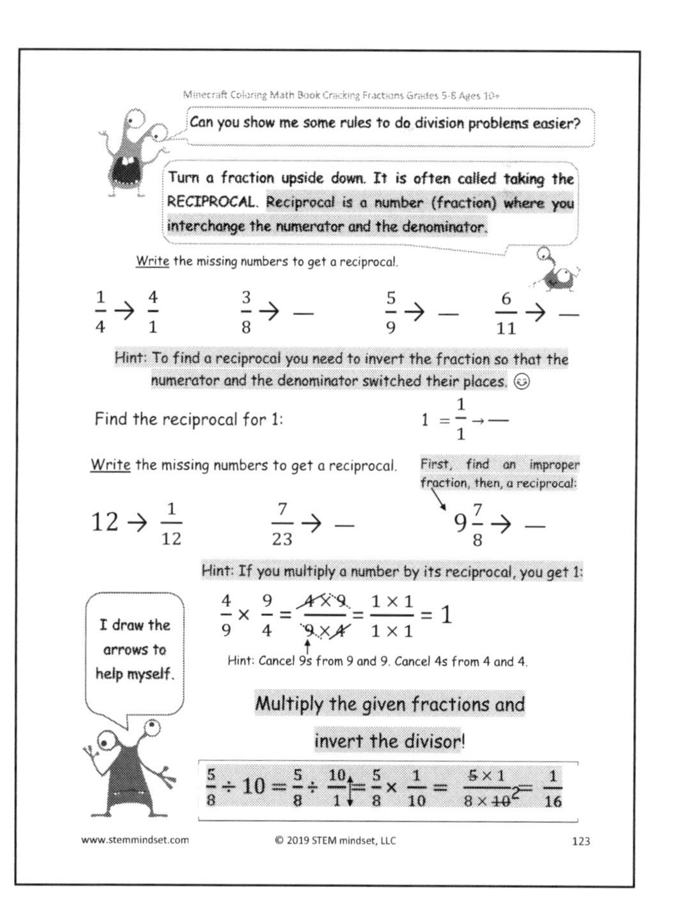

Can you show me some rules to do division problems easier?

Turn a fraction upside down. It is often called taking the RECIPROCAL. Reciprocal is a number (fraction) where you interchange the numerator and the denominator.

Write the missing numbers to get a reciprocal.

$\frac{1}{4} \rightarrow \frac{4}{1}$ $\frac{3}{8} \rightarrow$ — $\frac{5}{9} \rightarrow$ — $\frac{6}{11} \rightarrow$ —

Hint: To find a reciprocal you need to invert the fraction so that the numerator and the denominator switched their places. ☺

Find the reciprocal for 1: $1 = \frac{1}{1} \rightarrow$ —

Write the missing numbers to get a reciprocal.

First, find an improper fraction, then, a reciprocal:

$12 \rightarrow \frac{1}{12}$ $\frac{7}{23} \rightarrow$ — $9\frac{7}{8} \rightarrow$ —

Hint: If you multiply a number by its reciprocal, you get 1:

$$\frac{4}{9} \times \frac{9}{4} = \frac{\cancel{4} \times \cancel{9}}{\cancel{9} \times \cancel{4}} = \frac{1 \times 1}{1 \times 1} = 1$$

Hint: Cancel 9s from 9 and 9. Cancel 4s from 4 and 4.

I draw the arrows to help myself.

Multiply the given fractions and invert the divisor!

$$\frac{5}{8} \div 10 = \frac{5}{8} \div \frac{10}{1} = \frac{5}{8} \times \frac{1}{10} = \frac{\cancel{5} \times 1}{8 \times \cancel{10}_2} = \frac{1}{16}$$

1. Divide fractions. Write the missing numbers. Simplify if possible.

$\frac{10}{3} \div 2 = \frac{10}{3} \div \frac{2}{1} = \frac{10}{3} \times \frac{1}{2} = \frac{10 \times 1}{3 \times 2} = 1\frac{2}{3}$

$\frac{8}{15} \div 4 = \frac{8}{15} \div \frac{4}{1} = \frac{8}{15} \times \frac{1}{4} = \frac{8 \times 1}{15 \times 4} = \frac{2}{15}$

$\frac{2}{3} \div 4 = \frac{2}{3} \div \frac{4}{1} = \frac{2}{3} \times \frac{1}{4} = \frac{2 \times 1}{3 \times 4} = \frac{1}{6}$

$\frac{5}{6} \div 10 = \frac{1}{12}$ $\quad$ $\frac{13}{18} \div 26 = \frac{1}{36}$

2. My brother and my sister had a total of 56 swords at first. After she lost 8 of her swords and he lost $\frac{1}{2}$ of his swords, they had an equal number of swords. How many swords did she have at first?

Brother = B $\quad$ Sister = S

$B_1 + S_1 = 56$ $\quad$ $S_2 = S_1 - 8$ $\quad$ $B_2 = \frac{1}{2} B_1$

$\begin{cases} B_1 + S_1 = 56 \\ S_1 - 8 = \frac{1}{2} B_1 \end{cases} \rightarrow \begin{cases} S_1 = 56 - B_1 \\ S_1 - 8 = \frac{1}{2} B_1 \end{cases} \rightarrow \begin{matrix} B_1 = 32 \\ S_1 = 24 \end{matrix}$

Answer: 24 swords.

1. Divide fractions. Write the missing numbers. Simplify if possible.

$\frac{49}{50} \div 7 = \frac{49}{50} \div \frac{7}{1} = \frac{49}{50} \times \frac{1}{7} = \frac{49 \times 1}{50 \times 7} = \frac{7}{50}$

$\frac{14}{18} \div 2 = \frac{14}{18} \div \frac{2}{1} = \frac{14}{18} \times \frac{1}{2} = \frac{14 \times 1}{18 \times 2} = \frac{7}{18}$

$\frac{8}{27} \div 18 = \frac{4}{243}$ $\quad$ $\frac{6}{14} \div 21 = \frac{1}{49}$

$\frac{21}{25} \div 14 = \frac{3}{50}$ $\quad$ $\frac{12}{13} \div 24 = \frac{1}{26}$

2. My brother and my sister had a total of 20 wood planks at first. After she used $\frac{3}{4}$ of her planks and he used $\frac{5}{6}$ of his planks, they had an equal number of wood planks. How many wood planks did he have at first?

Brother = B $\quad$ Sister = S

$B_1 + S_1 = 20$ $\quad$ $B_2 = \frac{1}{6} B_1$ $\quad$ $S_2 = \frac{1}{4} S_1$

$\begin{cases} B_1 + S_1 = 20 \\ \frac{1}{4} S_1 = \frac{1}{6} B_1 \end{cases} \rightarrow \begin{cases} S_1 = 20 - B_1 \\ \frac{1}{4} S_1 = \frac{1}{6} B_1 \end{cases} \rightarrow B_1 = 12$

Answer: 12 wood planks.

1. Divide fractions. Write the missing numbers. Simplify if possible.

$\frac{9}{17} \div 18 = \frac{9}{17} \div \frac{18}{1} = \frac{9}{17} \times \frac{1}{18} = \frac{9 \times 1}{17 \times 18} = \frac{1}{34}$

$\frac{24}{35} \div 3 = \frac{24}{35} \div \frac{3}{1} = \frac{24}{35} \times \frac{1}{3} = \frac{24 \times 1}{35 \times 3} = \frac{8}{35}$

$\frac{49}{50} \div x = \frac{7}{50}$ $\quad$ $\frac{15}{16} \div x = \frac{3}{64}$

$x = \frac{49}{50} \div \frac{7}{50} = \frac{49}{50} \times \frac{50}{7} = 7$ $\quad$ $x = \frac{15}{16} \div \frac{3}{64} = \frac{15}{16} \times \frac{64}{3} = 20$

$\frac{36}{41} \div x = \frac{4}{41}$ $\quad$ $\frac{25}{50} \div x = \frac{1}{30}$

$x = \frac{36}{41} \div \frac{4}{41} = \frac{36}{41} \times \frac{41}{4} = 9$ $\quad$ $x = \frac{25}{50} \div \frac{1}{30} = \frac{25}{50} \times \frac{30}{1} = 15$

2. My brother destroyed $\frac{2}{5}$ as many spiders as my sister at first. After he destroyed 9 more spiders and she destroyed 6 more spiders, they had an equal number of spiders destroyed in the end. How many spiders did he destroy at first?

Brother = B $\quad$ Sister = S $\quad \rightarrow \quad \begin{cases} B_1 = \frac{2}{5} S_1 \\ B_1 + 9 = S_1 + 6 \end{cases}$

$B_2 = B_1 + 9$ $\quad$ $S_2 = S_1 + 6$ $\quad \rightarrow$

$B_1 = 2$

Answer: 2 spiders.

1. Find and circle or cross out the words to find out more about Minecraft.

A	C	T	I	V	A	T	O	R	O	D	B	TECHNIQUE
V	Z	D	A	M	A	G	E	P	P	O	O	OPPONENT
L	C	Y	V	A	J	H	O	G	P	X	O	DAMAGE
P	U	M	Y	J	M	Z	U	B	O	J	K	CHAMBER
A	G	G	R	A	V	A	T	I	N	G	S	GUARDIAN
O	L	A	O	C	R	A	H	C	E	R	H	AGGRAVATING
Q	Q	K	H	D	C	R	I	T	N	E	E	BOOKSHELVES
E	U	Q	I	N	H	C	E	T	T	B	L	CHARCOAL
H	K	A	F	E	X	C	Z	L	S	M	V	ACTIVATOR
G	N	D	P	U	S	T	X	N	M	A	E	
T	F	Z	J	M	B	P	F	I	S	H	S	
A	Z	Z	Y	H	S	K	E	G	Z	C	O	

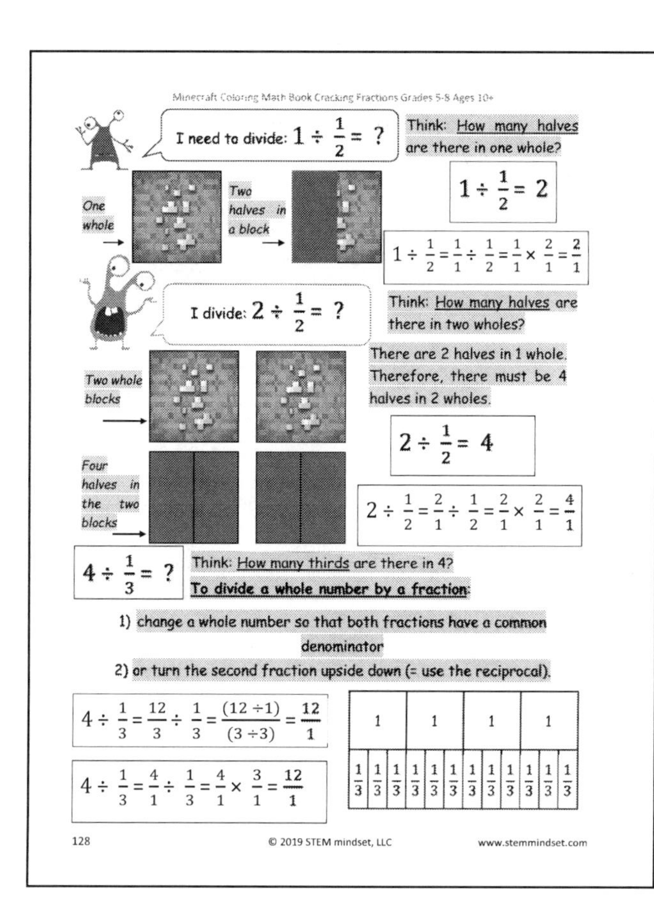

I need to divide: $1 \div \frac{1}{2} = ?$

Think: How many halves are there in one whole?

One whole

Two halves in a block

$1 \div \frac{1}{2} = 2$

$1 \div \frac{1}{2} = \frac{1}{1} \div \frac{1}{2} = \frac{1}{1} \times \frac{2}{1} = \frac{2}{1}$

I divide: $2 \div \frac{1}{2} = ?$

Think: How many halves are there in two wholes?

Two whole blocks

There are 2 halves in 1 whole. Therefore, there must be 4 halves in 2 wholes.

$2 \div \frac{1}{2} = 4$

Four halves in the two blocks

$2 \div \frac{1}{2} = \frac{2}{1} \div \frac{1}{2} = \frac{2}{1} \times \frac{2}{1} = \frac{4}{1}$

$4 \div \frac{1}{3} = ?$

Think: How many thirds are there in 4?

To divide a whole number by a fraction:

1) change a whole number so that both fractions have a common denominator
2) or turn the second fraction upside down (= use the reciprocal).

$4 \div \frac{1}{3} = \frac{12}{3} \div \frac{1}{3} = \frac{(12 \div 1)}{(3 \div 3)} = \frac{12}{1}$

$4 \div \frac{1}{3} = \frac{4}{1} \div \frac{1}{3} = \frac{4}{1} \times \frac{3}{1} = \frac{12}{1}$

1			1			1			1		
$\frac{1}{3}$	$\frac{1}{3}$	$\frac{1}{3}$	$\frac{1}{3}$	$\frac{1}{3}$	$\frac{1}{3}$	$\frac{1}{3}$	$\frac{1}{3}$	$\frac{1}{3}$	$\frac{1}{3}$	$\frac{1}{3}$	$\frac{1}{3}$

1. Divide fractions. Write the missing numbers. Simplify if possible.

$8 \div \frac{2}{7} = \frac{8}{1} \div \frac{2}{7} = \frac{8}{1} \times \frac{7}{2} = 28$ —

$6 \div \frac{12}{13} = \frac{6}{1} \div \frac{12}{13} = \frac{6}{1} \times \frac{13}{12} = 6\frac{1}{2}$

$5 \div \frac{10}{17} = \frac{5}{1} \div \frac{10}{17} = \frac{5}{1} \times \frac{17}{10} = 8\frac{1}{2}$

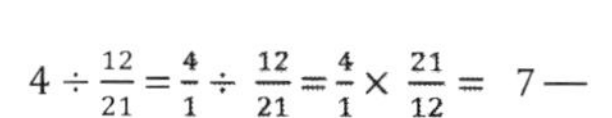

$4 \div \frac{12}{21} = \frac{4}{1} \div \frac{12}{21} = \frac{4}{1} \times \frac{21}{12} = 7$ —

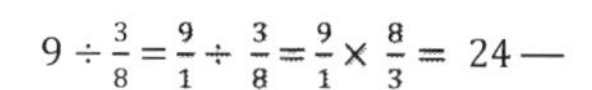

$9 \div \frac{3}{8} = \frac{9}{1} \div \frac{3}{8} = \frac{9}{1} \times \frac{8}{3} = 24$ —

2. I came across x spiders and destroyed $\frac{3}{4}$ of them with my iron sword. How many spiders did I come across if I destroyed 12 spiders?

$\frac{3}{4}x = 12$

$x = 12 \div \frac{3}{4} = \frac{12}{1} \div \frac{3}{4} = \frac{12}{1} \times \frac{4}{3} = \frac{12 \times 4}{1 \times 3} = 16$ —

Answer: 16 spiders.

1. Divide fractions. Write the missing numbers. Simplify if possible.

$3 \div \frac{21}{25} = \frac{3}{1} \div \frac{21}{25} = \frac{3}{1} \times \frac{25}{21} = 3\frac{4}{7}$

$7 \div \frac{14}{17} = \frac{7}{1} \div \frac{14}{17} = \frac{7}{1} \times \frac{17}{14} = 8\frac{1}{2}$

$3 \div \frac{6}{11} = \frac{3}{1} \div \frac{6}{11} = \frac{3}{1} \times \frac{11}{6} = 5\frac{1}{2}$

$6 \div \frac{24}{25} = 6\frac{1}{4}$ $\qquad 9 \div \frac{18}{22} = 11$

$3 \div \frac{15}{18} = 3\frac{3}{5}$ $\qquad 10 \div \frac{10}{15} = 15$

2. The villager planted x flowers on his farm and $\frac{3}{5}$ of them were beautiful 30 lilacs. How many flowers did he plant?

$\frac{3}{5}x = 30$

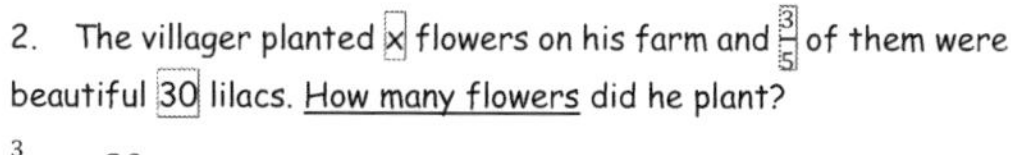

$x = 30 \div \frac{3}{5} = \frac{30}{1} \div \frac{3}{5} = \frac{30}{1} \times \frac{5}{3} = \frac{30 \times 5}{1 \times 3} = 50$ —

Answer: 50 flowers.

1. Divide or multiply. Write the missing numbers. Simplify if possible.

$x \div \frac{5}{7} = 7$ $\qquad x \div \frac{4}{15} = 22\frac{1}{2}$

Hint: Change the mixed number to an improper fraction, then divide or multiply!

$x = \frac{7}{1} \times \frac{5}{7} = \frac{7 \times 5}{1 \times 7} = 5$ $\qquad x = \frac{45 \times 4}{2 \times 15} = 6$

$x \div \frac{24}{25} = 8\frac{1}{3}$ $\qquad x \div \frac{18}{22} = 11$

$x = \frac{25 \times 24}{3 \times 25} = 8$ $\qquad x = \frac{11 \times 18}{1 \times 22} = 9$

$x \div \frac{24}{35} = 8\frac{3}{4}$ $\qquad x \div \frac{15}{36} = 28\frac{4}{5}$

$x = \frac{35 \times 24}{4 \times 35} = 6$ $\qquad x = \frac{144 \times 15}{5 \times 36} = 12$

2. I met x animals and $\frac{5}{12}$ of them were 35 black horses. How many animals did I meet?

$\frac{5}{12}x = 35$

$x = 35 \div \frac{5}{12} = \frac{35}{1} \div \frac{5}{12} = \frac{35}{1} \times \frac{12}{5} = 84$

Answer: 84 animals.

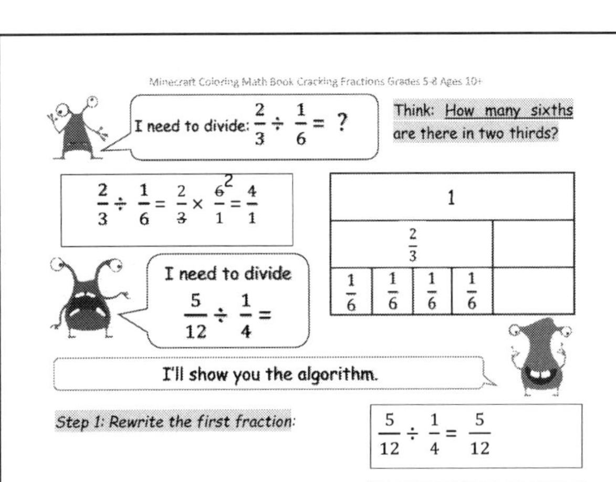
I need to divide: $\frac{2}{3} \div \frac{1}{6} = ?$
Think: How many sixths are there in two thirds?
$\frac{2}{3} \div \frac{1}{6} = \frac{2}{\not 3} \times \frac{\not 6^{2}}{1} = \frac{4}{1}$
1
$\frac{2}{3}$
$\frac{1}{6}$ $\frac{1}{6}$ $\frac{1}{6}$ $\frac{1}{6}$
I need to divide $\frac{5}{12} \div \frac{1}{4} =$

I'll show you the algorithm.

Step 1: Rewrite the first fraction:	$\frac{5}{12} \div \frac{1}{4} = \frac{5}{12}$
Step 2: Change the operation (÷ → ×):	$\frac{5}{12} \div \frac{1}{4} = \frac{5}{12} \times$
Step 3: Write the reciprocal of the second fraction – switch the numerator and the denominator:	$\frac{5}{12} \div \frac{1}{4} = \frac{5}{12} \times \frac{4}{1}$
Step 4: Multiply the numerators and denominators and simplify where possible:	$\frac{5}{12} \div \frac{1}{4} = \frac{5}{\not{12}\,3} \times \frac{4}{1}$
Step 5: Change to mixed fraction where possible:	$\frac{5}{12} \div \frac{1}{4} = \frac{5}{\not{12}\,3} \times \frac{4}{1} = 6\frac{2}{3}$

1. Divide fractions. Write the missing numbers. Simplify if possible.

$\frac{1}{12} \div \frac{5}{6} = \frac{1}{12} \times \frac{6}{5} = \frac{1}{10}$

$\frac{8}{9} \div \frac{24}{45} = \frac{8}{9} \times \frac{45}{24} = 1\frac{2}{3}$

$\frac{5}{18} \div \frac{8}{9} = \frac{5}{18} \times \frac{9}{8} = \frac{5}{16}$

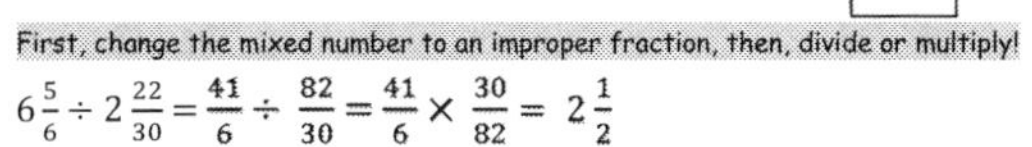

First, change the mixed number to an improper fraction, then, divide or multiply!

$6\frac{5}{6} \div 2\frac{22}{30} = \frac{41}{6} \div \frac{82}{30} = \frac{41}{6} \times \frac{30}{82} = 2\frac{1}{2}$

$1\frac{8}{9} \div 2\frac{14}{27} = \frac{3}{4}$ $\quad 4\frac{3}{7} \div 4\frac{6}{14} = 1$

2. I spent $\frac{1}{3}$ of an hour to find a cave. Then, I spent $\frac{5}{6}$ of an hour to dig out 2 diamonds. Then, I came back to my tiny house, found the recipe of a diamond sword, and spent $\frac{1}{4}$ of an hour to make it. How much time did I spend in all?

$\frac{1}{3} \times 60 = 20\ (min)$

$\frac{5}{6} \times 60 = 50\ (min)$

$\frac{1}{4} \times 60 = 15\ (min)$

Answer: 20 + 50 + 15 = 85 = 1h 25 min.

1. Divide fractions. Write the missing numbers. Simplify if possible.

$\frac{14}{18} \div \frac{2}{9} = \frac{14}{18} \times \frac{9}{2} = 3\frac{1}{2}$

$\frac{6}{7} \div \frac{24}{49} = 1\frac{3}{4}$ $\quad 3\frac{2}{6} \div 2\frac{2}{9} = 1\frac{1}{2}$

$\frac{21}{25} \div \frac{14}{15} = \frac{9}{10}$ $\quad 2\frac{5}{8} \div 2\frac{3}{12} = 1\frac{1}{6}$

$1\frac{8}{9} \div 2\frac{2}{33} = \frac{11}{12}$ $\quad 4\frac{2}{7} \div 3\frac{6}{14} = 1\frac{1}{4}$

$8\frac{3}{4} \div 3\frac{6}{8} = 2\frac{1}{3}$ $\quad 10\frac{2}{3} \div 5\frac{4}{6} = 1\frac{15}{17}$

2. I had $\frac{2}{7}$ as many emeralds as my brother. After my brother used 8 of his emeralds, and I used 6 emeralds, I have $\frac{1}{4}$ as many emeralds as my brother. How many emeralds do we have in the end?

$I_1 = \frac{2}{7}B_1$ $\quad B_2 = B_1 - 8$ $\quad I_2 = I_1 - 6$ $\quad I_2 = \frac{1}{4}B_2$

$I_2 = I_2$ $\quad \rightarrow \quad \frac{2}{7}B_1 - 6 = \frac{1}{4}(B_1 - 8)$

$B_1 = 112$ $\quad B_2 = 104$ $\quad I_2 = 26$ $\quad B_2 + I_2 = 130$

Answer: 130 emeralds.

1. Divide fractions. Write the missing numbers. Simplify if possible.

$\frac{16}{25} \div \frac{32}{55} = \frac{16}{25} \times \frac{55}{32} = 1\frac{1}{10}$

$\frac{6}{7} \div \frac{24}{49} = \frac{7}{4}$ $\quad 5\frac{1}{7} \div 4\frac{1}{2} = 1\frac{1}{7}$

$\frac{21}{25} \div \frac{14}{15} = \frac{9}{10}$ $\quad 11\frac{3}{7} \div 4\frac{10}{14} = 2\frac{6}{37}$

$15\frac{1}{8} \div 2\frac{1}{5} = 6\frac{7}{8}$ $\quad 4\frac{1}{5} \div 2\frac{1}{3} = 1\frac{4}{5}$

$x \div 3\frac{4}{8} = 2\frac{2}{7}$ $\quad x \div 7\frac{1}{7} = 3\frac{9}{25}$

$x = 8$ $\quad x = 24$

2. I crafted $\frac{3}{5}$ as many wooden sticks as my sister. After my sister used 8 of her wooden sticks to make weapons, and I crafted 12 more wooden sticks, I have $\frac{3}{4}$ as many wooden sticks as my sister. How many wooden sticks does she have in the end?

$I_1 = \frac{3}{5}S_1$ $\quad S_2 = S_1 - 8$ $\quad I_2 = I_1 + 12$ $\quad I_2 = \frac{3}{4}S_2$

$I_1 = I_1$ $\quad \rightarrow \quad \frac{3}{5}S_1 + 12 = \frac{3}{4}(S_1 - 8)$

$S_1 = 120$ $\quad S_2 = 112$

Answer: 112 sticks.

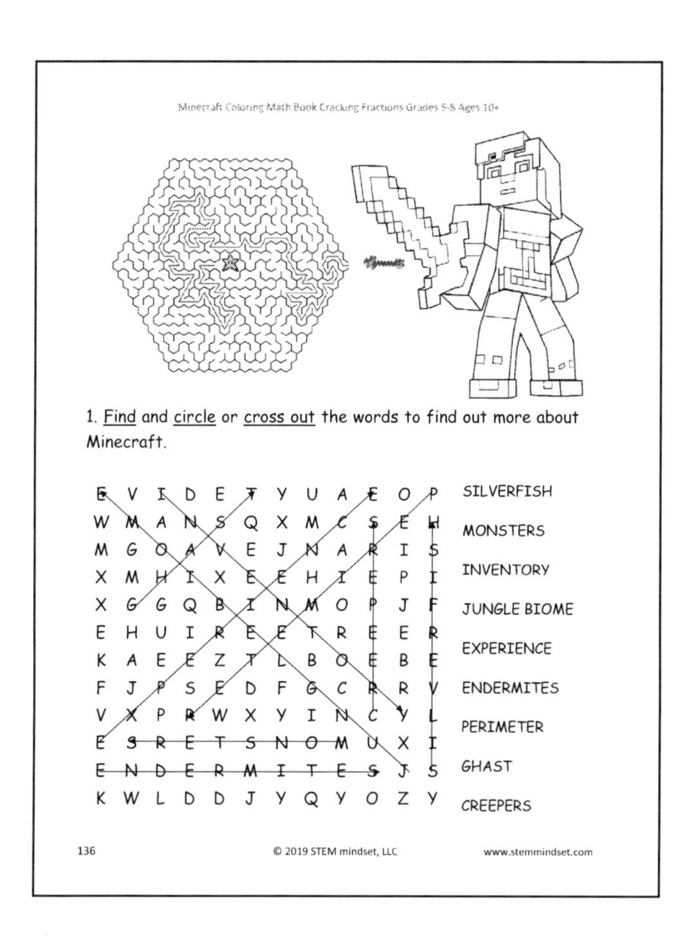

1. Find and circle or cross out the words to find out more about Minecraft.

E	V	I	D	E	T	Y	U	A	E	O	P	SILVERFISH
W	M	A	N	S	Q	X	M	C	S	E	H	MONSTERS
M	G	O	A	V	E	J	N	A	R	I	S	
X	M	H	I	X	E	E	H	I	E	P	I	INVENTORY
X	G	G	Q	B	I	N	M	O	P	J	F	JUNGLE BIOME
E	H	U	I	R	E	E	T	R	E	E	R	
K	A	E	E	Z	T	L	B	O	E	B	E	EXPERIENCE
F	J	P	S	E	D	F	G	C	R	R	V	ENDERMITES
V	X	P	R	W	X	Y	I	N	C	Y	L	PERIMETER
E	S	R	E	T	S	N	O	M	U	X	I	
E	N	D	E	R	M	I	T	E	S	J	S	GHAST
K	W	L	D	D	J	Y	Q	Y	O	Z	Y	CREEPERS

Made in the USA
Coppell, TX
06 January 2020

14170305R00094